This book provides a solid foundation and an extensive study for an important class of constrained optimization problems known as Mathematical Programs with Equilibrium Constraints (MPEC), which are extensions of bilevel optimization problems. The book begins with the description of many source problems arising from engineering and economics that are amenable to treatment by the MPEC methodology. Error bounds and parametric analysis are the main tools to establish a theory of exact penalization, a set of MPEC constraint qualifications and the first- and second-order optimality conditions. The book also describes several iterative algorithms such as a penalty-based interior point algorithm, an implicit programming algorithm and a piecewise sequential quadratic programming algorithm for MPECs. Results in the book will have significant impacts in such disciplines as engineering design, economics and game equilibria, and transportation planning, within all of which MPEC has a central role to play in the modeling of many practical problems.

A useful resource for applied mathematicians in general, this book will be a particularly valuable tool for operations researchers, transportation, industrial, and mechanical engineers, and mathematical programmers.

T0269195

Mathematical Programs
with
Equilibrium Constraints

Mathematical Programs with Equilibrium Constraints

ZHI-QUAN LUO

McMaster University

JONG-SHI PANG

The Johns Hopkins University

DANIEL RALPH

The University of Melbourne

CAMBRIDGE
UNIVERSITY PRESS

CAMBRIDGE UNIVERSITY PRESS
Cambridge, New York, Melbourne, Madrid, Cape Town, Singapore, São Paulo

Cambridge University Press
The Edinburgh Building, Cambridge CB2 8RU, UK

Published in the United States of America by Cambridge University Press, New York

www.cambridge.org
Information on this title: www.cambridge.org/9780521572903

First published 1996
This digitally printed version 2008

A catalogue record for this publication is available from the British Library

Library of Congress Cataloguing in Publication data
Luo, Zhi-Quan.
Mathematical programs with equilibrium constraints / Zhi-Quan Luo,
Jong-Shi Pang, Daniel Ralph.
p. cm.
Includes bibliographical references and index.
ISBN 0–521–57290–8
1. Mathematical optimization.   2. Nonlinear programming.
I. Pang, Jong-Shi.   II. Ralph, Daniel.   III. Title.
QA402.5.L83   1997
519.7′6–dc20   99–19428
                                                                      CIP

ISBN 978-0-521-57290-3 hardback
ISBN 978-0-521-06508-5 paperback

To our families

Contents

Numbering System

The six chapters of the book are numbered from 1 to 6, the sections are denoted by decimal numbers of the type **2.3** (meaning Section 3 of Chapter 2). Many sections are further divided into subsections, some subsections are numbered, others are not. The numbered subsections are by decimal numbers following the section numbers; e.g., Subsection **1.3.1** means Chapter 1, Section 3, Subsection 1.

All definitions, results, and miscellaneous items are numbered consecutively within each section in the form **1.3.5**, **1.3.6**, meaning Items 5 and 6 in Section 3 of Chapter 1. All items are also identified by their types (e.g., **1.4.1 Proposition.**, **1.4.2 Remark.**). When an item is referred to in the text, it is called out as Algorithm **5.2.1**, Theorem **4.1.7**, etc.

Equations are numbered consecutively in each section by (1), (2), etc. Any reference to an equation in the same section is by this number only, whereas equations in another section are identified by chapter, section, and equation. Thus (3.1.4) means Equation (4) in Section 1 of Chapter 3.

Acronyms

AVI	Affine Variational Inequality
BIF	B(ouligand)-Differentiable Implicit Function Condition
CQ	Constraint Qualification
C^r	Continuously differentiable of order r
CRCQ	Constant Rank Constraint Qualification
GBIF	Global BIF Condition
IMP	Implicit Programming
KKT	Karush-Kuhn-Tucker
LCP	Linear Complementarity Problem
LICQ	Linear Independence Constraint Qualification
MFCQ	Mangasarian-Fromovitz Constraint Qualification
MP	Mathematical Program
MPAEC	Mathematical Program with Affine Equilibrium Constraints
MPEC	Mathematical Program with Equilibrium Constraints
NCP	Nonlinear Complementarity Problem
NLP	Nonlinear Program
PCP	Piecewise Programming
PC^r	Piecewise smooth of order r
PIPA	Penalty Interior Point Algorithm
PSQP	Piecewise Sequential Quadratic Programming
SBCQ	Sequentially Bounded Constraint Qualification
SCOC	Strong Coherent Orientation Condition
SMFCQ	Strict Mangasarian-Fromovitz Constraint Qualification
SQP	Sequential Quadratic Programming
SRC	Strong Regularity Condition
VI	Variational Inequality

Glossary of Notation

Scalars

sgn t the sign, $1, -1, 0$, of a positive, negative, or zero scalar t

$t_+ \equiv \max(0, t)$ the nonnegative part of a scalar

$t_- \equiv \max(0, -t)$ the nonpositive part of a scalar

Spaces

$\Re^n$ real n-dimensional space

$\Re$ the real line

$\Re^{n \times m}$ the space of $n \times m$ real matrices

$\Re^n_+$ the nonnegative orthant of $\Re^n$

$\Re^n_{++}$ the positive orthant of $\Re^n$

Vectors

z^T the transpose of a vector z

$\{z^k\}$ a sequence of vectors $z^1, z^2, z^3, \ldots$

$x^T y$ the standard inner product of vectors in $\Re^n$

$\|x\| \equiv \sqrt{x^T x}$ the Euclidean norm of a vector $x \in \Re^n$

$x \geq y$ the (usual) partial ordering: $x_i \geq y_i, i = 1, \ldots n$

$x > y$ the strict ordering: $x_i > y_i, i = 1, \ldots, n$

$\min(x, y)$ the vector whose i-th component is $\min(x_i, y_i)$

$\max(x, y)$ the vector whose i-th component is $\max(x_i, y_i)$

$x \circ y \equiv (x_i y_i)$ the Hadamard product of x and y

$x \perp y$ x and y are perpendicular

$z^+ \equiv \max(0, z)$ the nonnegative part of a vector z

$z^- \equiv \max(0, -z)$ the nonpositive part of a vector z

Matrices

$\det A$	the determinant of a matrix A
A^{-1}	the inverse of a matrix A
$\|A\|$	the Euclidean norm of a matrix A
A^T	the transpose of a matrix A
A_α	the columns of A indexed by α
$A_{\alpha\cdot}$	the rows of A indexed by α
I	the identity matrix of appropriate order
I_k	the identity matrix of order k
$\mathrm{diag}(a)$	the diagonal matrix with diagonal elements equal to the components of the vector a

Functions

$f : \mathcal{D} \to \mathcal{R}$	a mapping with domain $\mathcal{D}$ and range $\mathcal{R}$
$f \circ g$	composition of two functions f and g
∇f	$(\partial f_i / \partial x_j)$, the $m \times n$ Jacobian of a mapping $f : \Re^n \to \Re^m$ $(m \geq 2)$
$\nabla_\beta f_\alpha$	$(\partial f_i / \partial x_j)_{i \in \alpha}^{j \in \beta}$, a submatrix of ∇f
$\nabla \theta$	$(\partial \theta / \partial x_j)$, the gradient of a function $\theta : \Re^n \to \Re$
$\nabla_y g(x, y)$	the partial Jacobian matrix of g with respect to y
$\nabla^2 \theta$	Hessian matrix of the scalar-valued function θ
$f'(\cdot\,;\,\cdot)$	directional derivative of the mapping f
f^{-1}	the inverse of f
$o(t)$	any function such that $\lim_{t \to 0} \frac{o(t)}{t} = 0$
$\Pi_K(x)$	the Euclidean projection of x on the set K
$\inf f(x)$	the infimum of the function f
$\sup f(x)$	the supremum of the function f
$\mathrm{dist}(x, W)$	distance function from vector x to set W
F_C	normal map associated with function F and set C

Sets

$\in$	element membership
$\notin$	not an element of
$\emptyset$	the empty set
$\subseteq$	set inclusion
$\subset$	proper set inclusion
$\cup, \cap, \times$	union, intersection, Cartesian product
$\prod S_i$	Cartesian product of sets S_i
$S_1 \setminus S_2$	the difference of sets S_1 and S_2
$|S|$	the cardinality of a finite set S
∂S	the (topological) boundary of a set S
$\mathrm{cl}\, S$	the (topological) closure of a set S
S^*	the dual cone of S
$0^+ S$	the cone of recession directions of S
$\mathrm{Gr}(\mathcal{A})$	the graph of a multifunction $\mathcal{A}$
$\mathrm{dom}(\mathcal{A})$	the domain of a multifunction $\mathcal{A}$
$\mathbb{B}(x, \delta)$	the closed ball with center at x with radius δ
$\mathrm{argmin}_x f(x)$	the set of x attaining the minimum of the real-valued function $f(x)$
$\mathrm{argmax}_x f(x)$	the set of x attaining the maximum of the real-valued function $f(x)$
$\mathrm{supp}(x)$	the support of vector x
$\mathcal{T}(x; S)$	tangent cone of set S at point $x \in S$
$\mathcal{C}(x; S)$	critical cone of set S at point $x \in S$ relative to an objective function
$\mathcal{N}(x; S)$	normal cone of set S at point $x \in S$
$[a, b]$	a closed interval in $\Re$
(a, b)	an open interval in $\Re$
$x^\perp$	the orthogonal complement of vector x

Problems

AVI (q, M, K)	AVI defined by vector q, matrix M and set K
LCP (q, M)	LCP defined by vector q and matrix M
SOL(F, K)	solution set of the VI (F, K)
SOL(q, M, K)	solution set of the AVI (q, M, K)
VI (F, K)	VI defined by mapping F and set K

MPEC symbols

$v \equiv (\zeta, \pi, \eta)$ MPEC multipliers

$w \equiv (x, y, \lambda)$ variable of MPEC in KKT form, $\lambda \in M(x, y)$

$y(x)$ implicit solution function of lower-level VI

$z \equiv (x, y)$ original MPEC variable

$\mathcal{B}(\bar{x})$ SCOC family of active index sets that define the C^1 pieces of $y(x)$ at $\bar{x}$

$\mathcal{C}(z; \mathcal{F})$ critical cone of MPEC at $z \in \mathcal{F}$ relative to the objective function f
$\equiv \bigcup_{\lambda \in M(z)} \mathcal{C}(z, \lambda)$ under full MPEC CQ

$\mathcal{C}(z, \lambda)$ a piece of $\mathcal{C}(z; \mathcal{F})$ corresponding to $\lambda \in M(z)$
$\equiv \mathcal{T}(z; Z) \cap \mathrm{Gr}(\mathcal{LS}_{(z,\lambda)}) \cap \nabla f(z)^{\perp}$

$\mathcal{F}$ MPEC's feasible region given by $Z \cap \mathrm{Gr}(\mathcal{S})$

$\mathcal{F}^{\mathrm{KKT}}$ feasible region of MPEC in KKT form

$\mathcal{I}(x, y)$ set of active indices at $(x, y) \in \mathrm{Gr}(\mathcal{S})$
$\equiv \{i : g_i(x, y) = 0\}$

$\mathcal{I}_0(x, y, \lambda)$ degenerate index set at $(x, y, \lambda) \in \mathcal{F}^{\mathrm{KKT}}$
$\equiv \{i : \lambda_i = g_i(x, y) = 0\}$

$\mathcal{I}_+(x, y, \lambda)$ nondegenerate index set at $(x, y, \lambda) \in \mathcal{F}^{\mathrm{KKT}}$
$\equiv \{i : \lambda_i > g_i(x, y) = 0\}$

$\mathcal{K}(z, \lambda)$ lifted critical cone at $(z, \lambda) \in \mathcal{F}^{\mathrm{KKT}}$

$\mathcal{K}(z, \lambda; dx)$ directional critical set at $(z, \lambda) \in \mathcal{F}^{\mathrm{KKT}}$ along direction dx

$L(x, y, \lambda)$ Lagrangean function for lower-level VI
$\equiv F(x, y) + \sum_{i=1}^{\ell} \lambda_i \nabla_y g_i(x, y)$

$\mathcal{L}(z; \mathcal{F})$ MPEC linearized cone at $z \in \mathcal{F}$
$\equiv \mathcal{T}(z, Z) \cap \left(\bigcup_{\lambda \in M(z)} \mathrm{Gr}(\mathcal{LS}_{(z,\lambda)}) \right)$

$\mathcal{L}^{\mathrm{MPEC}}(w, \zeta, \pi, \eta)$ MPEC Lagrangean function

$\mathcal{LS}_{(z,\lambda)}$ linearized solution map at $(z, \lambda) \in \mathcal{F}^{\mathrm{KKT}}$ for lower-level VI; $\mathcal{LS}_{(z,\lambda)}(dx)$ is defined as $\mathrm{SOL}(\nabla_x L(z, \lambda)dx, \nabla_y L(z, \lambda), \mathcal{K}(z, \lambda, dx))$

$\mathcal{LS}^{\mathrm{KKT}}_{(z,\lambda)}(dx)$ set of KKT pairs $(dy, d\lambda)$ of the AVI $(\nabla_x L(z, \lambda)dx, \nabla_y L(z, \lambda), \mathcal{K}(z, \lambda; dx))$

MPEC symbols
(continued)

$M(x,y)$	set of KKT multipliers of VI $(F(x,\cdot),C(x))$ at solution y
$M^c(z;dx)$	set of critical multipliers at $z \in \mathcal{F}$ along direction dx $\equiv \{\lambda \in M(z) : \mathcal{K}(z,\lambda;dx) \neq \emptyset\}$
$M^e(x,y)$	set of extreme points of $M(x,y)$
$\mathcal{S}(x)$	set of rational reactions of lower-level VI $\equiv \mathrm{SOL}(F(x,\cdot),C(x))$
Z	upper-level feasible region of (x,y)

Preface

This monograph deals with a class of constrained optimization problems which we call *Mathematical Programs with Equilibrium Constraints*, or simply, MPECs. Briefly, an MPEC is an optimization problem in which the essential constraints are defined by a parametric variational inequality or complementarity system. The terminology, MPEC, is believed to have been coined in [108]; the word "equilibrium" is adopted because the variational inequality constraints of the MPEC typically model certain equilibrium phenomena that arise from engineering and economic applications. The class of MPECs is an extension of the class of *bilevel programs*, also known as mathematical programs with optimization constraints, which was introduced in the operations research literature in the early 1970s by Bracken and McGill in a series of papers [34, 36, 37]. The MPEC is closely related to the economic problem of Stackelberg game [265] the origin of which predates the work of Bracken and McGill.

Our motivation for writing this monograph on MPEC stems from the practical significance of this class of mathematical programs and the lack of a solid basis for the treatment of these problems. Although there is a substantial amount of previous research on special cases of MPEC, no existing work provides such generality, depth, and rigor as the present study. Our intention in this monograph is to establish a sound foundation for MPEC that we hope will inspire further applications and research on this important problem.

This monograph consists of six chapters. Chapter **1** defines the MPEC, gives a brief description of several source problems, and presents various equivalent formulations of the equilibrium constraints in MPEC; the chapter concludes with some results of existence of optimal solutions. Chapter **2** presents an extensive theory of exact penalty functions for MPEC, using the theory of error bounds for inequality systems. This chapter ends with a brief discussion of how some exact penalty functions formulations of MPEC can be employed to obtain first-order optimality conditions; the latter topic and its extensions are treated in full in the next three chapters. Specifically, Chapter **3** presents the fundamental first-order optimality (i.e., stationarity) conditions of MPEC; Chapter **4** verifies in detail the hypotheses needed for the first-order conditions; Chapter **5** contains results on second-order optimality conditions. The sixth and last chapter presents several algorithms for solving MPECs including an interior point algorithm for MPECs with "monotone" inner problems, a conceptual iterative descent algorithm based on an implicit programming approach, and a locally superlinearly convergent Newton type (sequential quadratic programming) method based on a piecewise programming approach. Some preliminary computational results are reported. The monograph ends with an extensive list of references.

Due to the intrinsic complexity of the MPEC, a comprehensive study of this problem would inevitably require extensive tools from diverse disciplines. Besides a general knowledge of smooth (nonlinear) programming and multivariate analysis, which we assume as prerequisites for this work, such subjects as error bound theory for inequality systems, sensitivity and stability theory for parametric variational inequalities, piecewise smooth analysis, nonsmooth equations, the family of interior point methods, and some basic iterative descent methods for nonlinear programs are all important tools that will be used in this monograph. Since it is not possible for us to review in detail all the background material and keep the monograph within a reasonable length, we have chosen not to organize the preliminary results separately. Instead, we have included only the most useful background results relevant to the topics of discussion.

Throughout the monograph, we have taken several different points of view toward the MPEC, each of which is interesting by itself. Many results obtained herein are new and have not appeared in the literature before. For related approaches and results, we refer to [201, 214, 291, 292, 295]; see also the references in [5, 278].

The general MPEC is a highly nonconvex, nondifferentiable optimization problem that encompasses certain combinatorial features in its constraints. As such, it is computationally very difficult to solve, especially if one wishes to compute a globally optimal solution. Partly due to this pessimistic view, we have not attempted in this monograph to deal with the issue of finding a globally optimal solution to the general problem itself or to its special cases. The algorithms discussed in Chapter **6** are iterative schemes for computing a stationary point of the MPEC (and under mild conditions, a strict local minimum). We refer to [278] for references that discuss some global optimization approaches to solving bilevel programs.

Due to the broad applications of MPEC, this monograph is of interest to readers from diverse disciplines. In particular, operations researchers, economists, design and systems engineers, and applied mathematicians will likely find the subject matter interesting and challenging. We have written the monograph with these individuals in mind.

Acknowledgements

This monograph grew out of a research article [172] authored by us and Professor Shi-Quan Wu of the Academia Sinica in Beijing. We are indebted to Professor Wu's general interest in this subject which has provided the initial stimulus for us to undertake the project. We are also grateful to several funding agencies which have supported our research in the general optimization area. In particular, Tom Luo thankfully acknowledges the Natural Sciences and Engineering Research Council of Canada. Jong-Shi Pang gratefully acknowledges the generous support of the National Science Foundation (U.S.A.) over many years for his research in the general area of variational inequalities and complementarity problems; he is also grateful to the Office of Naval Research for its support on the work in the interior point methods. Danny Ralph wishes to thank the Australian Research Council for its support on this project.

During the preparation of the manuscript, we have presented its contents in several international conferences and received many valuable comments from colleagues. In particular, we have benefited from discussions with Professors Masao Fukushima, Olvi Mangasarian, Jiri Outrata, and Steve Robinson, and Drs. Yang Chen, Eric Ralph, and Houyuan Jiang. Dr. Alan Harvey from the Cambridge University Press has been very encouraging and supportive of this project. We thank them all.

The text of this monograph was typeset by the authors using Leslie Lamport's LaTeX, a document preparation system based on Donald Knuth's

TEX program. We have used the document style files of the book [54] that
were prepared by Richard Cottle and Richard Stone, based on the LATEX
book style.

Zhi-Quan (Tom) Luo,   McMaster, Ontario, Canada
Jong-Shi Pang,   Baltimore, Maryland, U.S.A.
Daniel Ralph,   Melbourne, Victoria, Australia

August 19, 1996

———

1

Introduction

This chapter introduces the main topic of this monograph, that is, the mathematical program with equilibrium constraints. Source problems from engineering and economics are described to justify the need for a thorough study of this important class of optimization problems. The rest of the chapter gives several useful formulations of the equilibrium constraints and presents sufficient conditions for the existence of optimal solutions to these problems.

1.1 Problem Formulation

A Mathematical Program with Equilibrium Constraints (MPEC) is an optimization problem with two sets of variables, $x \in \Re^n$ and $y \in \Re^m$, in which some or all of its constraints are defined by a parametric variational inequality or complementarity system with y as its primary variables and x the parameter vector. More specifically, this problem is defined as follows. Suppose that $f : \Re^{n+m} \to \Re$ and $F : \Re^{n+m} \to \Re^m$ are given functions, $Z \subseteq \Re^{n+m}$ is a nonempty closed set, and $C : \Re^n \to \Re^m$ is a set-valued map with (possibly empty) closed convex values; i.e., for each $x \in \Re^n$, $C(x)$ is a (possibly empty) closed convex subset of $\Re^m$. The set of all vectors $x \in \Re^n$

1

for which $C(x) \neq \emptyset$ is the *domain* of C and denoted $\text{dom}(C)$. Let X be the projection of Z onto $\Re^n$; i.e.,

$$X = \{x \in \Re^n : (x,y) \in Z \text{ for some } y \in \Re^m\}.$$

The function f is the overall objective function to be minimized; F is the *equilibrium* function of the *inner* problem, Z is a joint upper-level feasible region of the pair (x, y), and $C(x)$ defines the restriction of the variable y for each given $x \in X$. We shall make the blanket assumption that $X \subseteq \text{dom}(C)$. With this setup, the MPEC is:

$$
\begin{aligned}
\text{minimize} \quad & f(x,y) \\
\text{subject to} \quad & (x,y) \in Z, \text{ and} \\
& y \in S(x),
\end{aligned}
\tag{1}
$$

where for each $x \in X$, $S(x)$ is the solution set of the variational inequality (VI) defined by the pair $(F(x, \cdot), C(x))$; i.e., $y \in S(x)$ if and only if y is in $C(x)$ and satisfies the inequality:

$$(v - y)^T F(x,y) \geq 0, \quad \text{for all } v \in C(x).$$

In general, the graph of a set-valued map $\mathcal{A} : \Re^n \to \Re^m$ will be denoted $\text{Gr}(\mathcal{A})$; thus

$$\text{Gr}(\mathcal{A}) = \{(x,y) \in \Re^{n+m} : y \in \mathcal{A}(x)\}.$$

By considering the solution map S as a set-valued map from $\Re^n$ into $\Re^m$, we may write the problem (1) in the compact form:

$$
\begin{aligned}
\text{minimize} \quad & f(x,y) \\
\text{subject to} \quad & (x,y) \in Z \cap \text{Gr}(S).
\end{aligned}
$$

Let

$$\mathcal{F} \equiv Z \cap \text{Gr}(S) \tag{2}$$

denote the feasible region of (1). Throughout this monograph, we shall make the blanket assumption that this region is nonempty. We refer the reader to [220] for a comprehensive review of the VI and related problems and to [7] for the fundamental theory of set-valued maps.

The term "equilibrium constraints" in MPEC refers to the variational inequality constraint $y \in \mathcal{S}(x)$. Our usage of the word "equilibrium" reflects the fact that we are particularly interested in the case of MPEC where its constraints represent certain equilibrium applications in engineering and economics that are modeled by variational inequalities.

The formulation (1) of MPEC is very broad and encompasses a large number of interesting special cases. Foremost among these is the case where the mapping $F(x, \cdot)$ is the partial gradient map (with respect to the second argument) of a real-valued C^1 function $\theta : \Re^{n+m} \to \Re$; i.e., $F(x,y) = \nabla_y \theta(x,y)$ for all $(x,y) \in \text{Gr}(C)$ where ∇_y denotes the partial F(réchet)-differentiation with respect to the variable y. In this case, the VI $(F(x,\cdot), C(x))$ is, for each fixed $x \in X$, the set of stationarity conditions of the following optimization problem in the variable y:

$$\begin{aligned} \text{minimize} \quad & \theta(x,y) \\ \text{subject to} \quad & y \in C(x). \end{aligned} \tag{3}$$

This special case of the MPEC has traditionally been known as the *bilevel program* with (3) called its *inner program* or *lower-level program* for an obvious reason. In general, we shall use "argmin" to denote the optimal solution set of a minimization problem. Thus for a given vector x, $\text{argmin}\{\theta(x,y) : y \in C(x)\}$ denotes the optimal solution set of (3). For a convex set $C(x)$, we have

$$\text{argmin}\{\theta(x,y) : y \in C(x)\} \subseteq \mathcal{S}(x),$$

where $\mathcal{S}(x)$ is the solution set of the VI $(\nabla_y \theta(x,y), C(x))$; moreover, equality holds if in addition $\theta(x, \cdot)$ is convex in the second argument.

The MPEC (1) is a generalization of a bilevel program in which the inner problems are VIs. A bilevel program is in turn a special case of a hierarchical mathematical program which consists of multiple (possibly more than two) levels of optimization. Such multi-level mathematical programs have proven very useful in the modeling of hierarchical decision making processes and in the optimization of engineering designs.

A simple example of a two-level decision making process is as follows. Consider an economic planning process which involves several interacting agents (or individuals). Some agents, collectively called a *leader* or a *principal*, act as superiors who issue directives to the remaining agents,

collectively called a *follower* or simply an agent, who act as subordinates
to the leader. The leader's directives are described by the variable x and
the follower's decision variables are contained in the vector y. The variational inequality constraint $y \in \mathcal{S}(x)$ stipulates that for each of the leader's
directives x, the follower will choose a response vector y which is a solution
of a decision making problem modeled by the VI $(F(x, \cdot), C(x))$ that depends on x. Based on such *rational* responses from the subordinates, the
overall economic planning problem is to determine an optimal vector of the
leader's directives x^{opt} along with an equilibrium vector of the follower's
responses $y^{\mathrm{equ}} \in \mathcal{S}(x^{\mathrm{opt}})$ in order to minimize an economic performance
function modeled by $f(x, y)$ subject to the additional joint feasibility con-
dition $(x, y) \in Z$. With this economic interpretation, the solution set $\mathcal{S}(x)$
for the inner problem is sometimes called the set of *rational reactions* corresponding to x, the solution map $\mathcal{S}$ is called the *reaction map*, and the
graph $\mathrm{Gr}(\mathcal{S})$ is called the *rational set*. The objective function $f(x, y)$ and
the joint feasible region Z can be used to model additional anticipation of
the principal toward the (subordinate) agent's behavior. We will discuss
this modeling issue further in the context of the Stackelberg game; see
Section **1.2**.

Separately, in the modeling of many engineering design problems as
an MPEC, the first-level vector x typically contains the design variables
of an engineering process and the second-level vector y contains the state
variables of the system; each inner VI $(F(x, \cdot), C(x))$ corresponds to either
an optimization or equilibrium problem for a given admissible design x. The
overall optimization problem (1) is to determine an optimal pair of design
and state variables that will minimize the cost function $f(x, y)$ subject to
the joint feasibility condition $(x, y) \in Z$ and the design constraint $y \in \mathcal{S}(x)$.
Several design problems of this type will be presented in Section **1.2**.

An important special case of the MPEC (1) is where $C(x)$ is a convex
cone in $\Re^m$ for all $x \in X$. In this case, it is known from the theory of
variational inequalities [109] that the VI $(F(x, \cdot), C(x))$ is equivalent to a
generalized complementarity problem over the cone $C(x)$:

$$y \in C(x), \quad F(x, y) \in C(x)^*, \quad y^T F(x, y) = 0, \qquad (4)$$

where for an arbitrary subset S in an Euclidean space $\Re^N$,

$$S^* \equiv \{ z \in \Re^N : z^T v \ge 0 \text{ for all } v \in S \} \qquad (5)$$

is the *dual cone* of S. The case $C(x) = \Re^{m_1} \times \Re_+^{m_2}$ for some nonnegative integers m_1 and m_2 such that $m_1 + m_2 = m$ is particularly interesting. In this case, the vectors y and $F(x, y)$ can be partitioned into

$$y = \begin{pmatrix} y_1 \\ y_2 \end{pmatrix}, \quad F(x, y) = \begin{pmatrix} F_1(x, y) \\ F_2(x, y) \end{pmatrix},$$

where y_1, $F_1(x, y) \in \Re^{m_1}$ and y_2, $F_2(x, y) \in \Re^{m_2}$, and the problem (4) becomes

$$F_1(x, y) = 0,$$

$$(y_2, F_2(x, y)) \geq 0, \quad (y_2)^T F_2(x, y) = 0,$$

which is a mixed, nonlinear complementarity problem. When $m_1 = 0$, the latter problem is a standard nonlinear complementarity problem (NCP).

The LCP (linear complementarity problem) constrained MP (mathematical program) is a special case of the NCP constrained MP in which the function F is linear. The following mathematical program:

$$\begin{array}{ll} \text{minimize} & d^T x + e^T u + f^T v \\ \text{subject to} & Ax + Bu + Cv \geq g, \\ & (u, v) \geq 0, \quad u^T v = 0, \end{array} \quad (6)$$

whose constraints are in the form of a (nonstandard) linear complementarity problem, parametrized by x, has been called a *(linear) complementary program* in [118] and an LPEC in [187]. In turn, this complementary program is a special instance of a *disjunctive program* [14] which is an optimization problem with disjunctive (i.e., "or") constraints. To see that (6) is a disjunctive program, we note that the nonnegativity and complementarity constraint of this problem, $(u, v) \geq 0$, $u^T v = 0$, is equivalent to

$$(u, v) \geq 0, \quad \forall i \ (u_i = 0 \text{ or } v_i = 0),$$

which involves the disjunction "or". In essence, the NCP constrained MP, and more generally, the general MPEC (1) where the inner VIs are formulated in terms of their Karush-Kuhn-Tucker (KKT) conditions (see Subsection **1.3.2**), are special instances of a nonlinear, disjunctive program.

A special case of the linear complementary program is a linear (mixed) integer program in $\{0, 1\}$ variables. Due to this connection, the early study of the former program is closely tied to integer programming; in particular, cutting plane methods [121] and facial techniques [14] have been proposed for solving this problem.

The complexity of MPEC

The computational complexity of a (linear) disjunctive program is well known in the integer programming literature [14]. In essence, this complexity is caused by the disjunctive constraints which lead to some challenging combinatorial issues that typically are the main concern in the design of efficient solution algorithms. As these disjunctive constraints are also present implicitly in a general MPEC, the latter problem can be expected to be quite difficult.

Indeed, the general MPEC (1) is an extremely difficult optimization problem. Besides the intrinsic combinatorial curse of the constraints, the difficulty arises from several other sources that are more akin to this problem considered as a nonlinear program. One such source is the potential lack of convexity and/or closedness of the feasible region $\mathcal{F}$. The special case of the MPEC in which the inner problems are linear programs can be used to elucidate the lack of these useful properties. The following simple numerical example illustrates the possible nonconvexity of $\mathcal{F}$.

1.1.1 Example. Consider the bilevel program in $\Re^2$:

$$\text{minimize} \quad f(x,y)$$
$$\text{subject to} \quad x \geq 0,$$
$$\text{and} \quad y \in \text{argmin}\{y \; : \; y \in C(x)\},$$

where

$$C(x) \equiv \{y \in \Re_+ \; : \; x + 2y \geq 10, \; x - 2y \leq 6,$$
$$2x - y \leq 21, \; x + 2y \leq 38, \; -x + 2y \leq 18\}.$$

For $x \geq 0$, we have $C(x) \neq \emptyset$ if and only if $x \leq 16$. Since each inner problem is a linear program, we can solve for the optimal y for each given $x \in [0, 16]$, obtaining

$$\mathcal{S}(x) = \begin{cases} 5 - x/2 & \text{if } x \in [0,8] \\ -3 + x/2 & \text{if } x \in [8,12] \\ -21 + 2x & \text{if } x \in [12,16]. \end{cases}$$

The feasible region $\mathcal{F}$, which is equal to

$$\{(x, 5 - x/2) : x \in [0,8]\} \cup \{(x, -3 + x/2) : x \in [8,12]\} \cup$$
$$\{(x, -21 + 2x) : x \in [12, 16]\}$$

is the union of three noncollinear line segments in the plane $\Re^2$. Thus $\mathcal{F}$ is nonconvex.

In the above example, the set $\mathcal{S}(x)$ is a singleton for each x. However, this solution function also illustrates another difficulty with the MPEC in general: namely, the function $\mathcal{S}(x)$ (in the single-valued case) is in general not Fréchet differentiable. Thus nonsmoothness is also an intrinsic feature of an MPEC.

Though not convex, the region $\mathcal{F}$ in Example **1.1.1** is at least a closed connected set. Next we give an example to show that this set $\mathcal{F}$ could be disconnected and not closed.

1.1.2 Example. Consider the bilevel program in $\Re^2$:

$$\text{minimize} \quad f(x, y)$$
$$\text{subject to} \quad |x| \leq 1,$$
$$\text{and} \quad y \in \text{argmin}\{y : |y| \leq 1, xy \leq 0\}.$$

In the notation of the MPEC (1), we have $F(x, y) = 1$,

$$Z = \{(x, y) \in \Re^2 : |x| \leq 1\},$$

and

$$C(x) = \{y \in \Re : |y| \leq 1, xy \leq 0\}$$

is convex for each fixed x. It is not difficult to verify that

$$\mathcal{S}(x) = \begin{cases} \{-1\} & \text{if } x \in [0, 1] \\ \{0\} & \text{if } x \in [-1, 0). \end{cases}$$

Thus we have

$$\mathcal{F} = \{(x, -1) : x \in [0, 1]\} \cup \{(x, 0) : x \in [-1, 0)\}.$$

Clearly $\mathcal{F}$ is not closed.

The lack of closedness of $\mathcal{F}$ renders the MPEC more or less intractable. Subsequently, we shall impose some mild assumptions on the inner VIs that will allow us to circumvent this difficulty and focus on the case where $\mathcal{F}$ is indeed closed. Under these assumptions (that are satisfied by Example **1.1.1** but not by **1.1.2**), the feasible region $\mathcal{F}$ can be shown to be the union of finitely many closed sets. This structure of $\mathcal{F}$ brings out a combinatorial nature of the MPEC that adds to the difficulty of this problem. Indeed, the number of sets that constitute $\mathcal{F}$ could in general be large; they are the result of a complementarity condition implicit within the VIs.

Another difficulty with the MPEC is the multi-valued nature of the solution function $\mathcal{S}(x)$. This is illustrated by the following example.

1.1.3 Example. Consider the VI $(F(x,\cdot), C(x))$, where for all $(x,y) \in \Re^2$,

$$F(x,y) \equiv -y, \quad C(x) \equiv \{y \in \Re : |y| \le 1\}.$$

It can be verified that for all $x \in \Re$, $\mathcal{S}(x) = \{1, -1\}$ which is a discrete set.

A bilevel linear program is a special case of the MPEC in which f is a linear function in (x,y), Z is a polyhedron, F is a constant, and $C(x)$ is a polyhedron of special type:

$$
\begin{array}{lll}
\text{minimize} & c^T x + d^T y \\
\text{subject to} & A_1 x + A_2 y \ge a, & \qquad (7) \\
\text{and} & y \in \text{argmin}\{q^T y : B_1 x + B_2 y \ge b\},
\end{array}
$$

where the vectors a, b, c, d and matrices A_1, A_2, B_1, B_2 are of appropriate dimensions. In the notation of (1.1.1), we have

$$f(x,y) \equiv c^T x + d^T y, \quad Z \equiv \{(x,y) \in \Re^{n+m} : A_1 x + A_2 y \ge a\},$$

$$F(x,y) \equiv q, \qquad C(x) \equiv \{y \in \Re^m : B_1 x + B_2 y \ge b\}.$$

It has been shown [122, 106] that the bilevel linear program belongs to the class of strongly NP-hard problems. (See [92] for an introduction to the theory of computational complexity and the definitions for various complexity classes of problems, such as that of P, NP, and strong NP.) This implies that there can be no fully polynomial approximation scheme for solving (7) unless the classes P and NP are equal. (Roughly speaking, a fully polynomial approximation scheme is an algorithm for computing an "ε-optimal" solution to a given problem with running time which is a polynomial in terms of the problem size and $1/\varepsilon$.) In spite of its hardness, the bilevel linear program has been well researched and there are many algorithms of the enumerative, branch-and-bound, exact-penalty, decomposition type for solving this problem [23, 24, 25, 106, 129, 279, 284].

The intrinsic difficulties of MPEC are unfortunate since this optimization problem has a wide range of applications in engineering and economics (the next section outlines several of these applications). Partly due to these difficulties, many studies of bilevel programs in the past have not been based on very sound principles and are full of loose arguments and heuristic approaches. Added to the complication is the fact that some of the early results reported in the literature are in fact incorrect. To illustrate, the reference [50] gave a counterexample to demonstrate that the necessary optimality conditions for the bilevel programming problem obtained in [18] were not correct; the reference [20] gave examples to show that several known methods claimed by their authors to always yield a globally optimal solution of a bilevel linear program were flawed. MPEC, being defined formally only recently [108], deserves to be given a comprehensive investigation and put on a solid, rigorous ground. The present monograph is written with this as its main objective.

1.2 Source Problems

Although the origin of the MPEC can be traced to the economic notion of a Stackelberg game [265], in the operations research literature mathematical programming problems with optimization constraints, i.e., bilevel programs, were introduced in a series of papers by Bracken and McGill [34, 36, 37, 35, 38]. Applications of these programs to military defense and production and marketing decision making in a competitive environment were also discussed in these references. As we shall see, the Bracken-McGill bilevel programs are considerably easier than the general MPEC. In the Ph.D. thesis [61], de Silva discussed the application of "an implicitly defined optimization model" to U.S. crude oil production. These and other early applications of MPEC are mostly concerned with bilevel programs where the inner problems are optimization problems. Along with the advance of the theory and methods for variational inequalities and complementarity problems [109] comes the gradual broadening of MPEC's applications to equilibrium modeling. The term "mathematical programs with equilibrium constraints" was coined in [108].

The volume [4] contains a number of interesting articles describing the diverse applications of hierarchical optimization in engineering and economics. In the next few subsections, we discuss some selective applications of the MPEC.

The Bracken-McGill bilevel programs

For historical reasons, we begin our discussion of the applications of MPEC with the earliest models proposed by Bracken and McGill. Their first few papers addressed bilevel programming models of minimum-cost weapon mix and other defense problems; in what follows, we present their optimal production and marketing decision making model published in [38].

Consider a firm which produces several products labeled $i = 1, \ldots, m$ using a number of different resources labeled $j = 1, \ldots, n$. The firm wishes to maximize profit subject to resource and market share constraints. The products are manufactured within resource availabilities, and certain minimum market shares must be maintained in the face of competition from other firms.

We introduce some notation. Let x_i and y_i denote, respectively, the firm's production and marketing level of the i-th product; let $x \equiv (x_i)$ and $y \equiv (y_i)$ be the n-vectors of production and marketing levels respectively. The real-valued function $g_i(x, y, u, v)$ expresses the firm's market share for product i given the values of x, y, u and v, where u and v are vectors denoting the competitors' production and marketing levels of all the products. The resource utilization function $h_j(x, y)$ specifies the amount of resource j that is required for production level x and marketing level y. The minimum fraction of market share for product i that is required by the firm is denoted a_i, and the total amount of resource j available to the firm is denoted b_j. Finally, let W denote the set of all feasible production and marketing levels for the competitors.

For a given level (x, y) of production and marketing, the firm's minimum market share function for the i-th product in the face of competition is given by

$$\sigma_i(x, y) \equiv \min\{g_i(x, y, u, v) : (u, v) \in W\}.$$

This function σ_i is in general not Fréchet differentiable, regardless of how smooth g_i is. The firm's optimal production and marketing strategy can be obtained by solving the following optimization problem in the variables (x, y):

$$\begin{aligned}
\text{maximize} \quad & f(x, y) \\
\text{subject to} \quad & h_j(x, y) \leq b_j, \quad j = 1, \ldots, n, \\
\text{and} \quad & \sigma_i(x, y) \geq a_i, \quad i = 1, \ldots, m.
\end{aligned} \tag{1}$$

In other words, the firm chooses a strategy (x, y) to maximize its total profit, subject to the minimum specified level of market share a_i for each product i, and to the resource limitations. In this formulation, the firm behaves in a rather conservative manner: it takes into account the worst-case scenario on the part of the competitors in order to ensure its desired market share. This approach is related to the weak Stackelberg game which will be described shortly.

The above problem can be put into the form of a bilevel program in the variables x, y, and $\{(u^i, v^i)\}_{i=1}^m$, where the new (lower-level) variables (u^i, v^i) are minimizers of the function $\sigma_i(x, y)$. More precisely, the problem (1) is equivalent to

$$\begin{aligned}
\text{maximize} \quad & f(x, y) \\
\text{subject to} \quad & h_j(x, y) \leq b_j, \quad j = 1, \ldots, n, \\
& g_i(x, y, u^i, v^i) \geq a_i, \quad i = 1, \ldots, m; \\
\text{and} \quad & (u^i, v^i) \in \text{argmin}\{g_i(x, y, u, v) : (u, v) \in W\}.
\end{aligned} \tag{2}$$

Notice that the lower-level variables $\{(u^i, v^i)\}_{i=1}^m$ do not appear in the upper-level objective function f. Moreover, under the assumption that $g_i(x, y, u^i, v^i)$ is concave in (x, y) for fixed (u^i, v^i) and W is a convex set, the function $\sigma_i(x, y)$ is concave in (x, y); if in addition $f(x, y)$ and h_j are concave in (x, y), then (1) is a concave maximization program. Thus the bilevel program (2) is considerably easier than the general MPEC (1.1.1) which is not expected to possess much convexity or concavity property.

Stackelberg game

The MPEC is intimately related to the so-called leader-follower (or Stackelberg) game [265, 13]. This game problem has been studied extensively by economists and has found wide application in such areas as oligopolistic market analysis [210, 263], optimal product design [48], quality control in services [6], and pricing of electric transmission [114]. The usage of the terms "leader" and "follower" in our introduction of the MPEC was derived from this Stackelberg game problem; see Section **1.1**.

The Stackelberg game can be considered an extension of the renowned Nash game [207]. In the Nash game, there are a number of (say, M) players each of whom has a strategy set $Y_i \subseteq \Re^{m_i}$. The objective of player i is to minimize its economic cost $\theta_i(y_i, y_{\neq i}^{\text{given}})$ by selecting a strategy $y_i \in Y_i$ given that the other players have chosen their strategies $y_{\neq i}^{\text{given}}$, where $y_{\neq i}^{\text{given}}$ denotes the vector $(y_j^{\text{given}} \; : \; j \neq i)$. In other words, each player observes the actions of the remaining players and then reacts optimally, assuming that the other players' strategies remain unchanged. A strategy combination $y^* \in \prod_{j=1}^{m} Y_j$ is called a Nash equilibrium if no player has an incentive to deviate from his strategy y_i^* in the sense that

$$y_i^* \in \operatorname{argmin}\{\theta_i(y_i, y_{\neq i}^*) \; : \; y_i \in Y_i\}, \qquad \forall i.$$

It should be noted that the players in the Nash game are in a sense homogeneous since each of them has access to the same information regarding the other players' strategies and the strategy chosen is only dependent on this information.

In contrast, the Stackelberg game has a distinctive player (called the leader) who can *anticipate* the (re)actions of the remaining players (called followers) and use this knowledge in selecting his optimal strategy (see [242]). Specifically, the leader chooses a strategy from the strategy set $X \subseteq \Re^n$, while each follower (say i) has, corresponding to each of the leader's strategies $x \in X$, a strategy set $Y_i(x) \subseteq \Re^{m_i}$ that is closed and convex and a cost function $\theta_i(x, \cdot) : \prod_{j=1}^{M} \Re^{m_j} \to \Re$, where M is the number of followers in the Stackelberg game. Note that each follower's strategy is dependent on the particular strategy x of the leader and this follower's cost function is dependent on both the leader's and all followers' strategies. Let $p \equiv \sum_{i=1}^{M} m_i$. We assume that for any fixed but arbitrary $x^{\text{given}} \in X$ and $y_{\neq i}^{\text{given}} \equiv (y_j^{\text{given}} : j \neq i)$, the function

$$\theta_i(x^{\text{given}}, y_i, y_{\neq i}^{\text{given}}) \tag{3}$$

is convex and continuous differentiable in the variable $y_i \in Y_i(x^{\text{given}})$.

Collectively, the followers behave according to the Nash noncooperative principle described above. That is to say, they will choose, for each $x \in X$, a joint response vector

$$y^{\mathrm{opt}} \equiv (y_i^{\mathrm{opt}})_{i=1}^M \in \prod_{i=1}^M Y_i(x)$$

such that for each $i = 1, \ldots, M$

$$y_i^{\mathrm{opt}} \in \mathrm{argmin}\{\theta_i(x, y_i, y_{\neq i}^{\mathrm{opt}}) \, : \, y_i \in Y_i(x)\}. \tag{4}$$

By the convexity of the payoff functions (3) and the sets $Y_i(x)$, it is easy to show that (4) holds for all $i = 1, \ldots, M$ if and only if the vector y^{opt} is in $\mathrm{SOL}(F(x, \cdot), C(x))$ where for $y \in \Re^p$, $F(x, y) \equiv (F_i(x, y))_{i=1}^M$ with

$$F_i(x, y) \equiv \nabla_{y_i} \theta_i(x, y), \quad i = 1, \ldots, M,$$

and

$$C(x) \equiv \prod_{i=1}^M Y_i(x).$$

Let $f : \Re^{n+p} \to \Re$ be the leader's cost function which depends on both his own and the followers' strategies. The Stackelberg game problem is to determine a vector $(x, y) \in \Re^{n+p}$ in order to

$$
\begin{aligned}
&\text{minimize} \quad f(x, y) \\
&\text{subject to} \quad x \in X, \\
&\text{and} \quad\quad\quad y \in \mathcal{S}(x).
\end{aligned}
\tag{5}
$$

This is an MPEC.

As we can see, in the Stackelberg game the leader is more "powerful" than the followers in the sense that the leader is allowed to anticipate the reactions of the followers and select his strategy accordingly. Thus, the players of a Stackelberg game are no longer homogeneous as in the case of the Nash game. Recall that in the Nash game there is no distinction among the players since they can only observe but not anticipate the (re)actions of the other players. If the leader loses the advantage of anticipating the other players' reactions then the Stackelberg game reduces to the standard Nash game. The loss of leader's privilege will usually result in an increase in the leader's optimal cost and a decrease in the followers' optimal costs; see [94] for an example. Similarly, if the leader's ability of anticipation does not affect his objective function value (e.g., $f(x,y) \equiv 0$), then the Stackelberg game also reduces to a Nash game.

Refinements and variations of the above Stackelberg model have been proposed and studied extensively by Jacqueline Morgan and her collaborators; see [157, 158, 165, 179]. In the terminology used in these references, our model is a *strong* or *optimistic* Stackelberg game. To understand this terminology, note that the problem (5) is equivalent to

$$
\begin{aligned}
& \text{minimize} \quad f_{\text{strong}}(x) \equiv \inf_{y \in \mathcal{S}(x)} f(x,y) \\
& \text{subject to} \quad x \in X.
\end{aligned}
\tag{6}
$$

Thus the leader is optimistic in the sense that he assumes that the followers will act most favorably to his well-being by choosing, for each of his announced strategies $x \in X$, a reaction $y \in \mathcal{S}(x)$ among the rational reactions that will contribute to the minimization of the cost function f. This situation is to be contrasted with the *weak* or *pessimistic* Stackelberg game in which the leader assumes that the followers will choose their reactions from the rational set that will be least favorable to him; thus the leader will act conservatively to guard against the worst outcome. Mathematically, the leader will solve the following MPEC:

$$
\begin{aligned}
& \text{minimize} \quad f_{\text{weak}}(x) \equiv \sup_{y \in \mathcal{S}(x)} f(x,y) \\
& \text{subject to} \quad x \in X.
\end{aligned}
$$

When the reaction set $\mathcal{S}(x)$ is a singleton for each $x \in X$, there is no distinction between the above two situations. Nevertheless for Stackelberg games with a multivalued reaction map, the weak and strong versions are not necessarily equivalent. In [294], the terminology of "cooperative" and "noncooperative" has been used to mean "strong" and "weak" respectively. However, the former is confusing because a cooperative game normally has a somewhat different (and well established) meaning in game theory [216].

An application of the Stackelberg game model (5) in conjunction with the Cournot production model was discussed in [210, 263, 275]. The paper [48] proposes a Stackelberg game model for new product pricing and positioning in the face of price competition. Other studies on Stackelberg problems from an optimization point of view are [2, 5, 212, 213]. A numerical approach for computing a Stackelberg-Cournot-Nash equilibrium via nondifferentiable optimization is proposed in [215]. Incidentally, the bilevel programs introduced by Bracken and McGill, including the one discussed in the previous subsection, were developed to model a competitive environment similar to the leader-follower game discussed above. The difference is that in the Bracken-McGill model, there was no specification of the followers' (the competitors') behavior; also the way the leader (the firm) handled the followers' responses was different in the two approaches.

Misclassification minimization

Given two point sets $\mathcal{A}$ and $\mathcal{B}$ and a hyperplane H in an n-dimensional real Euclidean space, we nominate one side of H (a closed halfspace) as the A-side and the other as the B-side. We say that H *(linearly) separates* the two sets if the A-side can be chosen to contain all the points in $\mathcal{A}$ and the B-side all the points in $\mathcal{B}$. Such a separation is *strict* if no points of $\mathcal{A} \cup \mathcal{B}$ lie on H. When the two sets $\mathcal{A}$ and $\mathcal{B}$ are not linearly separated by a hyperplane H, then for any nomination of the A-side and B-side, there is some point in $\mathcal{A} \cup \mathcal{B}$ that lies in the wrong side of H (and not on H). Each of these misplaced points is called a *misclassification*. The problem of finding a hyperplane which separates $\mathcal{A}$ and $\mathcal{B}$ with a minimum number of misclassifications is of fundamental importance to the area of machine learning, pattern recognition, and artificial intelligence.

Mangasarian [187] has formulated the above misclassification problem as a bilevel linear program. Below is a brief description of this formulation. Suppose that the two sets of points $\mathcal{A}$, $\mathcal{B}$ are represented by the rows of two matrices A $(m \times n)$ and B $(k \times n)$ respectively, and suppose that the hyperplane H is given by

$$w^T x = \theta,$$

where $w \in \Re^n$ is the normal vector of the hyperplane and θ is a scalar; w and θ are to be determined. Clearly, the plane H separates the two sets $\mathcal{A}$ and $\mathcal{B}$ strictly if and only if

$$Aw > e\theta, \quad \text{and} \quad Bw < e\theta,$$

where e is a vector of ones of appropriate dimensions. By rescaling, the above equations are equivalent to

$$Aw > e\theta + e, \quad \text{and} \quad Bw < e\theta - e,$$

respectively. If the two sets $\mathcal{A}$, $\mathcal{B}$ are not linearly separable, then the last two equations cannot be satisfied by any choice of w and θ. In this case, we can count the total number of misclassifications as

$$c(w, \theta) = e^T(-Aw + e\theta + e)_* + e^T(Bw - e\theta + e)_*,$$

where $()_*$ denotes the step function

$$a_* \equiv \begin{cases} 1 & \text{if } a > 0 \\ 0 & \text{otherwise.} \end{cases}$$

Notice that if $\mathcal{A}$ and $\mathcal{B}$ are linearly separable, then $c(w, \theta) = 0$ for some w and θ. Thus, the misclassification minimization problem is to find (w, θ) in $\Re^{n+1}$ so as to

$$\text{minimize} \quad c(w, \theta). \tag{7}$$

This is a nonsmooth minimization problem; we can remove the nonsmoothness by reformulating the problem as a bilevel linear program. The key in the reformulation is the following observation: for any vector $a \in \Re^m$, $\text{argmin}\{-r^T a : 0 \le r \le e\}$ consists of all vectors $r \in \Re^m$ such that for all $i = 1, \ldots, m$,

$$r_i = \begin{cases} 1 & \text{if } a_i > 0 \\ \in [0, 1] & \text{if } a_i = 0 \\ 0 & \text{if } a_i < 0. \end{cases}$$

Thus by letting $r \equiv (-Aw + e\theta + e)_*$ and $s \equiv e^T(Bw - e\theta + e)_*$, the problem (7) is equivalent to

$$\text{minimize} \quad e^T r + e^T s$$

$$\text{subject to}$$

$$r \in \text{argmin}\{r^T(Aw - e\theta - e) : 0 \le r \le e\},$$

$$s \in \text{argmin}\{s^T(-Bw + e\theta - e) : 0 \le s \le e\},$$

which is a bilevel linear program with (w, θ) as the variables in the upper level and (r, s) as the variables in the lower level.

The reference [187] also suggested an iterative algorithm of the Frank-Wolfe type to solve the above misclassification problem, and some interesting numerical results were obtained. As a bilevel linear program, a host of other algorithms are also applicable; see [23, 24, 25, 129] and the references therein for algorithms of various kinds for solving general bilevel linear programs. Extensions of the misclassification problem are discussed in [39, 189].

Robotics

MPEC has recently been proposed as an approach to deal with the possible inconsistency of a certain motion planning model for robot hands (see Pang and Trinkle [228]). Specifically, consider the situation whereby a robot is to carry out several tasks involving frictional contacts with its surroundings. The friction forces satisfy a Coulomb-like law. The objective of the model is to determine, given the actuator torques and forces applied at the robotic joints, the instantaneous accelerations of the joints and the objects, the contact forces, and contact transitions at each contact point.

For simplicity, let us assume that there is only one object. Using the same notation as in the reference, we let n_c denote the number of contact points between the robot and the object. At each contact point $i = 1, ..., n_c$, we let $c_i = (c_{in}, c_{it}, c_{io})^T$ denote the unknown contact force under an appropriate coordinate frame, and we let $a_i = (a_{in}, a_{it}, a_{io})^T$ represent the relative linear acceleration at i. Furthermore, we use c_n, c_t, c_o and a_n, a_t, a_o to denote the n_c-dimensional vectors with components c_{in}, c_{it}, c_{io} and a_{in}, a_{it}, a_{io} respectively. Then Newton's equations of motion for the robot and the object can be set up as follows:

$$
\begin{pmatrix} a_n \\ a_t \\ a_o \end{pmatrix} = A \begin{pmatrix} c_n \\ c_t \\ c_o \end{pmatrix} + \begin{pmatrix} b_n \\ b_t \\ b_o \end{pmatrix}, \tag{8}
$$

where the matrix A is symmetric and positive semi-definite. Both A and (b_n, b_t, b_o) are known and they depend on such physical quantities as the joint torques and forces supplied by the joint actuators, the gravitational wrench and velocity product wrench experienced by the object, etc. The motions of the robot and the object are subject to kinematic constraints

and frictional laws. First, since the bodies do not interpenetrate, it follows
that
$$a_n \geq 0. \tag{9}$$
Second, the frictional force at each contact must obey Coulomb's law which
stipulates that the contact force must lie within a certain cone defined by
$$c_{it}^2 + c_{io}^2 \leq \mu_i^2 c_{in}^2, \qquad c_{in} \geq 0, \qquad i = 1, ..., n_c, \tag{10}$$
where μ_i is the effective frictional coefficient at the i-th contact. The third
type of constraint concerns the nonbreaking property of contacts. In particular, for each sliding contact point i that is characterized by $v_{it}^2 + v_{io}^2 > 0$,
we have the constraint
$$\mu_i c_{in}(v_{it}, v_{io}) + \sqrt{v_{it}^2 + v_{io}^2}(c_{it}, c_{io}) = 0, \qquad \forall i \in \mathcal{S}, \tag{11}$$
where (v_{it}, v_{io}) are the (known) tangential components of the relative veloc-
ity at the i-th contact and $\mathcal{S}$ denotes the set of sliding contacts. Similarly,
for each rolling contact i that satisfies $v_{it}^2 + v_{io}^2 = 0$, we have the following
constraint:
$$\mu_i c_{in}(a_{it}, a_{io}) + \sqrt{a_{it}^2 + a_{io}^2}(c_{it}, c_{io}) = 0, \qquad \forall i \in \mathcal{R}, \tag{12}$$
where $\mathcal{R}$ denotes the set of rolling contacts. The last set of constraints are
the following complementarity conditions
$$c_{in} a_{in} = 0, \qquad \forall i = 1, ..., n_c. \tag{13}$$

The dynamic 3-dimensional multi-body motion planning (subject to
frictional contacts) problem can now be stated as finding vectors
$$c_n, c_t, c_o, a_n, a_t, a_o$$
satisfying the conditions (8)–(13). It is well known that this problem does
not always have a solution [16, 166]. As a remedy, Pang and Trinkle [228]
proposed the following relaxed model:

$$
\begin{aligned}
&\text{minimize} \quad f(c_{\mathcal{S}n}, c_{\mathcal{S}t}, c_{\mathcal{S}o}) \\
&\text{subject to} \quad \text{(8)–(10), (12), (13)} \\
&\text{and} \quad v_{it} c_{it} + v_{io} c_{io} \leq 0, \quad \forall i \in \mathcal{S},
\end{aligned}
\tag{14}
$$

where

$$f(c_{\mathcal{S}n}, c_{\mathcal{S}t}, c_{\mathcal{S}o}) \equiv$$

$$\sum_{i \in \mathcal{S}} \left((\mu_i v_{it} c_{in} + \sqrt{v_{it}^2 + v_{io}^2} c_{it})^2 + (\mu_i v_{io} c_{in} + \sqrt{v_{it}^2 + v_{io}^2} c_{io})^2 \right)$$

is the residual of the sliding constraints (11). This relaxed formulation differs from the original formulation of the multi-body motion planning problem in that the original sliding constraint (11) has been replaced by the minimization of its residual and by a new set of inequalities $v_{it}c_{it} + v_{io}c_{io} \leq 0$, for all $i \in \mathcal{S}$. Physically, these new inequalities ensure that the friction forces do not produce energy.

The relaxed model (14) of multi-body motion planning can be seen to be a special case of MPEC. In particular, the first-level variables are the tangential forces at the sliding contacts, $(c_{\mathcal{S}t}, c_{\mathcal{S}o})$; and the second-level variables are the normal forces and accelerations (c_n, a_n) as well as the tangential forces and accelerations at the rolling contacts $(c_{\mathcal{R}t}, c_{\mathcal{R}o})$ and $(a_{\mathcal{R}t}, a_{\mathcal{R}o})$. As shown in [228], for fixed $(c_{\mathcal{S}t}, c_{\mathcal{S}o})$, the problem of determining $(c_n, a_n), (c_{\mathcal{R}t}, c_{\mathcal{R}o})$, and $(a_{\mathcal{R}t}, a_{\mathcal{R}o})$ satisfying (8), (9), (10) for $i \in \mathcal{R}$ only, (12), and (13) is equivalent to a mixed nonlinear complementarity problem which defines the inner VI of MPEC. In [228], a variation of the model (14) was also proposed whereby the nonlinear frictional constraints (10) and (12) were replaced with a set of pyramidal cone constraints. With these modifications, (14) becomes a quadratic minimization problem subject to linear complementarity constraints.

Residual minimization of complementarity systems

The model (14) was built based on the notion of residual minimization of a complementarity system. The original formulation of the robotic problem consisted of conditions (8)–(13). In order to deal with the situation where these conditions might be too restrictive and therefore no solution could exist, we proposed to do the best possible by minimizing the residual of the constraints which we believed might be the culprit for the inconsistency of the full model, subject to the satisfaction of the remaining constraints. In what follows, we place this discussion in a somewhat broader context and explore the role of MPEC in the study of general, infeasible complementarity systems.

To simplify the notation, we consider the standard NCP:

$$x \geq 0, \quad F(x) \geq 0, \quad x^T F(x) = 0, \tag{15}$$

where $F : \Re^n_+ \to \Re^n$ is a given continuous mapping. Associated with this problem, we define the mapping

$$H : x \in \Re^n \mapsto \min(x, F(x)) \in \Re^n,$$

where "min" is the componentwise minimum operator. Clearly the NCP (15) is equivalent to the optimization problem:

$$\begin{aligned} \text{minimize} \quad & \theta(x) \equiv \tfrac{1}{2} H(x)^T H(x) \\ \text{subject to} \quad & x \geq 0, \end{aligned} \tag{16}$$

in the sense that a solution of (15) is an optimal solution of (16) with zero objective value. The function θ is an example of a *residual* for the NCP (15); indeed θ has the following two defining properties of a residual function: (a) it is nonnegative everywhere, and (b) it is equal to zero at a vector if and only if this vector solves the NCP (15).

With $y \equiv x - \min(x, f(x))$, it is not hard to see that the problem (16) is equivalent to:

$$\begin{aligned} \text{minimize} \quad & \tfrac{1}{2}(x - y)^T (x - y) \\ \text{subject to} \quad & x \geq 0, \\ & y \geq 0, \quad F(x) - x + y \geq 0, \\ & y^T (F(x) - x + y) = 0, \end{aligned} \tag{17}$$

which is in the form of the MPEC (1.1.1) with $Z \equiv \Re^n_+$, $C(x) \equiv \Re^n_+$ for all $x \in \Re^n_+$, and

$$f(x, y) \equiv \tfrac{1}{2}(x - y)^T (x - y), \quad F(x, y) \equiv F(x) - x + y.$$

The minimization problem (16) is the cornerstone to the NE/SQP algorithm and its variants for solving the NCP (15); see [224, 90, 91, 27]. More generally, using the normal equation formulation of the VI (F, C) where C is a closed convex set in $\Re^n$ (see Proposition **1.3.3** for this formulation), it is possible to derive an MPEC formulation for the residual minimization problem of a VI. See [237, 63] for a Newton-like approach to minimizing the normal residual, and [72] for a Gauss-Newton approach that is more directly comparable to the NE/SQP method [224].

Besides being useful for solving the NCP (15), the residual function $\theta(x)$ plays an important role in the theory of error bounds; in turn these error bounds are instrumental for the derivation of an exact penalization theory for MPEC. The latter topic will be covered in full detail in the next chapter.

The parametric feasibility problem

A class of parametric feasibility problems that have important applications in chemical engineering can be formulated as a bilevel nonlinear program [79, 99]. These problems can be described as follows. Let $F : \Re^{n+m} \to \Re^{\ell}$ be a given vector-valued function; let X and Y be given subsets of $\Re^n$ and $\Re^m$ respectively. The *feasibility test problem*, as it is called in the references, is to determine if for every $x \in X$, there exists a $y \in Y$ such that $F(x,y) \leq 0$. It is easy to see that this problem has an equivalent formulation as the following max-min-max problem:

$$\max_{x \in X} \min_{y \in Y} \max_{1 \leq j \leq \ell} F_j(x,y).$$

More precisely, we let

$$\xi_{\text{opt}} \equiv \max_{x \in X} \xi(x), \tag{18}$$

where

$$\xi(x) \equiv \min_{y \in Y} \max_{1 \leq j \leq \ell} F_j(x,y);$$

then the feasibility test problem has an affirmative answer if and only if $\xi_{\text{opt}} \leq 0$. In principle, due to the nonsmoothness of the function $\xi(x)$ in general, the problem of determining ξ_{opt} is a nonsmooth optimization problem. In Grossman and Floudas [99], a bilevel program was proposed as a reformulation of (18) and an algorithm based on a mixed integer programming technique was presented for solving the feasibility test problem. The bilevel program is:

maximize ξ

subject to $x \in X$,

and $(\xi, y) \in \text{argmin}\{\xi : F_j(x,y) \leq \xi, j = 1, \ldots, \ell; \ y \in Y\}.$

Here x is the first-level variable and the pair $(\xi, y) \in \Re \times \Re^m$ is the second-level variable.

Besides the above feasibility test problem, the cited references also discuss a *flexibility index problem* which leads to a three-level nonlinear program; see also [270, 271].

The continuous network design problem

Many transportation planning and design problems can be formulated as MPECs. Typically, the outer objective function of the resulting MPEC represents the combined system and design costs, whereas each inner problem represents either an optimization or an equilibrium problem that describes the behavior of the users of the transportation network corresponding to a particular configuration of the network dictated by a given set of the design variables. In this and the next two subsections, we shall discuss several transportation planning problems and present their MPEC formulations.

The network design problem is concerned with the modification of a transportation infrastructure by adding new links or improving existing ones in order to maximize social welfare and/or minimize design and other costs [56]. The continuous version of this problem arises when the design variables are represented by real numbers (representing divisible capacity enhancements) [82]. The early work of the continuous network design problem dealt with a bilevel program in which the lower-level problems were optimization formulations of the traffic assignment problem [1, 56]. Inspired by Smith's renowned observation [264] that the traffic equilibrium problem under Wardrop's user equilibrium principle [282] could be formulated as a variational inequality, Marcotte [199] formulated the network design problem as an MPEC (although he did not call it that). We refer the reader to [78] for a summary of various VI and NCP formulations and solution methods of the traffic equilibrium problem as well as the diverse applications of this problem to national, regional, and urban planning.

To simplify the discussion and notation, we consider the fixed-demand model and use the arc-flow formulation of the traffic user equilibrium problem which defines the inner VI. The following are given:

- a network $\mathcal{G}$ with node set $\mathcal{N}$ and link set $\mathcal{A}$;

- a subset $\Theta \subseteq \mathcal{N} \times \mathcal{N}$ denoting the origin-destination (OD) pairs;

- the known vector $d \equiv (d_{ij})$ of traffic demands for all OD pairs (i, j) in Θ;

- for each OD pair $(i, j) \in \Theta$, the set of paths P_{ij} using links in $\mathcal{A}$ that join i to j; let

$$P \equiv \bigcup_{(i,j) \in \Theta} P_{ij}$$

denote the set of all paths in the network joining all the OD pairs.

The variables of the model are:

- the vector of link flows $f \equiv (f_a)$, where f_a represents the flow on link $a \in \mathcal{A}$; this is the lower-level variable;

- the vector of link capacity enhancements $x = (x_a)$, where x_a represents the capacity increase on link $a \in \mathcal{A}$; this is the upper-level variable.

Let $X \subseteq \Re^{|\mathcal{A}|}$ be the feasible set of the link capacity enhancements x. The set of *feasible* link flows, denoted Y, consists of all flow vectors f that are nonnegative and for which there exists, for each OD pair $(i, j) \in \Theta$ and each path $p \in P_{ij}$, a path flow amount $h_p \geq 0$ such that

$$\begin{aligned}
\sum_{p \in P_{ij}} h_p &= d_{ij}, \quad \text{for all } (i, j) \in \Theta \\
f_a &= \sum_{p \in P} \delta_{ap} h_p, \quad \text{for all } a \in \mathcal{A},
\end{aligned} \tag{19}$$

where

$$\delta_{ap} \equiv \begin{cases} 1 & \text{if path } p \text{ uses link } a \\ 0 & \text{if path } p \text{ does not use link } a. \end{cases}$$

It is not difficult to see that Y is a bounded polyhedron in $\Re_+^{|\mathcal{A}|}$.

The following (real-valued) functions are also given:

- $c_a(x, f)$, the perceived unit cost of travel on link a as a function of the flow vector f and link capacity vector x;

- $C_a(x, f)$, the total travel cost on link a; one form of this function is $f_a c_a(x, f)$;

- $g(x)$, the capital investment and operating costs associated with the capacity vector x.

Under the above specifications, the continuous network design problem can be formulated as the MPEC:

$$\text{minimize} \quad \sum_{a \in \mathcal{A}} C_a(x, f) + g(x)$$

$$\text{subject to} \quad x \in X, \tag{20}$$

$$\text{and} \quad f \text{ solves the VI } (c(x, \cdot), Y),$$

where $c(x, f)$ is the vector-valued function with components $c_a(x, f)$ for all $a \in \mathcal{A}$.

There have been several descent-type algorithms for solving the continuous network design problem [199, 268, 269]. Kim and Suh [130] have proposed a national transportation planning model for Korea based on a bilevel programming approach.

Origin-destination demand adjustment problem

The problem of estimation and/or adjustment of travel demands between OD pairs is fundamental to transportation planning and has a long history. For instance, the demand quantities d_{ij} are basic inputs to the design problem (20); therefore the availability of these quantities is essential to the treatment of the problem. Many proposals exist for dealing with the estimation/adjustment problem of travel demands. In what follows we describe an MPEC approach to a demand adjustment problem that is based on the work of Chen and Florian [47], which also contains many references to this and related problems. Another paper by the same authors [46] discusses the general nonlinear bilevel program and provides optimality conditions of the Fritz-John type.

In the problem under consideration, a set of observed link counts $\hat{v}_a$ is available for a subset of arcs $\hat{\mathcal{A}} \subseteq \mathcal{A}$; also available is a set of known OD demands $\hat{d}_{ij}$ which correspond to some previous values that require adjustment (for instance, because they are out of date). Analogous to the notation used above, let $c \equiv (c_a(f))$ where $c_a(f)$ denotes the perceived unit cost of travel on link a as a function of the vector of link flows $f \in \Re_+^{|\mathcal{A}|}$. Let $\mathcal{Y}(d)$ be the set of all admissible link flow vectors f corresponding to a given demand vector d; that is $\mathcal{Y}(d)$ consists of all link flow vectors $f \geq 0$

for which there exists a vector of path flows $h = (h_p) \geq 0$, for $p \in P_{ij}$ and $(i,j) \in \Theta$, such that the conditions (19) hold. Let $\mathcal{D}$ be a given set of admissible OD demands; for instance, $\mathcal{D}$ consists of given upper and lower bounds on the demands. Then the OD demand adjustment problem is

$$\text{minimize} \quad \textstyle\sum_{a \in \hat{A}}(f_a - \hat{v}_a)^2 + \sum_{(i,j) \in \Theta}(d_{ij} - \hat{d}_{ij})^2$$

$$\text{subject to} \quad d \in \mathcal{D}, \tag{21}$$

$$\text{and} \quad f \text{ solves VI } (c, \mathcal{Y}(d)).$$

This is an MPEC with d as the first-level variable and f the lower-level variable. In contrast to the network design problem (20), the mapping c in the lower-level VI of the present MPEC (21) is independent of the first-level variable d but the feasible set $\mathcal{Y}(d)$ is dependent on d. These two features distinguish the two MPECs (20) and (21). The objective functions of these two problems are also different in nature.

A discrete transit planning problem

Many network design problems involve discrete variables which correspond to either-or decisions. In what follows we consider a transit decision making problem discussed in [55].

Consider a generalized transportation network $\mathcal{G} \equiv (\mathcal{N}, \mathcal{A})$ where the set of links $\mathcal{A}$ corresponds to the set of transit lines (e.g., bus or subway) and the set of nodes $\mathcal{N}$ signifies the transit stops. Let $\mathcal{A}_i^+$ and $\mathcal{A}_i^-$ denote respectively the set of outgoing and incoming links at node $i \in \mathcal{N}$. A strategy is, by definition, a set of links $\bar{A} \subseteq \mathcal{A}$ to be used by a traveler to reach a given destination; at each stop i, the traveler waits and considers a subset of attractive links $\bar{A}_i^+ \subseteq \bar{A}$ outgoing from node i. We shall use w_i to denote the waiting time at node i, and use g_i to denote the demand from i to the destination. It should be noted that w_i is a variable to be determined and g_i is a constant assumed to be known.

For each link $a \in \mathcal{A}$, we define several variables:

- f_a: the frequency of transit service on a;

- v_a: the traffic volume on link a;

- x_a: a 0-1 variable given by

$$x_a = \begin{cases} 1 & \text{if } a \in \bar{A} \\ 0 & \text{otherwise.} \end{cases}$$

Also, we will denote the required travel time (or generalized cost) on link a, which is assumed to be a known constant, by t_a. Notice that the traffic volume, the service frequency and the waiting time satisfy the following simple relation:

$$v_a = x_a f_a w_i, \qquad \forall a \in \mathcal{A}_i^+, \ i \in \mathcal{N}.$$

The total waiting time for a given strategy $\bar{A}$ and a given set of service frequencies $\{f_a : a \in A\}$ is given by

$$\sum_{a \in \mathcal{A}} t_a v_a + \sum_{i \in \mathcal{N}} w_i$$

where the first sum corresponds to the in-vehicle travel time and the second sum signifies the waiting time. The traveler's objective is to determine an optimal strategy that has the least total travel time. This is formulated as the following bilevel programming problem:

$$\begin{array}{ll} \text{minimize} & \sum_{a \in \mathcal{A}} t_a v_a + \sum_{i \in \mathcal{N}} w_i \\ \text{subject to} & \sum_{a \in \mathcal{A}} t_a f_a \leq N, \\ & f_a \geq \underline{f}_a, \quad \forall a \in \mathcal{A} \end{array}$$

and (f, x) solves

$$\begin{array}{lll} \text{minimize} & \sum_{a \in \mathcal{A}} t_a v_a + \sum_{i \in \mathcal{N}} w_i & (22) \\ \text{subject to} & v_a = x_a f_a w_i, \quad a \in \mathcal{A}_i^+, \ i \in \mathcal{N}, \\ & \sum_{a \in \mathcal{A}_i^+} v_a - \sum_{a \in \mathcal{A}_i^-} v_a = g_i, \quad i \in \mathcal{N}, \\ & v_a \geq 0, \quad a \in \mathcal{A}, \\ & x_a = 0 \text{ or } 1, \quad a \in \mathcal{A}. \end{array}$$

Notice that in the upper-level problem, the variable is the service frequency f, and the constraints $\sum_{a \in \mathcal{A}} t_a f_a \leq N$ and $f_a \geq \underline{f}_a$ correspond to the fleet size requirement and the minimum level of service requirement respectively.

In the lower-level problem, the service frequency is held fixed, and the problem is a simple linear program (actually a shortest path problem) with the standard flow conservation constraints.

There are many other applications of bilevel linear programs to the qualitative study of transportation. In particular, we cite the study [21] which introduced a bilevel linear programming model for the optimization of investments in interregional highway networks in developing countries; this model was subsequently applied to the Tunisian network using real data.

A facility location and production problem

MPEC can also be used to model a certain facility location and production problem [206]. Specifically, consider a large firm which plans to enter an industry to compete with a number of small firms for a certain homogeneous commodity. The entering firm needs to determine its production facility locations and the production levels at these locations, with the objective of maximizing its profits while taking into account the impact of its policy on the market price. In other words, the entering firm anticipates the reaction of the existing firms before choosing its optimal policy. As an MPEC, the inner problem is a spatial price equilibrium model [84, 83, 107] which determines the prices as a function of a particular production and shipment plan of the firm, whereas the outer problem is the firm's profit maximization problem subject to the constraints of facility choices and production and shipment restrictions.

Let $\mathcal{N}$ denote the set of all locations relevant to the firm's production and distribution; $\mathcal{N}$ is the node set of a transportation network whose link set is denoted $\mathcal{A}$. We make the simplifying assumption that each node $i \in \mathcal{N}$ is a potential supply and demand market. The notation $(\mathcal{A}_i^+, \mathcal{A}_i^-)$ is the same as in the previous transit model. Let $\mathcal{N}_0 \subseteq \mathcal{N}$ denote the set of eligible locations from which the entering firm can choose to locate its production facilities. Then the firm's profit as a result of production at node $i \in \mathcal{N}_0$ is given by

$$Z_i(\pi, Q, s) \equiv$$

$$\pi_i \left(Q_i - \sum_{a \in \mathcal{A}_i^+} s_a \right) - V_i(Q) - F_i + \sum_{a=(i,j) \in \mathcal{A}} \pi_j s_a - \sum_{a \in \mathcal{A}_i^+} t_a(s),$$

where for each such location i, π_i is the market price of the commodity

(with $\pi \equiv (\pi_j)_{j\in\mathcal{N}}$ being the vector of all prices), Q_i is the production
level, $V_i(Q)$ is the cost of production as a function of the total output
$Q \equiv (Q_j)_{j\in\mathcal{N}_0}$, F_i is the fixed cost of setting up the facility, s_a is the
amount the firm's production transported on link a, and $t_a(s)$ is the cost
to the firm of shipping on link a as a function of the total shipping pattern
$s \equiv (s_a)$. The objective of the firm is to select a set of locations from $\mathcal{N}_0$
and to determine the vector of production levels Q at these locations and
the vector of shipment amount s, so as to maximize its total profits. Notice
that the vector of market prices $\pi = (\pi_i)$ which has direct impact on profit,
depends implicitly on the full production vector Q and shipment vector s
through the spatial price equilibrium model to be described shortly. For
each $i \in \mathcal{N}_0$, let

$$y_i \equiv \begin{cases} 1 & \text{if site } i \text{ is selected} \\ 0 & \text{otherwise} \end{cases}$$

be the decision variable expressing whether or not a site $i \in \mathcal{N}_0$ is selected
by the firm. In terms of these discrete variables, the firm's total profit is
given by

$$Z(\pi,y,Q,s) \equiv \sum_{i\in\mathcal{N}_0} (Z_i(\pi,Q,s)y_i + F_i(1-y_i)).$$

The profit maximization problem is then

maximize $Z(\pi,y,Q,s)$

subject to $Q_i \leq \bar{Q}_i y_i, \quad i \in \mathcal{N}_0,$

$$Q_i + \sum_{j\in\mathcal{N}_0}\sum_{a=(j,i)\in\mathcal{A}} s_a - \sum_{a\in\mathcal{A}_i^+} s_a \geq 0, \quad i \in \mathcal{N}_0,$$

$$\sum_{j\in\mathcal{N}_0}\sum_{a=(j,i)\in\mathcal{A}} s_a - \sum_{a\in\mathcal{A}_i^+} s_a \geq 0, \quad i \notin \mathcal{N}_0, \qquad (23)$$

$$\sum_{i\in\mathcal{N}_0} Q_i \leq \bar{Q},$$

$$Q_i \geq 0, \ s_{ij} \geq 0, \quad i \in \mathcal{N}_0, \ (i,j) \in \mathcal{A},$$

$$y_i = 0, 1, \quad i \in \mathcal{N}_0,$$

where $\bar{Q}$ is the total production limit and $\bar{Q}_i$ is the production capacity at
node $i \in \mathcal{N}_0$.

It remains to describe the implicit relationship between the market price
π and the production level Q and the shipping pattern s. We need to fix

some notation. For each node $i \in \mathcal{N}$, we denote the total demand and supply at i, as a function of the prevailing price π, by $D_i(\pi)$ and $S_i(\pi)$ respectively. For each link $a \in \mathcal{A}$, let f_a be the flow of the commodity on a, and let $f \equiv (f_a)$ be the vector of these link flows. Let $c_a(f)$ denote the unit cost of transportation on link $a \in \mathcal{A}$ as a function of all link flows.

With the above notation, the conditions defining the spatial price equilibrium model are as follows.

(a) The flows are nonnegative:

$$f_a \geq 0, \quad \forall a \in \mathcal{A}.$$

(b) The local prices cannot exceed delivered prices:

$$\pi_i + c_a(f) \geq \pi_j, \quad \forall a = (i,j) \in \mathcal{A}.$$

(c) No flow should occur when delivered price exceeds local price:

$$\pi_i + c_a(f) > \pi_j \text{ implies } f_a = 0, \quad \forall a = (i,j) \in \mathcal{A}.$$

(d) Flow occurs only when the local price and delivered price are equal:

$$f_a > 0 \text{ implies } \pi_i + c_a(f) = \pi_j, \quad \forall a = (i,j) \in \mathcal{A}.$$

(e) The prices are nonnegative and the flows are conserved: for each $i \in \mathcal{N}$, $\pi_i \geq 0$ and

$$D_i(\pi) - S_i(\pi) + \sum_{a \in A_i^+} f_a - \sum_{a \in A_i^-} f_a =$$

$$\begin{cases} Q_i - \sum_{a \in A_i^+} s_a + \sum_{j \in \mathcal{N}_0} \sum_{a=(j,i) \in A} s_a & \text{if } i \in \mathcal{N}_0 \\ \sum_{j \in \mathcal{N}_0} \sum_{a=(j,i) \in A} s_a & \text{if } i \notin \mathcal{N}_0. \end{cases} \quad (24)$$

Note that conditions (a)–(d) can be combined into the complementarity condition:

$$\pi_i + c_a(f) \geq \pi_j, \quad f_a \geq 0, \quad f_a(\pi_i + c_a(f) - \pi_j) = 0.$$

For each given pair (Q, S), it can be shown [84] that under the assumption that the link cost functions $c_a(f)$ are all positive and the supply and demand functions $S_i(\pi)$ and $D_i(\pi)$ satisfy a technical condition, the above conditions (a)–(e) are equivalent to a nonlinear complementarity problem in the flow and price variables (f, π). Specifically, let $h_i(f, S, D)$ and $R_i(Q, s)$ denote, respectively, the left- and right-hand expression in (24); then (e) can be replaced by

$$\left.\begin{array}{c} \pi_i \geq 0, \quad h_i(f, S, D) \leq R_i(Q, s) \\[2mm] \pi_i\left(h_i(f, S, D) - R_i(Q, s)\right) = 0 \end{array}\right\} \quad \forall i \in \mathcal{N}. \qquad (25)$$

The firm's overall optimal facility location and production problem is now formulated as the MPEC with (y, Q, s) as the outer-level variables and (f, π) as the inner-level variables; the inner problem is the NCP defined by conditions (a)–(d) and (25), and the outer problem is the profit maximization problem (23). Due to the presence of the 0-1 variables y_i, this MPEC is expected to be particularly difficult. Some numerical results for solving this problem are discussed in [206].

There are numerous other applications of MPEC and/or bilevel programs to economic planning. For example, the paper [28] introduced the Indus Basin model for the study of surface and groundwater related policies in Pakistan; it turns out that this model is a special bilevel linear program. Other related applications of MPEC include that of Cassidy, Kirby, and Raike [44] who studied the resource distribution problems, that of Fortuny-Amart and McCarl [80], and that of McCarl and Spreen [203] who studied the problem of economic decisions and sector analysis; the last named reference also has an extensive bibliography on the applications of MPEC to economics. The papers [208, 209] describe a nonlinear bilevel programming approach for aluminum production modeling via the smelting process; the model was developed for management use at Portland Aluminum located at Portland Victoria, Australia. Another nonlinear bilevel model was developed for the analysis of electric utility demands [115]. The paper [62] presents an MPEC formulation for the gas transmission problem to minimize the total supply cost subject to satisfaction of nodal demands at minimal guaranteed pressure.

Optimal design problems in mechanical structures

A large number of engineering design problems are of the MPEC type. In what follows, we describe a frictionless, minimum weight design problem of structural optimization involving unilateral constraints [135, 136]. The dissertation [231] documents an extensive study of optimization problems of this type and presents numerical methods for solving them. The physical objects of these problems are discrete mechanical structures, such as trusses. For the latter, the unilateral constraints are of the contact type and those related to cables that do not admit compressive stresses. The stiffness equation for a truss is

$$F = K(t)u, \qquad (26)$$

where F is the nodal forces which are composed of external forces f (given) and forces due to unilateral constraints; u is the vector of nodal displacements; and $K(t)$ is the stiffness matrix, regarded as a function of a vector $t \in \Re^m$ of bar volumes, $K(t)$ is symmetric positive definite for $t > 0$. A common form of $K(t)$ is

$$K(t) \equiv \sum_{i=1}^{m} t_i K_i,$$

where each K_i is a symmetric positive definite matrix. A node that may come into frictionless contact with a rigid obstacle is considered. The kinematic condition that nodes of the structure cannot penetrate the obstacle is expressed by

$$Cu \geq g, \qquad (27)$$

where C is a kinematic transformation matrix and g is a vector of initial distances between nodes and the rigid obstacle. In terms of the matrix C, we have

$$F = f + C^T p, \qquad (28)$$

where p is the vector of contact forces, work conjugate to the vector Cu of contact displacements. Adhesionless contact requires that

$$p \leq 0; \qquad (29)$$

and ruling out action at a distance gives the complementarity condition:

$$p^T(Cu - g) = 0. \qquad (30)$$

The four conditions (26)–(30) define a mixed linear complementarity problem in the variables f and p. By the symmetry and positive definiteness of the matrix $K(t)$, these conditions are precisely the optimality conditions of the strictly convex quadratic program in the variable u:

$$\text{minimize} \quad -f^T u + \tfrac{1}{2} u^T K(t) u$$
$$\text{subject to} \quad Cu \geq g; \tag{31}$$

the vector p in (28)–(30) is the multiplier of the constraint $Cu \geq g$.

A general minimum weight design problem can now be defined as follows:

$$\text{minimize} \quad w(u, p, t)$$
$$\text{subject to} \quad t \in \mathcal{T},$$
$$h(u, p, t) \leq 0,$$
$$\text{and} \quad (26)\text{–}(30),$$

where w is a generalized weight function, $\mathcal{T}$ is the constraint set of the variable t of bar volumes (examples of $\mathcal{T}$ include $\{t \in \Re^m_+ : \sum_{i=1}^m t_i = V\}$, where $V > 0$ is the given total volume of the bars, and $\{t \in \Re^m : \bar{t} \geq t \geq \underline{t}\}$, where $\bar{t} > \underline{t} > 0$ are upper and lower bounds on the volume of the bars respectively), and usual instances of the constraint function h are bounds on displacements and stresses (in which case h will be independent of p), and/or bounds on contact forces. Clearly, this problem is an instance of the MPEC with mixed linear complementarity constraints; t is the upper-level variable and (u, p) are the lower-level variables.

Note that for a given value of t, there is a unique u vector satisfying (26)–(30), although the p vector need not be unique. When p is not unique, it creates a source for the multi-valuedness of the solution map of the lower-level problems. Nevertheless, since the complementarity system (26)–(30) is derived from a strictly convex quadratic program, it is likely that this feature can be profitably exploited in the future study of the weight design problem as an MPEC.

In [138], a shape optimization problem of a mechanical structure was formulated as an MPEC with nonlinear complementarity constraints. This problem was the result of a finite element discretization of an infinite-dimensional control problem with complementarity constraints; hence the size of the resulting MPEC could potentially be very large. A bundle method from nondifferentiable optimization coupled with a nonsmooth variant of Newton's method was proposed in the reference. A related optimum design problem is discussed in [139].

Optimal prestress of cracked structures

In two papers [266, 267], Stavroulakis studied the problem of optimal prestress stabilization of cracked structures by formulating the problem as an optimal control problem for variational inequalities. In the first paper, the frictionless case was considered and the resulting mathematical model led to an MPEC similar to the minimum weight design problem discussed previously. In the second paper, frictional unilateral contact surfaces were considered and this led to an MPEC whose inner problems are asymmetric mixed LCPs. In essence the inner LCPs model the contact problem of deformable bodies under Coulomb friction for given prestress forces; the objective function of the MPEC is a cost function that is an additive measure of the cost of controlling the prestress forces and that of the crack opening and frictional slip. We refer the reader to [131, 132, 133, 134] for various LCP formulations of a deformable-body contact problem subject to Coulomb friction. In what follows, we give the MPEC formulation of the optimal prestress problem as described in [267].

As before, let u denote the nodal displacement of a structure which is subject to an external force vector f and a variable vector z of prestress forces acting on the structure via a transformation matrix A; also, let $p_t \equiv (p_{it})$ denote the tangential force and C_t be a (given) static force transformation matrix. With these new ingredients, the kinematic and contact conditions can be stated in terms of the quadratic program in the variable u for given (p_t, z):

$$
\begin{aligned}
&\text{minimize} \quad (C_t^T p_t - Az - f)^T u + \tfrac{1}{2} u^T K u \\
&\text{subject to} \quad C_n u \geq g,
\end{aligned}
\tag{32}
$$

with the multiplier $p_n \equiv (p_{in})$ of the constraints $C_n u \geq g$ yielding the normal force. Expressing this quadratic program equivalently in terms of its KKT system, we have

$$
Ku + C_n^T p_n + C_t^T p_t = Az + f,
\tag{33}
$$

$$
p_n \leq 0, \quad C_n u \geq g, \quad p_n^T (C_n u - g) = 0.
\tag{34}
$$

For this planar model, the Coulomb friction law stipulates that

$$|p_{it}| \leq \mu p_{in}, \quad \text{for all } i, \tag{35}$$

where $\mu > 0$ is the friction coefficient assumed for simplicity to be the same at all contact points. Furthermore, the maximum work principle states that there exist $r_{it}^{\pm}$ and $s_{it}^{\pm}$ such that for all i,

$$p_{it} + s_{it}^{+} = \mu p_{in}, \quad -p_{it} + s_{it}^{-} = \mu p_{in}, \tag{36}$$

(this is simply a restatement of the Coulomb friction law (35) with slack variables $s_{it}^{\pm}$ added);

$$C_{it}u + r_{it}^{+} - r_{it}^{-} = 0, \tag{37}$$

$$(s_{it}^{+}, s_{it}^{-}) \geq 0, \quad (r_{it}^{+}, r_{it}^{-}) \geq 0, \tag{38}$$

$$s_{it}^{+}r_{it}^{+} = s_{it}^{-}r_{it}^{-} = 0, \tag{39}$$

where C_{it} denote the i-th row of the matrix C_t. Letting $f(z, r_t, v_n)$ denote the cost function, then the MPEC is

$$\begin{aligned} \text{minimize} \quad & f(z, r_t, C_n u - g) \\ \text{subject to} \quad & z \in \mathcal{Z} \\ \text{and} \quad & \text{(33), (34), (36)--(39)}, \end{aligned}$$

where $\mathcal{Z}$ is the set of admissible controls; e.g., a rectangular box. This problem is an MPEC with z as the first-level variable and (u, p_n, p_t, r_t, s_t) as the second-level variable; each lower-level problem is a mixed LCP.

1.3 Equivalent Constraint Formulations

Though compact, the formulation (1.1.1) is not very practical for analysis and other purposes. We shall introduce several equivalent formulations of the inner VI $(F(x, \cdot), C(x))$ that will form the basis of our subsequent study. The first formulation involves the casting of this VI as a system of equations defined by the projection operator; see (1). The second formulation expresses the same VI in terms of its so-called Karush-Kuhn-Tucker (KKT) system under an inequality representation of the set $C(x)$; see (8). The third formulation is a compact representation of the second and expresses the VI as a mixed nonlinear complementarity system; see (11). The fourth formulation expresses the complementarity (or VI) constraints in an MPEC using the notion of a merit function for a complementarity problem (or VI); see (25) and (27). The fifth and last formulation deals with the case where for each $x \in X$, the set $\mathcal{S}(x)$ consists of the unique solution of the VI $(F(x, \cdot), C(x))$; this leads to the formulation of the MPEC as an "implicit program".

1.3.1 MPEC in normal form

In order to introduce the first equivalent formulation of the inner VIs of the MPEC, we need to introduce some familiar concepts and review some known results in convex analysis. Since the proofs of these results are not difficult and can be found in a standard textbook on convex analysis such as [256], they are omitted.

Projection and related concepts

For a nonempty closed convex set $C \subseteq \Re^n$, let $\Pi_C(z)$, called the *projection* of z onto C, denote the vector in C that is closest to the (given) vector $z \in \Re^n$ under the Euclidean norm; that is, $\Pi_C(z)$ is the optimal solution of the following minimization problem in the variable x:

$$\text{minimize} \quad \tfrac{1}{2}(x - z)^T(x - z)$$
$$\text{subject to} \quad x \in C.$$

The associated mapping $\Pi_C : z \mapsto \Pi_C(z)$ is called the *projection map* onto C. The result below summarizes some key properties of this mapping: existence, uniqueness, variational characterization, and nonexpansiveness.

1.3.1 Proposition. Let C be a closed convex subset of $\Re^n$.

(a) For each $z \in \Re^n, \Pi_C(z)$ exists and is unique; $\Pi_C(z)$ is the unique vector $\bar{z} \in C$ that satisfies

$$(x - \bar{z})^T(\bar{z} - z) \geq 0, \quad \text{for all } x \in C.$$

(b) For any $z, z' \in \Re^n$,

$$\|\Pi_C(z) - \Pi_C(z')\| \leq \|z - z'\|.$$

In particular, Π_C is a Lipschitz continuous function on $\Re^n$.

Let C be a nonempty subset of $\Re^n$. The *tangent cone* of C at a given vector $x \in C$, denoted $\mathcal{T}(x; C)$, is the set of all vectors $v \in \Re^n$, called *tangent vectors*, for which there exist sequences of vectors $\{x^k\} \subset C$ converging to x and positive scalars $\{\tau_k\}$ converging to zero such that

$$v = \lim_{k \to \infty} \frac{x^k - x}{\tau_k}.$$

Clearly $\mathcal{T}(x; C)$ is a cone in the sense that $\lambda v \in \mathcal{T}(x; C)$ for all scalars $\lambda \geq 0$ and vectors $v \in \mathcal{T}(x; C)$. In general $\mathcal{T}(x; C)$ is closed but not necessarily convex; but if C is closed and convex, then so is this tangent cone.

The *normal cone* of C at $x \in C$, denoted $\mathcal{N}(x; C)$, is defined as the polar of the tangent cone $\mathcal{T}(x; C)$. In general, the *polar*, denoted S^p, of a subset S of $\Re^n$ is defined to be

$$S^p \equiv \{y \in \Re^n : y^T x \leq 0 \text{ for all } x \in S\}.$$

It is easy to verify that S^p is always a closed convex cone. The set $-S^p$ is precisely the dual cone S^* of S defined in (1.1.5). Unlike the tangent cone, the normal cone $\mathcal{N}(x; C)$ is always closed and convex for any set C and vector $x \in C$. Moreover if C is convex, we have the following characterization of the normal cone.

1.3.2 Proposition. Let C be a nonempty subset of $\Re^n$ and $x \in C$. Then

$$\{u \in \Re^n : u^T(y - x) \leq 0 \text{ for all } y \in C\} \subseteq \mathcal{N}(x; C);$$

moreover equality holds if C is convex.

Let C be a closed convex subset of $\Re^n$ and $F : C \to \Re^n$ be a given mapping. The *normal map* associated with this pair (F, C) is the mapping $F_C : \Re^n \to \Re^n$ defined by:

$$F_C(z) \equiv F \circ \Pi_C(z) + z - \Pi_C(z), \quad \text{for all } z \in \Re^n;$$

the associated equation

$$F_C(z) = 0$$

is called a *normal equation*. This terminology is motivated by the concept of the normal cone defined above. In fact combining Propositions **1.3.1** and **1.3.2**, we deduce that $\Pi_C(z)$ is the unique vector $\bar{z} \in C$ such that $z - \bar{z}$ is in $\mathcal{N}(\bar{z}; C)$. Using this observation, we have the following characterization.

1.3.3 Proposition. Let C be a nonempty closed convex subset of $\Re^n$ and let $F : C \to \Re^n$ be a given mapping. The following statements are equivalent for $x, z \in \Re^n$.

(a) The vector x solves the VI (F, C), and $z = x - F(x)$.

(b) The vector x belongs to C, $-F(x) \in \mathcal{N}(x; C)$, and $z = x - F(x)$.

(c) The vector z satisfies $F_C(z) = 0$, and $x = \Pi_C(z)$.

This result implies that the VI (F, C) is equivalent to the normal equation $F_C(z) = 0$ in the following sense: if x solves the VI (F, C), then $z \equiv x - F(x)$ is a zero of F_C; conversely, if z is a zero of F_C, then $x \equiv \Pi_C(z)$ solves the VI (F, C). Moreover, the VI (F, C) has a unique solution if and only if the normal map F_C has a unique zero.

When C is the nonnegative orthant $\Re^n_+$, the projection map $\Pi_{\Re^n_+}$ takes on a very simple form; specifically,

$$\Pi_{\Re^n_+}(z) = \max(z, 0),$$

where "max" is the componentwise maximum operator. We write

$$z^+ \equiv \max(z, 0), \quad \text{and} \quad z^- \equiv \max(0, -z).$$

In this notation, the nonlinear complementarity problem:

$$x \geq 0, \quad F(x) \geq 0, \quad x^T F(x) = 0,$$

which is equivalent to the VI $(F, \Re^n_+)$, has the following equivalent normal equation formulation:

$$F(z^+) - z^- = 0.$$

Historically, Eaves [65] seemed to have been the first person to have used this equation for the study of complementarity problems; Kojima [140] was among the pioneers who used a related nonsmooth equation formulation for the sensitivity analysis of stationary points of parametric nonlinear programs.

The normal formulation

Returning to the MPEC (1.1.1) and utilizing the equivalence between a VI and its normal equation, we obtain an equivalent formulation of (1.1.1) via a change of variables. Specifically, for each $x \in X$, let $F_{C(x)}(x, \cdot)$ denote the normal map associated with the pair $(F(x, \cdot), C(x))$. This map is well defined provided that $C(x)$ is a nonempty closed convex subset of $\Re^m$. Under this assumption, the problem (1.1.1) is then equivalent to the following optimization problem in the variables (x, v):

$$\begin{aligned} \text{minimize} \quad & f(x, \Pi_{C(x)}(v)) \\ \text{subject to} \quad & (x, \Pi_{C(x)}(v)) \in Z, \\ & F_{C(x)}(x, \Pi_{C(x)}(v)) = 0. \end{aligned} \tag{1}$$

Let $\tilde{Z}$ be the subset of $\Re^{n+m}$ consisting of all vectors (x, v) such that $(x, \Pi_{C(x)}(v)) \in Z$ and define

$$\tilde{f}(x, v) \equiv f(x, \Pi_{C(x)}(v)), \quad H(x, v) \equiv F_{C(x)}(x, \Pi_{C(x)}(v)) = 0.$$

The formulation (1) can be written in the following compact form:

$$\begin{aligned} \text{minimize} \quad & \tilde{f}(x, v) \\ \text{subject to} \quad & (x, v) \in \tilde{Z}, \\ & H(x, v) = 0. \end{aligned} \tag{2}$$

With the exception of the set $\tilde{Z}$, the last formulation is an equality constrained, nondifferentiable optimization problem. The nonsmoothness is due to the projection operator $\Pi_{C(x)}$ which, even for a fixed x, is typically not Fréchet or Gâteaux differentiable (see [211, Chapter 2] for a review of these basic differentiability concepts). Thus, although the given functions f and F could be fairly smooth, the resulting formulation (2) would destroy the smoothness property. Nevertheless the simplicity of the formulation (2) is in fact an advantage in the derivation of optimality conditions for the original MPEC (1.1.1). This is the main approach taken in the paper [172].

1.3.2 MPEC in KKT form

When the set $C(x)$ is represented by finitely many convex inequalities, we can derive an equivalent formulation for the VI $(F(x, \cdot), C(x))$ that will be useful for deriving optimality conditions for MPEC and designing solution methods.

Specifically, assume that for each $x \in X$,

$$C(x) \equiv \{y \in \Re^m : g(x,y) \leq 0\}, \tag{3}$$

where $g : \Re^{n+m} \to \Re^\ell$ is continuously differentiable and for each $x \in X$ and $i = 1, \ldots, \ell$, $g_i(x, \cdot)$ is a convex function in the second argument. An immediate consequence of this representation of $C(x)$ is that it renders the set-valued map $C : x \mapsto C(x)$ closed: that is, the graph $\mathrm{Gr}(C)$ is a closed set in $\Re^{n \times m}$. Note that the representation (3) allows for the presence of linear equality constraints defining $C(x)$.

Associated with the representation (3) of $C(x)$, we define the set-valued map $M : \Re^{n+m} \to \Re^\ell$, where for each $(x,y) \in \Re^{n+m}$, $M(x,y)$ is the (possibly empty) set of multipliers $\lambda \in \Re_+^\ell$ satisfying

$$
\begin{aligned}
F(x,y) + \sum_{i=1}^\ell \lambda_i \nabla_y g_i(x,y) &= 0, \\
\lambda_i \geq 0, \quad g_i(x,y) \leq 0, \quad \lambda_i g_i(x,y) &= 0, \quad \text{for } i = 1, \ldots, \ell.
\end{aligned}
\tag{4}
$$

Note that $\mathrm{dom}(M) \subseteq \mathrm{Gr}(C)$. Moreover, for each $(x,y) \in \mathrm{dom}(M)$, $M(x,y)$ is a nonempty convex polyhedron; indeed, (4) are just the Karush-Kuhn-Tucker (KKT) conditions for the VI $(F(x, \cdot), C(x))$ at the vector $y \in C(x)$; see [220]. Let

$$\mathcal{I}(x,y) \equiv \{i : g_i(x,y) = 0\}$$

be the index set of active constraints at the pair $(x,y) \in \mathrm{Gr}(C)$. In terms of this notation, we have the following explicit representation of the multiplier set $M(x,y)$:

$$
\begin{aligned}
M(x,y) = \{\lambda \in \Re_+^\ell : F(x,y) + \sum_{i \in \mathcal{I}(x,y)} \lambda_i \nabla_y g_i(x,y) &= 0, \\
\lambda_j = 0, \ \forall j \notin \mathcal{I}(x,y)\}.
\end{aligned}
\tag{5}
$$

As a set-valued map, the multiplier map M is closed. This can be seen easily by applying a limit argument to the KKT system (4).

We shall impose an important assumption on the map M. Implicit in this assumption is the inclusion: $\mathcal{F} \subseteq \mathrm{dom}(M)$, which simply says that for each vector (x,y) feasible to (1.1.1), KKT multipliers exist for the solution y to the VI $(F(x, \cdot), C(x))$. The full assumption is a technical condition which will be used to show that some (and not necessarily all) multipliers in the set $M(x,y)$ will be uniformly bounded for all $(x,y) \in \mathcal{F}$. More precisely, this assumption, which we shall refer to as the *sequentially bounded constraint qualification* (SBCQ) for the MPEC (1.1.1), is

(SBCQ): for any convergent sequence $\{(x^k, y^k)\} \subseteq \mathcal{F}$, there exists for each k a multiplier vector $\lambda^k \in M(x^k, y^k)$ and $\{\lambda^k\}$ is bounded.

The SBCQ is actually quite mild and is satisfied under a number of very common constraint qualifications (CQs) in nonlinear programming. We shall postpone the discussion of these other CQs. For now, we give several important consequences of the SBCQ on the inner VIs in the context of the MPEC.

1.3.4 Theorem. Let F and each g_i be continuous; let Z be closed. Suppose that each $g_i(x, \cdot)$ is convex for all $x \in X$ and $\nabla_y g_i(x, y)$ exists and is continuous at all points (x, y) in an open set containing $\mathcal{F}$. Assume that the SBCQ holds on $\mathcal{F}$ for the set-valued map M defined above, then the following two statements hold.

(a) Let $(x, y) \in Z$ be given. Then $y \in \mathcal{S}(x)$ if and only if $M(x, y) \neq \emptyset$.

(b) If Z is closed, then so is the feasible region $\mathcal{F}$.

Proof. Clearly, a vector y lies in $\mathcal{S}(x)$ if and only if y is an optimal solution of the following convex program in the variable v:

$$\begin{aligned} \text{minimize} \quad & v^T F(x, y) \\ \text{subject to} \quad & v \in C(x). \end{aligned} \tag{6}$$

Since this is a convex program, it follows that its KKT conditions, which are exactly those given in (4), are well defined by SBCQ and are both necessary and sufficient for the vector $y \in \mathcal{S}(x)$. Thus (a) holds.

To prove (b), let $\{(x^k, y^k)\} \subset \mathcal{F}$ be a sequence converging to the limit $(\bar{x}, \bar{y})$ which must necessarily belong to $Z \cap \mathrm{Gr}(C)$. It remains to show that $\bar{y} \in \mathcal{S}(\bar{x})$. By SBCQ, there exists a bounded sequence of multipliers $\{\lambda^k\}$ such that $\lambda^k \in M(x^k, y^k)$ for each k. Thus, we have

$$\begin{aligned} F(x^k, y^k) + \sum_{i=1}^{\ell} \lambda_i^k \nabla_y g_i(x^k, y^k) &= 0, \\ \lambda_i^k \geq 0, \quad g_i(x^k, y^k) \leq 0, \quad \lambda_i^k g_i(x^k, y^k) &= 0. \end{aligned} \tag{7}$$

Without loss of generality, we may assume that $\{\lambda^k\}$ converges to $\bar{\lambda}$. Taking the limit $k \to \infty$ in the above KKT conditions, we easily deduce that

$\bar{\lambda} \in M(\bar{x}, \bar{y})$. Hence, by (a), it follows that $\bar{y} \in \mathcal{S}(\bar{x})$. Thus $(\bar{x}, \bar{y}) \in \mathcal{F}$ and $\mathcal{F}$ is closed. **Q.E.D.**

The SBCQ is equivalent to two conditions taken together: (i) $M(x, y)$ is nonempty for all $(x, y) \in \mathcal{F}$, and (ii) for any bounded subset $S \subseteq \mathcal{F}$, there exists $\rho > 0$ such that $M(x, y) \cap \mathbb{B}(0, \rho) \neq \emptyset$ for all $(x, y) \in S$. Clearly, (i) and (ii) imply the SBCQ. The converse can be proved by contradiction. Roughly speaking, the SBCQ amounts to saying that "KKT conditions hold with bounded multipliers on bounded sets".

Example 1.1.2 revisited. We illustrate that the SBCQ fails for this example. As noted before, for $x < 0$, the only solution to the inner-level program is $y = 0$. It is easy to show that at $(\bar{x}, 0)$, where $\bar{x} < 0$, the multiplier corresponding to the constraint $xy \leq 0$ is equal to $-1/\bar{x}$ which tends to $-\infty$ as $\bar{x}$ approaches zero. The failure of the SBCQ provides one reason why the feasible region $\mathcal{F}$ is not closed in this example.

The KKT formulation

Returning to the MPEC (1.1.1) and utilizing the equivalence established in part (a) of Theorem **1.3.4**, we obtain an equivalent formulation of (1.1.1) via the introduction of the multiplier variables. Specifically, we have the following result.

1.3.5 Theorem. Under the assumptions of Theorem **1.3.4**, the following two statements hold.

(a) The problem (1.1.1) is equivalent to the following minimization problem in the variables (x, y, λ):

$$
\begin{aligned}
\text{minimize} \quad & f(x, y) \\
\text{subject to} \quad & (x, y) \in Z, \\
& F(x, y) + \sum_{i=1}^{\ell} \lambda_i \nabla_y g_i(x, y) = 0, \\
& \lambda \geq 0, \quad g(x, y) \leq 0, \quad \lambda^T g(x, y) = 0.
\end{aligned}
\tag{8}
$$

More precisely, if the pair (x^*, y^*) is a global minimizer of (1.1.1) then for any $\lambda^* \in M(x^*, y^*)$, the triple (x^*, y^*, λ^*) is a global minimizer of (8); conversely, if there exists a $\lambda^* \in M(x^*, y^*)$ such that the triple

(x^*, y^*, λ^*) is a global minimizer of (8), then the pair (x^*, y^*) is a global minimizer of (1.1.1).

(b) If in addition Z is compact, then there exists some $c > 0$ such that (1.1.1) is further equivalent, in the same sense, to:

$$
\begin{aligned}
\text{minimize} \quad & f(x, y) \\
\text{subject to} \quad & (x, y, \lambda) \in Z \times (\mathbb{B}(0, c) \cap \Re_+^\ell), \\
& F(x, y) + \sum_{i=1}^{\ell} \lambda_i \nabla_y g_i(x, y) = 0, \\
& g(x, y) \leq 0, \quad \lambda^T g(x, y) = 0.
\end{aligned}
\tag{9}
$$

Theorem **1.3.5** requires no proof: the existence of c is a simple consequence of the SBCQ condition and the compactness of Z. In Section **2.2** we use the formulation (9) to derive exact penalty functions for (1.1.1). We call this formulation the KKT constrained MP.

The equivalence of the problems (1.1.1) and (8) is with respect to the global minima. The matter is slightly different as far as the local minima of the two problems are concerned. This difference is due to the presence of the multipliers λ in (8). See Section **3.4** for more discussion of this issue; in particular, Proposition **3.4.1** summarizes the connection between the local minima of the problems (1.1.1) and (8).

There are several differences between the formulation (8) (or (9)) and (1.3.2). One is that in (1.3.2) there is no assumption on the form of the set $C(x)$. The other difference is that (1.3.2) is defined by a modified set of variables, (x, v) instead of (x, y), whereas (8) involves the additional multiplier vector λ. The latter formulation remains a smooth optimization problem (provided that f, F and g_i are smooth) whereas the former involves the nondifferentiable projection map.

Note that (8) is a well-defined optimization problem even without the convexity of the functions $g_i(x, \cdot)$. Such a convexity property is needed to ensure the equivalence of (8) with the MPEC (1.1.1). Although many results in this monograph assume the convexity of $g_i(x, \cdot)$, most of them remain valid without this assumption; in such cases, the conclusions apply to (8) but not necessarily to (1.1.1). For an example of a result of this kind, see Theorem **1.4.1**.

Define the (vector-valued) VI-Lagrangean function $L : Z \times \Re_+^\ell \to \Re^m$:

for $(x, y, \lambda) \in Z \times \Re^\ell_+$,

$$L(x, y, \lambda) \equiv F(x, y) + \sum_{i=1}^{\ell} \lambda_i \nabla_y g_i(x, y). \tag{10}$$

Introducing the slack variable u for the inequality constraint $g(x, y) \leq 0$, we define the function $G : Z \times \Re^{2\ell}_+ \to \Re^{m+\ell}$ as follows:

$$G(x, y, \lambda, u) \equiv \left(\begin{array}{c} L(x, y, \lambda) \\ u + g(x, y) \end{array} \right).$$

In terms of this G, the formulation (8) can be written in the following compact form:

$$\begin{aligned} \text{minimize} \quad & f(x, y) \\ \text{subject to} \quad & (x, y) \in Z \\ & G(x, y, \lambda, u) = 0 \\ & (\lambda, u) \geq 0, \quad \lambda \circ u = 0, \end{aligned} \tag{11}$$

where $a \circ b$ denotes the Hadamard product of two vectors; that is, the i-th component of $a \circ b$ is equal to $a_i b_i$. The problem (11) is the MPEC (1.1.1) expressed in mixed, nonlinear complementarity constraints. For arbitrary functions G, the formulation (11) can be used to model an optimization problem with general complementarity systems as constraints. With an abuse of terminology, we will call (11) an MPEC also.

The formulation (8) reveals an interesting structure of the feasible region $\mathcal{F}$ of the MPEC when $C(x)$ is given by (3) with each $g_i(x, \cdot)$ being convex. Indeed, for each subset $\mathcal{I}$ of $\{1, \ldots, \ell\}$ let

$$\begin{aligned} S_{\mathcal{I}} \equiv \{ (x, y, \lambda) \in Z \times \Re^\ell \; : \; & L(x, y, \lambda) = 0, \\ & \lambda_i = 0, \; g_i(x, y) \leq 0, \; i \in \mathcal{I}, \\ & \lambda_i \geq 0, \; g_i(x, y) = 0, \; i \notin \mathcal{I} \}; \end{aligned} \tag{12}$$

also let

$$S'_{\mathcal{I}} \equiv \{ (x, y) \in \Re^{n+m} \; : \; (x, y, \lambda) \in S_{\mathcal{I}} \text{ for some } \lambda \in \Re^\ell_+ \}.$$

It is then easy to see, assuming a constraint qualification such as the SBCQ on $\mathcal{F}$, that

$$\mathcal{F} = \bigcup_{\mathcal{I}} S'_{\mathcal{I}} \tag{13}$$

which expresses the feasible region $\mathcal{F}$ as a finite union of the sets $S'_{\mathcal{I}}$, called the *pieces* of $\mathcal{F}$. With an abuse of language, we also call each $S_{\mathcal{I}}$ a piece of $\mathcal{F}$. Note that some $S_{\mathcal{I}}$ (and thus $S'_{\mathcal{I}}$) might be empty. The expression (13) suggests that $\mathcal{F}$ is not likely to be convex in general, which is no surprise given Example **1.1.1**. Moreover, the number of pieces of $\mathcal{F}$, though finite, could be very large; in fact, it is often impossible to have a complete knowledge of all these pieces. Thus the representation (13) is seldom used for computational purposes; nevertheless it does have some theoretical importance in the study of the MPEC.

An important special case of the MPEC occurs when F and g are affine functions. In this case, the VI $(F(x,\cdot), C(x))$ becomes an affine variational inequality (AVI). Due to its significance, we temporarily depart from our discussion of MPEC and briefly review some basic properties of the AVI.

Let $M \in \Re^{m \times m}$, $q \in \Re^m$, and X be a polyhedral set in $\Re^m$. The *affine variational inequality*, which we denote AVI (q, M, X), is to find a vector $x^* \in X$ such that

$$(x - x^*)^T (q + Mx^*) \geq 0, \qquad \forall x \in X.$$

The solution set of AVI (q, M, X) is denoted $\mathrm{SOL}(q, M, X)$. When X is equal to the nonnegative orthant $\Re^m_+$, then the AVI (q, M, X) is equivalent to the linear complementarity problem defined by the vector q and matrix M, which we denote by LCP (q, M). We write $\mathrm{SOL}(q, M)$ for $\mathrm{SOL}(q, M, \Re^n_+)$. By using Proposition **1.3.3**, it is easy to show that

$$x \in \mathrm{SOL}(q, M, X) \iff x = \Pi_X(x - Mx - q)$$

where $\Pi_X(\cdot)$ is the Euclidean projector onto X. In the case where $X = \Re^n_+$, we have $\Pi_X(z) = \max(z, 0)$.

When X has an explicit representation of the form:

$$X \equiv \{x \in \Re^n : Ax \geq b\},$$

where $A \in \Re^{\ell \times m}$ and $b \in \Re^\ell$ are given, a vector $x \in \mathrm{SOL}(q, M, X)$ if and only if there exists a multiplier $\lambda \in \Re^\ell$ such that

$$q + Mx - A^T \lambda = 0,$$
$$Ax - b \geq 0, \quad \lambda \geq 0, \tag{14}$$
$$\lambda^T(Ax - b) = 0;$$

the latter KKT system is the equivalent mixed linear complementarity formulation of the AVI (q, M, X). The feasible region of the problem (14) is the set of vectors $(x, \lambda) \in \Re^{m+\ell}$ satisfying the defining conditions of this problem except for the complementarity condition $\lambda^T(Ax - b) = 0$.

We now return to the discussion of the MPEC where F and g are affine functions of the form:

$$F(x, y) \equiv q + Nx + My, \quad g(x, y) \equiv b + Ax + By, \qquad (15)$$

for some matrices $N \in \Re^{m \times n}$, $M \in \Re^{m \times m}$, $A \in \Re^{\ell \times n}$, $B \in \Re^{\ell \times m}$ and vectors $q \in \Re^m$, and $b \in \Re^\ell$, each piece S_I is a (possibly empty) polyhedron. Hence if Z is polyhedral, then from (13) it follows that $\mathcal{F}$ is the union of finitely many polyhedra. As we shall see subsequently, an affine function g of the above form will satisfy the SBCQ.

When F and g are given by (15), the formulation (8) has the following form:

$$\begin{aligned}
\text{minimize} \quad & f(x, y) \\
\text{subject to} \quad & (x, y) \in Z, \\
& q + Nx + My + B^T\lambda = 0, \qquad (16) \\
& \lambda \geq 0, \quad b + Ax + By \leq 0, \\
& \lambda^T(b + Ax + By) = 0.
\end{aligned}$$

With the exception of the joint constraint $(x, y) \in Z$, the remaining constraints of (16) are now in a mixed linear complementarity form. We call an AVI constrained mathematical program a Mathematical Program with Affine Equilibrium Constraints (MPAEC). As we shall see from the results in the subsequent chapters, the MPAEC shares much resemblance with a linearly constrained nonlinear program in standard NLP theory. In particular, we are able to obtain a fairly complete optimality theory for an MPAEC; it is also considerably easier to develop solution algorithms for this special class of MPECs.

All bilevel linear programs are MPAECs. In fact, they are LPECs; i.e., MPAECs whose first-level objective functions are linear. Another example of an MPAEC is the residual minimization problem of a linear complementarity problem (LCP), that is, the problem:

$$\text{minimize} \quad \tfrac{1}{2} \| \min(x, q + Mx) \|^2, \qquad (17)$$

for a given vector $q \in \Re^n$ and matrix $M \in \Re^{n \times n}$. Indeed, this problem is
equivalent to (1.2.17) with $f(x) \equiv q + Mx$. The latter problem is clearly an
instance of an MPAEC with a convex objective function. More generally,
the residual minimization problem of an AVI is an MPAEC of the same
type. Indeed let C be a polyhedron in $\Re^n$ and M an $n \times n$ matrix. Let
M_C be the normal map associated with this pair (M, C); that is

$$M_C(z) \equiv M\Pi_C(z) + z - \Pi_C(z), \quad \text{for } z \in \Re^n.$$

Let $q \in \Re^n$; consider the unconstrained least-squares problem:

$$\text{minimize} \quad \tfrac{1}{2} \|q + M_C(z)\|^2. \tag{18}$$

We claim that this problem can be formulated as an MPAEC. Specifically,
given the equivalence of solutions z of the normal equation

$$q + M_C(z) - x = 0$$

and solutions y of the AVI $(q - x, M, C)$ for each fixed x, the least-squares
problem (18) is equivalent to

$$\text{minimize} \quad \tfrac{1}{2} x^T x$$
$$\text{subject to} \quad y \in \text{SOL}(q - x, M, C).$$

This approach was taken for an NCP constrained MP in [98]; see also [72].
An alternative formulation can be made using the explicit representation
of C:

$$C \equiv \{x \in \Re^n : Ax \geq b\}$$

for some matrix A and vector b of appropriate dimensions. Then $\bar{z} = \Pi_C(z)$
if and only if there exists a multiplier vector λ such that

$$\bar{z} = z + A^T \lambda$$
$$A\bar{z} \geq b, \quad \lambda \geq 0, \quad \lambda^T(A\bar{z} - b) = 0.$$

Using the first equation to eliminate $\Pi_C(z)$, we deduce that the problem
(18) is equivalent to

$$\text{minimize} \quad \tfrac{1}{2} \|q + Mz + (M - I)A^T\lambda\|^2$$
$$\text{subject to} \quad -b + Az + AA^T\lambda \geq 0, \quad \lambda \geq 0 \tag{19}$$
$$\lambda^T(-b + Az + AA^T\lambda) = 0,$$

which is an MPAEC with linear complementarity constraints and a convex quadratic objective function; the upper-level variable is z and the lower-level variable is λ. Note also that for a fixed z, the lower-level LCP is defined by the symmetric positive semidefinite matrix AA^T.

It should be pointed out that when $A = I$ and $b = 0$, the problem (18) is the residual minimization problem of an LCP defined by the matrix M and vector q. Moreover, with the following identification of variables:

$$x \equiv z + \lambda, \quad y = x + \lambda - q - Mx,$$

this problem is equivalent to (1.2.17) with $f(x) = q + Mx$, or equivalently, to (17). In other words, for $C = \Re^n_+$, the two problems (17) and (18) are equivalent under a change of variables. This is consistent with the normal equation formulation of a VI which also involves a change of variables; cf. Proposition **1.3.3**.

Alternatively, one could use a polyhedral (such as the max or sum) norm to define the residual minimization problem (18) of the AVI (q, M, C). This will then lead to an LPEC which unlike (19) will have a linear first-level objective function. This alternative formulation was essentially the approach Mangasarian [188] took for the study of the "ill-posed" LCP.

Other constraint qualifications

We discuss how the SBCQ is implied by some renowned constraint qualifications in nonlinear programming. First is the Mangasarian-Fromovitz constraint qualification (MFCQ) [191, 93] on the set $Z \cap \mathrm{Gr}(C)$. Specifically, the MFCQ is said to hold at a vector $y \in C(x)$ if there exists a vector $v \in \Re^\ell$ such that

$$v^T \nabla_y g_i(x, y) < 0, \quad \text{for all } i \in \mathcal{I}(x, y). \tag{20}$$

By a theorem of the alternatives [180], it can be shown that the MFCQ holds at the pair $(x, y) \in \mathrm{Gr}(C)$ if and only if the implication below is valid:

$$\left. \begin{array}{c} \sum_{i \in \mathcal{I}(x,y)} \lambda_i \nabla_y g_i(x, y) = 0 \\ \lambda_i \geq 0, \ i \in \mathcal{I}(x, y) \end{array} \right\} \Rightarrow \lambda_i = 0, \ \forall i \in \mathcal{I}(x, y). \tag{21}$$

In turn, the latter implication holds if and only if the set $M(x, y)$ is nonempty and bounded [93].

In the case where $g_i(x, \cdot)$ is convex for each i (as we have assumed in much of our discussion), the above MFCQ is equivalent to the Slater CQ, that is, to the existence of a vector $\bar{y}$ such that $g(x, \bar{y}) < 0$. Indeed if such vector $\bar{y}$ exists, then for all $i \in \mathcal{I}(x, y)$, we have, by the gradient inequality for convex functions [256],

$$0 > g_i(x, \bar{y}) \geq g_i(x, y) + \nabla_y g_i(x, y)^T (\bar{y} - y).$$

Thus $v \equiv \bar{y} - y$ satisfies (20).   Conversely if a vector v satisfying (20) exists, then with $\bar{y} \equiv y + \tau v$ where $\tau > 0$ is sufficiently small, we have, by continuity, $g_i(x, \bar{y}) < 0$ for all $i \notin \mathcal{I}(x, y)$, and by (20), $g_i(x, \bar{y}) < 0$ for all $i \in \mathcal{I}(x, y)$.

Roughly speaking, the MFCQ implies that *all* multipliers are bounded on bounded sets, which is a stronger form of SBCQ. This conclusion is made precise in the following result.

1.3.6 Proposition. Let F and each g_i be continuous; let Z be closed. Suppose that $\nabla_y g_i(x, y)$ exists and is continuous at all points (x, y) in an open set containing $\mathcal{F}$. If the MFCQ holds at all pairs $(x, y) \in Z \cap \mathrm{Gr}(C)$, then the SBCQ holds on $\mathcal{F}$.

Proof. Let $\{(x^k, y^k)\} \subseteq \mathcal{F}$ be a sequence converging to some limit $(\bar{x}, \bar{y})$ which must necessarily belong to $Z \cap \mathrm{Gr}(C)$ by the closedness of the these sets. For each k, let $\lambda^k \in M(x^k, y^k)$ be arbitrary. If the sequence $\{\lambda^k\}$ is not bounded, then without loss of generality, we may assume that

$$\lim_{k \to \infty} \|\lambda^k\| = \infty, \quad \text{and} \quad \lim_{k \to \infty} \frac{\lambda^k}{\|\lambda^k\|} = \lambda^*,$$

for some $\lambda^* \in \Re_+^\ell \setminus \{0\}$.   Dividing (7) by $\|\lambda^k\|$ and letting $k \to \infty$, we deduce

$$\sum_{i=1}^{\ell} \lambda_i^* \nabla_y g_i(\bar{x}, \bar{y}) = 0.$$

But this contradicts (21), and therefore contradicts the MFCQ at $(\bar{x}, \bar{y})$. Hence the SBCQ holds on $\mathcal{F}$.   **Q.E.D.**

Another condition we shall consider is the constant-rank constraint qualification (CRCQ) [120, 227] on $Z \cap \mathrm{Gr}(C)$, which requires that for each $(x, y) \in Z \cap \mathrm{Gr}(C)$, there exists a neighborhood V of (x, y) such that for each subset $\mathcal{J}$ of $\mathcal{I}(x, y) \equiv \{i : g_i(x, y) = 0\}$, the family of gradients

$$\{\nabla_y g_i(u, v) : i \in \mathcal{J}\} \tag{22}$$

has the same rank (which depends on $\mathcal{J}$) for all vectors $(u, v) \in V$. The CRCQ is different from the MFCQ. For instance, an affine function g of the form:

$$g(x, y) = b + Ax + By, \quad \text{for } (x, y) \in \Re^{n+m}$$

for arbitrary matrices $A \in \Re^{\ell \times n}$ and $B \in \Re^{\ell \times m}$ and vector $b \in \Re^{\ell}$, easily satisfies the CRCQ but could violate the MFCQ; conversely, the following example shows that the MFCQ can hold whereas the CRCQ fails to hold.

1.3.7 Example. Let $Z = \Re^4$ and for $x, y \in \Re^2$,

$$g_1(x, y) \equiv y_1 + (y_2)^2 - x_1, \quad \text{and} \quad g_2(x, y) \equiv y_1 - x_2.$$

It is easy to see that with $\bar{x} = \bar{y} = (0, 0)$, $(\bar{x}, \bar{y}) \in Z \cap \mathrm{Gr}(C)$. We have

$$\nabla_y g_1(x, y) = \begin{pmatrix} 1 \\ 2y_2 \end{pmatrix}, \quad \text{and} \quad \nabla_y g_2(x, y) = \begin{pmatrix} 1 \\ 0 \end{pmatrix}.$$

Therefore the MFCQ holds at $(\bar{x}, \bar{y})$ but the CRCQ fails at this vector.

By a result of Janin [120], we deduce that under the CRCQ on $\mathcal{F}$, $M(x, y) \neq \emptyset$ for all $(x, y) \in \mathcal{F}$. The following result shows that CRCQ implies SBCQ; this result justifies, in particular, that the formulation (16) is completely equivalent to the MPAEC.

1.3.8 Proposition. Let F and each g_i be continuous; let Z be closed. Suppose that $\nabla_y g_i(x, y)$ exists and is continuous at all points (x, y) in an open set containing $\mathcal{F}$. If the CRCQ holds at all pairs $(x, y) \in Z \cap \mathrm{Gr}(C)$, then the SBCQ holds on $\mathcal{F}$.

Proof. Let $\{(x^k, y^k)\} \subseteq \mathcal{F}$ be any convergent sequence with limit $(\bar{x}, \bar{y})$ belonging to $Z \cap \mathrm{Gr}(C)$. We will show the existence of a bounded sequence of multipliers $\{\lambda^k\}$ such that $\lambda^k \in M(x^k, y^k)$ for all k. For each k, let $\mathcal{B}_k$ be the collection of index sets $\mathcal{J} \subseteq \mathcal{I}(x^k, y^k)$ such that the gradient vectors

$$\{\nabla_y g_i(x^k, y^k) : i \in \mathcal{J}\}$$

are linearly independent, and let $\nabla_y g_{\mathcal{J}}(x^k, y^k)^T$ be the corresponding matrix of order $m \times |\mathcal{J}|$ with these vectors as its columns. Also let

$$B_k(\mathcal{J}) \equiv \nabla_y g_{\mathcal{J}}(x^k, y^k) \nabla_y g_{\mathcal{J}}(x^k, y^k)^T, \quad \text{for } \mathcal{J} \in \mathcal{B}_k.$$

Clearly, each such matrix $B_k(\mathcal{J})$ is nonsingular. We claim that

$$\sup_k \max\{ \|B_k(\mathcal{J})^{-1}\| : \mathcal{J} \in \mathcal{B}_k\} < \infty. \tag{23}$$

Assume this is false; then there exist a subsequence $\{(x^k, y^k) : k \in \kappa\}$ and a sequence of index sets $\{\mathcal{J}_k : k \in \kappa\}$ such that $\mathcal{J}_k \in \mathcal{B}_k$ for each $k \in \kappa$ and $\{ \|B_k(\mathcal{J}_k)^{-1}\| \}$ is unbounded. Since there are only finitely many index subsets, we may assume without loss of generality that $\mathcal{J}_k$ is equal to a common index set, say $\mathcal{J}$, for all $k \in \kappa$. Clearly, $\mathcal{J} \subseteq \mathcal{I}(\bar{x}, \bar{y})$. Hence, by the CRCQ, it follows that the gradient vectors

$$\{\nabla_y g_i(\bar{x}, \bar{y}) : i \in \mathcal{J}\}$$

are linearly independent; the sequence of matrices $\{B_k(\mathcal{J}) : k \in \kappa\}$ obviously converges to the nonsingular matrix $\nabla_y g_{\mathcal{J}}(\bar{x}, \bar{y}) \nabla_y g_{\mathcal{J}}(\bar{x}, \bar{y})^T$. Hence, $\{B_k(\mathcal{J})^{-1} : k \in \kappa\}$ converges to the inverse of the latter matrix. This is a contradiction. Thus, (23) holds.

Now for each k, $M(x^k, y^k)$ is a nonempty polyhedron in $\Re_+^\ell$; thus it has at least one extreme point $\lambda^k \in M(x^k, y^k)$ with the property that the gradient vectors

$$\{\nabla_y g_i(x^k, y^k) : \lambda_i^k > 0\}$$

are linearly independent; moreover, $\mathcal{I} \equiv \{i : \lambda_i^k > 0\} \subseteq \mathcal{I}(x^k, y^k)$. Thus $\mathcal{I} \in \mathcal{B}_k$; since

$$F(x^k, y^k) + \nabla_y g_{\mathcal{I}}(x^k, y^k)^T \lambda_I^k = 0,$$

we have

$$\lambda_{\mathcal{I}}^k = -B_k(\mathcal{I})^{-1} \nabla_y g_{\mathcal{I}}(x^k, y^k) F(x^k, y^k).$$

Consequently, by (23), the convergence of $\{(x^k, y^k)\}$, and the continuity of the functions $\nabla_y g$ and F, we deduce that $\{\lambda^k\}$ must be bounded. **Q.E.D.**

Both the MFCQ and the CRCQ on $Z \cap \text{Gr}(C)$ are implied by the linear independence constraint qualification (LICQ) on $Z \cap \text{Gr}(C)$, which states that for all vectors $(x, y) \in Z \cap \text{Gr}(C)$, the gradient vectors

$$\{\nabla_y g_i(x, y) : i \in \mathcal{I}(x, y)\}$$

are linearly independent. Moreover, if the LICQ is valid, then $M(x,y)$ is a singleton for all $(x,y) \in Z \cap \mathrm{Gr}(C)$.

1.3.3 Merit functions for CP/VI

The complementarity conditions:

$$0 \le u \perp v \ge 0,$$

where for two vectors $u, v \in \Re^n$, $u \perp v$ means $u^T v = 0$, have appeared in several equivalent formulations of the constraints in the MPEC. It is possible to express these conditions as an equivalent system of equations, or even more simply, as a single differentiable equation. Loosely defined, any function defining such an equation is called a *merit function* for the above complementarity conditions. In what follows we describe several equivalent formulations of the complementarity and/or VI constraints in an MPEC using some renowned merit functions.

We define the convex functional $\varphi : \Re^2 \to \Re$:

$$\varphi(a,b) \equiv \sqrt{a^2 + b^2} - (a+b), \quad (a,b) \in \Re^2.$$

Fischer [76] credited W. Burmeister with the introduction of this functional which was used for the first time by Fischer [75] to transform the KKT conditions of a nonlinear program into a system of equations. The basic property of the functional φ is that

$$\min(a,b) = 0 \iff \varphi(a,b) = 0.$$

With this observation, the problem (11) admits the equivalent formulation:

$$\begin{aligned}
\text{minimize} \quad & f(x,y) \\
\text{subject to} \quad & (x,y) \in Z, \\
& G(x,y,\lambda,u) = 0, \\
& \varphi(\lambda_i, u_i) = 0, \quad \forall i,
\end{aligned} \tag{24}$$

where the lower-level constraints are all cast as equations. By introducing the aggregate functional:

$$\Phi(u,v) \equiv \sum_{i=1}^{\ell} \varphi(u_i, v_i)^2, \quad \text{for } (u,v) \in \Re^{2\ell},$$

the problem (24) can be written more simply as

$$
\begin{aligned}
\text{minimize} \quad & f(x, y) \\
\text{subject to} \quad & (x, y) \in Z, \\
& G(x, y, \lambda, u) = 0, \\
& \Phi(\lambda, u) = 0.
\end{aligned}
\tag{25}
$$

Properties of the functionals φ and Φ have been obtained by Fischer [75]. In particular, φ is convex, subadditive, positively homogeneous, globally Lipschitz continuous, and continuously differentiable everywhere in $\Re^2$ except at the origin; moreover, Ψ is continuously differentiable and nonnegative everywhere on $\Re^n$. Based on these two functionals, highly effective iterative descent algorithms have been proposed for solving nonlinear programs and linear and nonlinear complementarity problems [75, 76, 77, 123, 58, 289]. See [277] for interesting growth properties of the function Φ.

Functions similar to Φ have been introduced for a variational inequality. Specifically, for the VI (G, K) where $G : \Re^n \to \Re^n$ is C^1 and K is a closed convex subset of $\Re^n$, consider the functional:

$$
g_\alpha(x) \equiv \max_{y \in X} \left(G(x)^T (x - y) - \frac{\alpha}{2} \| y - x \|^2 \right), \quad x \in \Re^n,
$$

where $\alpha \geq 0$ is a fixed scalar and $\| \cdot \|$ denotes the Euclidean norm of vectors. For $\alpha = 0$, this is the *gap function* introduced by Auslender [10] for the VI and revisited by Hearn [110] for a convex program. In general, the gap function g_0 is not F-differentiable; nevertheless, for $\alpha > 0$, g_α is continuously differentiable (recall, by our assumption, G is C^1). This *regularized gap function*, as g_α is called for $\alpha > 0$, was discovered independently by Auchmuty [8] and Fukushima [86]. Observing that $g_\alpha(x) \geq 0$ for all $x \in K$, we may conclude that

$$
[u \in \operatorname{argmin} \{ g_\alpha(x) : x \in K \} \text{ and } g_\alpha(u) = 0]
$$

$$
\Longleftrightarrow [g_\alpha(u) = 0,\ u \in K] \Longleftrightarrow [u \in \mathrm{SOL}(G, K)],
$$

where $\mathrm{SOL}(G, K)$ denotes the solution set of the VI (G, K). Thus g_α is an example of a *constrained merit function* for the VI (G, K). By exploiting the above equivalences, descent algorithms for solving this VI have been proposed that are based on decreasing the function g_α; see [198, 200, 86, 274].

There are many other merit functions for CPs and VIs; the survey [87] presents an excellent summary of efforts in this area. In particular, we mention the one introduced by Peng [230] and extended by Yamashita, Taji, and Fukushima [290]. This *unconstrained merit function* for the VI (G, K) is defined to be

$$h_{\alpha\beta}(x) \equiv g_\beta(x) - g_\alpha(x), \quad \text{for } x \in \Re^n,$$

where $\beta > \alpha > 0$ are given scalars. It has been shown in the cited references that an unconstrained global minimum $u \in \Re^n$ of $h_{\alpha\beta}$ on $\Re^n$ with $h_{\alpha\beta}(u) = 0$ is a solution of the VI (G, K), and vice versa; thus the term "unconstrained merit function". When $\beta \equiv 1/\alpha$ and $K \equiv \Re^n_+$, $h_{\alpha\beta}$ reduces to the *implicit Lagrangean function* for the NCP introduced by Mangasarian and Solodov [196] and further studied in [170, 288, 67, 89].

In principle, we could use any one of the above merit functions to restate the VI/NCP constraints in the MPEC (1.1.1) as a single equation. We illustrate such a formulation for the NCP constrained MP:

$$\begin{aligned}
&\text{minimize} \quad f(x, y) \\
&\text{subject to} \quad (x, y) \in Z, \\
&\text{and} \quad 0 \le y \perp F(x, y) \ge 0.
\end{aligned} \tag{26}$$

Using the implicit Lagrangean function: for $\alpha > 0$,

$$m_\alpha(x, y) \equiv y^T F(x, y) +$$

$$\tfrac{\alpha}{2} \left(\| [y - \alpha^{-1} F(x, y)]^+ \| + \| [F(x, y) - \alpha^{-1} y]^+ \| - \| y \|^2 - \| F(x, y) \|^2 \right),$$

we obtain the following equivalent formulation of (26):

$$\begin{aligned}
&\text{minimize} \quad f(x, y) \\
&\text{subject to} \quad (x, y) \in Z, \\
&\text{and} \quad m_\alpha(x, y) = 0.
\end{aligned} \tag{27}$$

The extent to which this and other related formulations can be used for the study of the MPEC is an area that has not been investigated. When this manuscript was completed, there had been only two references [88, 68] which discussed the use of the formulation (24) in the design of globally convergent iterative methods for solving the MPEC. These reformulations seem promising and deserve further investigation.

1.3.4 *Monotonicity and the implicit form*

If the upper-level feasible region Z is of the form

$$Z = X \times \Re^m$$

for some closed convex set $X \subseteq \Re^n$ and if the solution map $\mathcal{S}$ is single-valued on X, that is, if $\mathcal{S}(x)$ is a singleton, which we write as $\{y(x)\}$, for all $x \in X$, then the MPEC (1.1.1) can be written as an (implicit) optimization problem in the variable x alone:

$$\begin{aligned} \text{minimize} \quad & \tilde{f}(x) \\ \text{subject to} \quad & x \in X, \end{aligned} \tag{28}$$

where

$$\tilde{f}(x) \equiv f(x, y(x)).$$

With a strong monotonicity assumption on the equilibrium function $F(x, \cdot)$, the set $\mathcal{S}(x)$ can be shown to be a singleton.

1.3.9 Definition. A mapping $G : K \subseteq \Re^m \to \Re^m$ is said to be

(a) *monotone* on K if for all pairs $(u, v) \in K \times K$,

$$(u - v)^T (G(u) - G(v)) \geq 0;$$

(b) *strictly monotone* on K if for all pairs $(u, v) \in K \times K$ with $u \neq v$,

$$(u - v)^T (G(u) - G(v)) > 0;$$

(c) *strongly monotone* on K if there exists a constant $c > 0$, called the *modulus*, such that for all pairs (u, v) in $K \times K$,

$$(u - v)^T (G(u) - G(v)) \geq c \|u - v\|^2.$$

If K is convex and G is continuously differentiable on an open set containing K, then G is monotone on K if and only if the Jacobian matrix $\nabla G(x)$ is positive semidefinite for all $x \in K$; if $\nabla G(x)$ is positive definite for all $x \in K$, then G is strictly monotone on K; G is strongly monotone on K if and only if $\nabla G(x)$ is uniformly positive definite for all $x \in K$; that is, there exists a constant $c' > 0$ such that $y^T \nabla G(x) y \geq c' y^T y$ for all $x \in K$ and $y \in \Re^m$.

Monotonicity plays an important role in the existence and uniqueness of solutions to VIs. The following proposition summarizes some key results for a monotone VI.

1.3.10 Proposition. Let K be a closed convex set in $\Re^m$ and $G : K \to \Re^m$ be a continuous mapping. Let $\mathrm{SOL}(G, K)$ denote the (possibly empty) solution set of the VI (G, K).

(a) If G is monotone on K, then $\mathrm{SOL}(G, K)$, if nonempty, is a closed convex set.

(b) If G is strictly monotone on K, then $\mathrm{SOL}(G, K)$ consists of at most one element.

(c) If G is strongly monotone on K, then $\mathrm{SOL}(G, K)$ consists of exactly one element.

In many special cases, the set K is equal to the Cartesian product of finitely many sets of lower dimensions; that is,

$$K \equiv \prod_{i=1}^{p} K_i, \tag{29}$$

where $K_i \subseteq \Re^{m_i}$ and $\sum_{i=1}^{p} m_i = m$. An example of such a set K is the nonnegative orthant $\Re^m_+$. If K is given by (29), then the above monotonicity concepts can be weakened and an existence and uniqueness result analogous to Proposition **1.3.10** can be established.

Specifically, we say that the mapping G is *uniform P* on the set K given by (29) if there exists a constant $c > 0$, called a *modulus*, such that

$$\max_{1 \leq i \leq p} (u_i - v_i)^T (G_i(u) - G_i(v)) \geq c \|u - v\|^2,$$

for all pairs $(u, v) \in K \times K$, where $u \equiv (u_i)$, $v \equiv (v_i)$, $G(u) \equiv (G_i(u))$, and $G(v) \equiv (G_i(v))$, and for each i, $(u_i, v_i) \in K_i \times K_i$ and $(G_i(u), G_i(v)) \in \Re^{2m_i}$. Since for all vectors u and v,

$$\max_{1 \leq i \leq p} (u_i - v_i)^T (G_i(u) - G_i(v)) \geq \frac{1}{p} (u - v)^T (G(u) - G(v)),$$

it follows that a strongly monotone mapping must be uniform P; the converse is not true in general.

Analogous to part (c) of Proposition **1.3.10**, we have the following result.

1.3.11 Proposition. Let K_i be a closed convex set in $\Re^{m_i}$ for all i, and $G : K \to \Re^m$ be a continuous mapping, where K is given by (29). If G is uniform P on K, then SOL(G, K) is a singleton.

When G is an affine mapping, say $G(x) \equiv q + Mx$ for some vector $q \in \Re^m$ and matrix $M \in \Re^{m \times m}$, and each K_i is the interval $[0, \infty)$ for $i = 1, \ldots, m$, then the VI (G, K) is equivalent to the LCP:

$$x \geq 0, \quad q + Mx \geq 0, \quad x^T(q + Mx) = 0,$$

which we denote by the pair (q, M). In this case, G is a uniform P function on K, which is the nonnegative orthant $\Re^n_+$ of $\Re^n$, if and only if M satisfies

$$\max_{1 \leq i \leq m} x_i(Mx)_i > 0, \quad \text{for all } 0 \neq x \in \Re^n. \tag{30}$$

A real square matrix M with the latter property is called a *P matrix*. There are many interesting characterizations of such a matrix; see Section 3.3 in [54]. Here we mention two of these that will be useful in this monograph.

1.3.12 Proposition. For a given $M \in \Re^{m \times m}$, the following three statements are equivalent.

(a) M satisfies (30).

(b) All principal minors of M are positive.

(c) The LCP (q, M) has a unique solution for all vectors $q \in \Re^m$.

Returning to the MPEC (1.1.1), we conclude that if for each $x \in X$, $F(x, \cdot)$ is strongly monotone on the closed convex set $C(x)$, then (28) is a valid, equivalent formulation of (1.1.1). For the NCP constrained MP (26) where $Z \equiv X \times \Re^m$, the equivalent implicit formulation (28) is valid if for each $x \in X$, $F(x, \cdot)$ is a uniform P function on $\Re^m_+$.

Admittedly, the assumption that $F(x, \cdot)$ is strongly monotone (or uniform P) for all $x \in X$ is rather restrictive. There are several relaxations of this assumption under which an implicit solution function $y(\cdot)$ will exist near any given $x \in X$ (though S may no longer be a single-valued map). Such a function $y(\cdot)$ will typically have certain interesting properties that can be put to use for a local analysis (such as the characterization of stationarity) and also for computational purposes; see Sections **4.2** and **6.3**.

Alternatively, without assuming the single-valuedness of the reaction map S, another implicit program formulation of the MPEC (1.1.1) with $Z = X \times \Re^m$ is possible by recalling the discussion about the weak Stackelberg game model; see Section **1.2**. There it was noted that (1.1.1) is equivalent to (1.2.6) which is a nonlinear program in the variable x only. Although it would be valuable to explore the formulation (1.2.6) for the study of the MPEC, we will not do so in this monograph.

There is a common deficiency with the implicit formulation (28) and the alternative (1.2.6); namely, the upper-level constraint region Z is restricted to be of the form $X \times \Re^m$; in other words, no joint constraints among the variables x and y are allowed except those in the lower level. For (28), this restriction is needed in view of the fact that it is rather unlikely for the implicit (single-valued) solution function $y(x)$ to exist when there are additional constraints on the variable y. As we shall see later, as far as the optimality conditions for MPEC are concerned, such a restriction is not necessary; see Subsection **3.3.1** and Sections **4.3** and **5.6**. Also, a certain family of iterative algorithms can be developed without requiring Z to be of the special form; see Section **6.4**.

1.4 Existence of Optimal Solutions

As an optimization problem, a general result can be stated for the existence of an optimal solution to the MPEC (1.1.1). We say that a real-valued function $\theta : S \subseteq \Re^N \to R$ is *coercive* on the set S if

$$\lim_{\|x\| \to \infty, x \in S} \theta(x) = \infty.$$

It is easy to show that if S is a closed subset of $\Re^N$ and θ is a continuous, coercive function on S, then θ achieves its minimum on S. More generally, if S is closed, θ is continuous and there exists a vector $x^0 \in S$ such that the *level set*

$$\{x \in S : \theta(x) \le \theta(x^0)\}$$

is bounded (and thus compact), then θ achieves its minimum on S. Applying these remarks to the MPEC in its KKT formulation (1.3.8), we obtain the following existence result for the problem (1.3.8) in which the convexity of the functions $g_i(x, \cdot)$ is not essential.

1.4.1 Theorem. Let Z be a closed set, and let f, g, F, $\nabla_y g_i$ be continuous. Let $\mathcal{F}^{\text{KKT}} \subseteq \Re^{n+m} \times \Re^\ell_+$ denote the feasible region of (1.3.8).

(a) If there exists $(x^0, y^0, \lambda^0) \in \mathcal{F}^{\text{KKT}}$ such that the set

$$\{(x, y, \lambda) \in \mathcal{F}^{\text{KKT}} : f(x, y) \le f(x^0, y^0)\}$$

is bounded, then the problem (1.3.8) has an optimal solution.

(b) If the SBCQ holds on $\mathcal{F}$, each $g_i(x, \cdot)$ is convex, and there exists $(x^0, y^0) \in \mathcal{F}$ such that the set

$$\{(x, y) \in \mathcal{F} : f(x, y) \le f(x^0, y^0)\}$$

is bounded, then (1.1.1) has an optimal solution.

Proof. The assumptions imply that $\mathcal{F}^{\text{KKT}}$ is a closed set. Thus part (a) follows from the above discussion. By Theorem **1.3.4**(b), $\mathcal{F}$ is a closed set under the assumptions in part (b) of the present theorem. Hence the desired conclusion follows similarly.   **Q.E.D.**

In [294], Zhang has obtained an existence result for the MPEC when the lower-level problem is an optimization problem; in essence he assumed a certain continuity condition on the "value function" of the lower-level optimization problem under which the closedness of the feasible region $\mathcal{F}$ can be established. As we have explained above, the closedness property of $\mathcal{F}$ together with an "inf-compactness" condition on the objective function f (i.e., the level sets of f are bounded) will yield the existence of an optimal solution to the MPEC.

When F and g are affine functions (see (1.3.15)), and f is a quadratic function, we can use the renowned Frank-Wolfe theorem in quadratic programming [81] to obtain a necessary and sufficient condition for the existence of an optimal solution to the associated MPEC. We say that a function $\theta : S \subseteq \Re^N \to \Re$ is *bounded below* on S if there exists a scalar α such that $\theta(x) \geq \alpha$ for all $x \in S$. Clearly, the property of being bounded below on S is a necessary condition for θ to attain its minimum on S. The Frank-Wolfe theorem states that this condition is also sufficient if θ is a quadratic function and S is polyhedral. A refinement of this theorem was obtained by Eaves [64] who showed that a quadratic function is bounded below on a polyhedral set if and only if it is bounded below on every feasible ray of the set. These results can be extended to the MPEC with a quadratic objective function f and AVI constraints.

In general, given a subset S of $\Re^N$, we say that a vector $v \in \Re^N$ is a *recession direction* of S [256] if there exists a vector $x \in S$ such that $x + \tau v \in S$ for all $\tau \geq 0$. The set of all recession directions of S forms a cone in $\Re^N$ and is denoted $0^+ S$; for a nonconvex set S, such as the feasible region $\mathcal{F}$ of the MPEC (1.1.1), $0^+ S$ is not necessarily convex. A feasible ray of S is a set of the form $\{x + \tau v : \tau \geq 0\}$ for some $x \in S$ and $v \in 0^+ S$.

The following result, Proposition **1.4.2**, identifies a superset of $0^+ \mathcal{F}$ for the feasible region of an MPAEC. (Note that $\mathcal{F}$, and thus $0^+ \mathcal{F}$, is nonconvex even for an MPAEC.) Although it is possible to extend this result to yield a complete characterization of a feasible ray of $\mathcal{F}$ in this case, using the representation (1.3.13) of $\mathcal{F}$ in terms of its polyhedral pieces, we will omit this extension because it is not essential to the rest of the book.

1.4.2 Proposition. Let F and g be given by (1.3.15). If $(\bar{u}, \bar{v}) \in 0^+ \mathcal{F}$ then $(\bar{u}, \bar{v}) \in 0^+ Z$ and $\bar{v}$ is a solution of the AVI defined by the affine map $v \mapsto N\bar{u} + Mv$ and the polyhedral set $\{v \in \Re^m : A\bar{u} + Bv \leq 0\}$.

Proof. By assumption, there exists $(\bar{x}, \bar{y}) \in \mathcal{F}$ such that $(\bar{x}, \bar{y}) + \tau(\bar{u}, \bar{v}) \in \mathcal{F}$ for all $\tau \geq 0$. Hence $(\bar{u}, \bar{v}) \in 0^+ Z$ and $\bar{v}$ satisfies $A\bar{u} + B\bar{v} \leq 0$. To show that $\bar{v}$ is a solution of the AVI as stated, let v satisfy $A\bar{u} + Bv \leq 0$. Then for all $\tau \geq 0$, $\bar{y} + \tau v \in C(\bar{x} + \tau \bar{u})$. Since $\bar{y} + \tau \bar{v} \in \mathcal{S}(\bar{x} + \tau \bar{u})$, we deduce

$$(v - \bar{v})^T \left(q + N(\bar{x} + \tau \bar{u}) + M(\bar{y} + \tau \bar{v}) \right) \geq 0.$$

Dividing by $\tau > 0$ and passing to the limit $\tau \to \infty$, we obtain

$$(v - \bar{v})^T (N\bar{u} + M\bar{v}) \geq 0,$$

as desired. **Q.E.D.**

The following result gives necessary and sufficient conditions for the MPAEC (1.3.16) to have an optimal solution; it extends the aforementioned result due to Frank-Wolfe and Eaves for quadratic programs.

1.4.3 Theorem. Let Z be a polyhedron and let f be a quadratic function. Suppose $\mathcal{F} \neq \emptyset$. The following statements are equivalent.

(a) The MPAEC (1.3.16) has a globally optimal solution.

(b) f is bounded below on $\mathcal{F}$.

(c) f is bounded below on feasible rays of $\mathcal{F}$.

Proof. Clearly (a) $\Rightarrow$ (b) $\Rightarrow$ (c). It remains to show (c) $\Rightarrow$ (a). Suppose that f is bounded below on feasible rays of $\mathcal{F}$. Then f must be bounded below on feasible rays of the nonempty pieces $S_{\mathcal{I}}$ for all subsets $\mathcal{I}$ of $\{1, \ldots, \ell\}$, where $S_{\mathcal{I}}$ is defined in (1.3.12). Since each piece $S_{\mathcal{I}}$ is polyhedral, it follows that f attains its minimum on each of the nonempty pieces $S_{\mathcal{I}}$. Since the number of pieces is finite, (a) follows.   **Q.E.D.**

As an application of Theorem **1.4.3**, we show that the residual minimization problem of an AVI (1.3.18) always has an optimal solution, provided that the set C is nonempty. In particular, the residual minimization problem of an LCP (1.3.17) always has an optimal solution.

1.4.4 Corollary. Let C be a nonempty polyhedron in $\Re^n$. The least-squares problem (1.3.18) has an optimal solution.

Proof. Consider the equivalent MPAEC formulation (1.3.19) of (1.3.18). Let $\bar{z} \in C$; then this vector $\bar{z}$ and $\lambda = 0$ constitute a feasible solution to (1.3.19). Moreover, the objective function of this problem is nonnegative, thus bounded below by zero. The existence of an optimal solution to (1.3.19) thus follows from Theorem **1.4.3**.   **Q.E.D.**

In the above existence results, an inequality representation has been assumed for the sets $C(x)$. In general, the explicit representation can be replaced by a "continuity" assumption on the set-valued map C under which the MPEC (1.1.1) will continue to have an optimal solution. (This includes the case where $C(x)$ is an abstract closed convex set that is independent of x.) Since this monograph deals almost exclusively with the case of $C(x)$ with a finite inequality representation, we omit the details of the more general existence result; see [108].

2

Exact Penalization of MPEC

For reasons given in Section **1.1**, we have seen that the general MPEC (1.1.1) is a very difficult constrained optimization problem to deal with. One reason is that the feasible region of the MPEC is defined implicitly as the solution set of a parametric variational inequality. To facilitate the design of solution procedures for MPEC, we need to represent its constraints in terms of a finite system of (nonlinear) equalities and inequalities, thus casting MPEC in the form of a standard nonlinear program. In Section **1.3**, we have given various equivalent formulations of the constraints of the MPEC; in particular, the KKT forms (1.3.8) and (1.3.9) will play a major role in the study of the MPEC (1.1.1).

Once an MPEC has been formulated as a standard constrained optimization problem, such as (1.3.9), we may attempt to use any one of a number of nonlinear programming approaches to deal with it. Traditionally, penalty functions provide a powerful approach, both as a theoretical tool and as a computational vehicle, for the study of mathematical programs. Based on a recent exact penalty function theory of subanalytic optimization problems first obtained by Warga [283] and subsequently extended by Dedieu [57], this chapter develops a general exact penalization

theory for the MPEC (1.1.1) under some mild continuity and subanalyticity assumptions on the objective and constraint functions of MPEC. We
also obtain variations and improvements of the basic exact penalty function results. The principal tool that enables us to develop this theory is
the theory of error bounds for systems of analytic inequalities, particularly
those for quadratic inequalities. In turn, the latter theory has in recent
years received a great deal of attention in the mathematical programming
literature, due to its prevalence in optimization and equilibrium problems.

A brief motivation

Before embarking on the derivation of the exact penalty results, we
find it useful to motivate the approach to be taken herein. The theory
of penalty functions for constrained optimization problems has a long history; the recent article by Burke [41] provides a comprehensive review of
the theory of exact penalization and gives an elegant, unified approach for
deriving many known results in the literature. We elicit two particularly
important messages from Burke's theory. One is the fact that well-known
optimality conditions for nonlinear programs are often obtained as a consequence of what we would call *exact penalization of order* 1; we explain
the latter concept later. The other important point is that a constraint
regularity condition is needed for such exact penalization to hold. On close
inspection of the regularity condition, one realizes that this condition can
be considered as a certain error bound on perturbations of the constraint
set of the nonlinear program. This point of view, which connects the theory of exact penalization with that of error bounds for constraint sets, is
consistent with the recent theory of Warga [283] and Dedieu [57] which
is based on Lojasiewicz' inequality for "subanalytic" sets [113, 164]; the
definitions of these sets and related concepts are reviewed in the next subsection. The main difference between this novel theory and the classical
regularity theory (as explained in [41]) is that the regularity assumption in
the latter theory is replaced by some generalized analytic properties of the
constraint set of the nonlinear program. These analytic properties are very
broad and often can be satisfied by sets that fail the regularity assumption.
Thus, the results of Warga and Dedieu have broadened the classical exact
penalty function theory considerably. The results obtained under the subanalytic assumption could be called *exact penalization results of order* γ,

where γ is a scalar in the interval $(0, 1]$. The case where $\gamma < 1$ occurs when the regularity assumption fails to hold.

The development of this chapter starts from an extension of the work by Warga and Dedieu for general nonlinear programs with subanalytic constraints. Our approach is based on Lojasiewicz' error bound for the subanalytic functions. We believe that this approach is appropriate for the MPEC because the approach allows us to establish exact penalty functions under some broad analytic conditions; more importantly, the derived results will highlight the essential features of the MPEC as a difficult nonlinear program and provide the motivation for the subsequent development on error bounds. In order to apply this theory of exact penalization to MPEC, we rely on the formulation of MPEC as a standard constrained optimization problem by using the KKT representation of the inner variational inequality. Then we apply Lojasiewicz' error bound to this KKT formulation to obtain an exact penalty function for MPEC. As will be seen, this penalty function is equal to the residual function of the KKT system raised to a fractional power γ. Typically, γ is only guaranteed to exist theoretically and the theory provides no clue as to how this exponent can be calculated (or just estimated) efficiently in practice. The final two sections of this chapter are devoted to the derivation of improved penalty functions for some special classes of MPEC. The latter penalty functions all have the exponent γ equal to either 1 or $\frac{1}{2}$, and they are based on the various sharpened error bound results developed in Section **2.3**. The chapter ends with a brief, preliminary discussion of optimality conditions for MPEC.

2.1 General Exact Penalty Results

We recall that a real-valued function f defined on open subset U of $\Re^n$ is *analytic* if it can be represented by a convergent power series in the neighborhood of any point of U; a vector-valued function F from the open set U into $\Re^m$ is *analytic* if each of its component functions is analytic. A subset X of $\Re^n$ is *semianalytic* if for each vector $a \in \Re^n$ there are a neighborhood U of a and a finite family of sets $X_{ij} \subseteq \Re^n, i = 1, \ldots, p$, and $j = 1, \ldots, q$, each of the form

$$X_{ij} \equiv \{x \in U : f_{ij}(x) = 0\} \quad \text{or} \quad \{x \in \Re^n : f_{ij}(x) < 0\}$$

for some real analytic function f_{ij} on U, such that

$$X \cap U = \bigcup_{i=1}^{p} \bigcap_{j=1}^{q} X_{ij}.$$

A subset X of $\Re^n$ is *subanalytic* if for each vector $a \in \Re^n$ there exists a neighborhood U of a such that $X \cap U$ is the projection of a bounded semianalytic set $A \subset \Re^{n+p}$ for some nonnegative integer p; i.e.,

$$X \cap U = \Pi(A),$$

where $\Pi : \Re^{n+p} \to \Re^n$ is given by $\Pi(x, y) = x$ for all $(x, y) \in \Re^{n+p}$. A vector-valued function is *semianalytic (subanalytic)* if its graph is semianalytic (subanalytic).

The reference [26] provides an extensive study of semianalytic and subanalytic sets, and [57, Section 2] summarizes several important examples of these sets and functions. In the former reference, credit was given to M.S. Lojasiewicz [163, 164] for the origin of the theory of semianalytic and subanalytic sets. Of interest to us are the following facts:

(a) piecewise analytic functions defined over a semianalytic partition are semianalytic (thus subanalytic);

(b) the pointwise supremum of a finite family of continuous subanalytic functions is subanalytic;

(c) the class of continuous subanalytic functions defined on a compact subanalytic set is closed under algebraic operations;

(d) the image of a bounded subanalytic set by a subanalytic function is subanalytic but this property is not valid if "subanalytic" is replaced by "semianalytic".

In particular, semianalytic and subanalytic functions need not be smooth.

Unlike the class of analytic functions (which we assume is familiar to a general reader of this monograph), the class of subanalytic functions is rather abstract and much less known to researchers in the mathematical programming community. Nevertheless, since the latter class of functions is considerably broader than the former class (and more importantly, contains functions that need not be differentiable), we feel that it is useful to develop our results using the subanalytic functions and sets. As an aid to those readers uncomfortable with the notion of subanalyticity, we recommend that they replace subanalyticity by (piecewise) analyticity everywhere. Consider now the following nonlinear program (NLP):

$$\text{minimize} \quad \theta(x)$$

$$\text{subject to} \quad x \in X, \tag{1}$$

$$g(x) \le 0, \quad h(x) = 0,$$

where $\theta : \Re^n \to \Re$, $g : \Re^n \to \Re^m$, and $h : \Re^n \to \Re^q$ are continuous functions and X is a subset of $\Re^n$. Let

$$r(x) = \sum_{i=1}^{m} (g_i(x))_+ + \sum_{j=1}^{q} |h_j(x)|$$

denote the residual of the functional constraints in (1) at a point x in X, where the subscript "+" denotes the nonnegative part of a scalar. In this and the next section, unless otherwise specified an optimal solution (or a minimum) of a minimization problem (such as (1), (1.1.1), and all their penalty function formulations) always refers to a global solution (or minimum).

Adding to the results in [283, 57], we prove a theorem which gives an exact penalty function formulation for the problem (1) under the assumption that θ is Lipschitz continuous, the constraint functions are continuous subanalytic, and X is a compact subanalytic set in $\Re^n$. Similar to the proof in [57], our proof also uses the following result known as Lojasiewicz' inequality [26, Theorem 6.4].

2.1.1 Theorem. Let $\phi, \psi : S \to \Re$ be two continuous subanalytic functions defined on the compact subanalytic set $S \subset \Re^n$. If $\phi^{-1}(0) \subseteq \psi^{-1}(0)$, then there exist a scalar $\rho > 0$ and an integer $N^* > 0$ such that for all $x \in S$,

$$\rho |\psi(x)|^{N^*} \le |\phi(x)|.$$

Thus, if the reverse inclusion $\psi^{-1}(0) \subseteq \phi^{-1}(0)$ also holds, then the two functions ϕ and ψ are "equivalent" on S in the sense that there exist scalars ρ_1 and ρ_2, both positive, and positive integers N_1 and N_2, such that for all $x \in S$,

$$\rho_1 |\psi(x)|^{N_1} \le |\phi(x)| \le \rho_2 |\psi(x)|^{1/N_2}.$$

Below is the promised exact penalty function result for the nonlinear program (1). Some exact penalty terminology is given after the proof.

2.1.2 Theorem. Let $X \subset \Re^n$ be a compact subanalytic set, θ be Lipschitz continuous on X, and g_i and h_j be continuous subanalytic. Suppose (1) is feasible. Then there exist a scalar $\alpha^* > 0$ and a positive integer N^* such that

(a) for all scalars $\alpha \geq \alpha^*$ and $\gamma \geq N^*$, if $\bar{x}$ solves the problem (1), then $\bar{x}$ is an optimal solution of the problem:

$$\begin{aligned} \text{minimize} \quad & \theta(x) + \alpha r(x)^{1/\gamma} \\ \text{subject to} \quad & x \in X; \end{aligned} \tag{2}$$

(b) conversely, if $\bar{x}$ solves the problem (2) for some scalars $\gamma \geq N^*$ and $\alpha \geq \alpha^*$, then $r(\bar{x}) = 0$ and $\bar{x}$ solves (1).

Proof. Let W be the feasible set of (1); i.e.,

$$W = \{x \in X \; : \; g(x) \leq 0, \; h(x) = 0\}.$$

Then W is a subanalytic set; moreover, the residual function r is continuous and subanalytic. (These facts can be found in [57].) Let $\psi(x) \equiv \text{dist}(x, W)$ be the distance function from the point x to the set W. Then ψ is continuous and subanalytic. Let $r|_X$ denote the restriction of the function r to the set X. Clearly, $(r|_X)^{-1}(0) \subseteq \psi^{-1}(0)$. Hence, by the Lojasiewicz' inequality for subanalytic functions, it follows that there exist a scalar $\rho > 0$ and an integer $N^* > 0$ such that for all $x \in X$,

$$\rho \, \text{dist}(x, W) \leq r(x)^{1/N^*}. \tag{3}$$

Let $L > 0$ be the Lipschitzian modulus of θ on X; i.e., for all $x, y \in X$,

$$|\theta(x) - \theta(y)| \leq L \, \|x - y\|.$$

Also let $\mu \equiv \max(1, \sup\{r(x) : x \in X\})$ and α^* be an arbitrary scalar greater than $\rho^{-1} L \mu^{1/N^*}$.

We are now ready to prove the two assertions (a) and (b). To prove (a), let $\bar{x}$ solve (1); let $\alpha \geq \alpha^*$ and $N \geq N^*$ be arbitrary scalars and $x \in X$ an arbitrary vector. Let $z \in W$ be such that $\text{dist}(x, W) = \|x - z\|$. We have $\theta(z) \geq \theta(\bar{x})$ and $\theta(x) - \theta(z) \geq -L\|x - z\|$. Thus,

$$
\begin{aligned}
\theta(x) + \alpha r(x)^{1/N} &= \theta(z) + (\theta(x) - \theta(z)) + \alpha r(x)^{1/N^*} r(x)^{1/N - 1/N^*} \\
&\geq \theta(\bar{x}) - L\|x - z\| + \alpha \rho \|x - z\| \mu^{1/N - 1/N^*} \\
&\geq \theta(\bar{x}) + (\alpha \rho (1/\mu)^{1/N^*} - L)\|x - z\| \\
&\geq \theta(\bar{x}) = \theta(\bar{x}) + \alpha r(\bar{x})^{1/N}.
\end{aligned}
$$

Consequently, (a) holds. Conversely, suppose x' is an optimal solution of (2). Since W is compact and nonempty, (1) has an optimal solution $\bar{x}$. By the above inequalities, we deduce

$$
\begin{aligned}
\theta(\bar{x}) &\geq \theta(x') + \alpha r(x')^{1/N} \\
&\geq \theta(\bar{x}) + \left(\alpha \rho (1/\mu)^{1/N^*} - L\right) \|x' - z\| \\
&\geq \theta(\bar{x}),
\end{aligned}
$$

where z is the vector in W closest to x'. Consequently, equalities hold throughout and $x' \in W$ by the choice of α. Thus (b) follows.   **Q.E.D.**

Given the above result, we call the minimization problem (2) an *exact penalty equivalent (of order $1/\gamma$)* of the NLP (1), and call its objective function an *exact penalty function (of order $1/\gamma$)*. An inequality of the type (3) is known as an *error bound (of order $1/N^*$)* for the set W. As we have seen from the proof of Theorem **2.1.2**, the validity of such an error bound is the key to obtaining the penalty equivalent (2) of (1). In essence, this line of proof is quite standard and has been adopted by many authors in their work on exact penalty functions, e.g. [41, 57, 182, 283, 292]. Several subsequent results in this monograph will also be based on related error bounds for the feasible set of the problem in question.

2.1.3 Remark. The two statements (a) and (b) in the above theorem can be summarized by saying that for all $\alpha \geq \alpha^*$ and all $\gamma \geq N^*$,

$$
\text{argmin}\{\theta(x) + \alpha r(x)^{1/\gamma} : x \in X\} = \text{argmin}\{\theta(x) : x \in \mathcal{F}^{\text{NLP}}\},
$$

where $\mathcal{F}^{\text{NLP}}$ denotes the feasible region of (1). It is interesting to note that in Theorem **2.1.2**, the objective function θ is assumed to be Lipschitz continuous. Incidentally, we are not certain at this time whether a Lipschitz continuous function must be subanalytic, or vice versa.

The proof of Theorem **2.1.2** illustrates a general principle which will be used repeatedly in this chapter. Namely, consider a constrained optimization problem posed in the abstract form:

$$\text{minimize} \quad \theta(x)$$

$$\text{subject to} \quad x \in W \subseteq X,$$

where W and X are two nonempty closed subsets of $\Re^n$ and θ is Lipschitz continuous on X. Assume that this problem has a global minimum. If $\psi(x)$ is a real-valued function majorizing the distance function $\text{dist}(x, W)$ on X, that is, if there is a constant $c > 0$ such that

$$\text{dist}(x, W) \leq c\,\psi(x), \quad \text{for all } x \in X,$$

then there exists a scalar $\alpha^* > 0$ such that for all $\alpha \geq \alpha^*$,

$$\text{argmin}\{\theta(x) + \alpha\psi(x) : x \in X\} = \text{argmin}\{\theta(x) : x \in W \cap X\}.$$

When $\psi(x)$ is equal to the distance function itself, this principle reduces to an important result of Clarke [51, Proposition 2.4.3] that has provided the basis for much of the classical exact penalization theory of constrained optimization [41].

Using Theorem **2.1.2**, we can derive a necessary and sufficient condition for a feasible vector of the program (1) to be its local minimum in terms of two penalty functions:

$$\theta(x) + r(x)^{1/N}, \quad \text{for all } N \geq N^*, \text{ for some } N^*,$$

$$\theta(x) - 1/\log(r(x)). \tag{4}$$

The latter function is particularly interesting because it does not involve the integer N which in general is very difficult, if not impossible, to estimate. The proof of the characterization of a local minimum is based on the observation that a local minimum x^* of (1) is a global minimum of the following restricted program

$$\text{minimize} \quad \theta(x)$$

$$\text{subject to} \quad x \in X \cap \mathbb{B}(x^*, \varepsilon),$$

$$g(x) \leq 0, \quad h(x) = 0,$$

for some closed ball $\mathbb{B}(x^*, \varepsilon)$ with center at x^* and radius $\varepsilon > 0$. We state the following corollary without proof (see [57]). (Note: since this is a local result, the boundedness of X is not needed and the Lipschitzian property of θ can be assumed to hold only locally. Also note that there is no penalty parameter α in (4) because this parameter can be "absorbed" into the residual term, for x sufficiently close to x^*.)

2.1.4 Corollary. Let $X \subseteq \Re^n$ be a closed subanalytic set, and g_i and h_j be continuous subanalytic. Let $x^* \in W$. Suppose that θ is Lipschitz continuous in a neighborhood of x^*. Then x^* is a local minimum of (1) if and only if x^* locally minimizes on X either one of the functions in (4).

2.1.5 Remark. The above corollary assumes that $x^* \in W$. Without this assumption, the "if" part of the conclusion is not necessarily valid. This remark applies to Corollary **2.4.5** as well.

Unlike the second function in (4) which involves no parameter, the constant N in the function

$$\psi_N(x) \equiv \theta(x) + r(x)^{1/N}$$

is dependent on the given vector x^*. This makes ψ_N not very useful in practice if one does not know x^* in advance. Thus it would be nice to know whether it is possible to obtain a constant N such that the local minimizers of (1) coincide with that of ψ_N on X. The following proposition addresses this question.

2.1.6 Proposition. Let X, g_i, h_j and θ be given by Corollary **2.1.4**. Suppose that the set of local minimizers of (1) is compact and has a finite number of connected components. Then, there exists some $N > 0$ such that the set of local minimizers of (1) is contained in that of the function ψ_N on X.

Proof. Let S denote the set of local minimizers of the problem (1). Then, for each $x^* \in S$, there exists some neighborhood $\mathbb{B}(x^*, \varepsilon^*)$ of x^* such that x^* is a global minimizer over $\mathbb{B}(x^*, \varepsilon^*)$, and this neighborhood intersects with only one of the connected components of S. By slightly extending Corollary **2.1.4**, we deduce the existence of an $N(x^*) > 0$ such that for all

$N \geq N(x^*)$, $\psi_N(x)$ is an exact penalty function for $\theta(x)$ over $\mathbb{B}(x^*, \varepsilon^*)$. In other words, a vector x' minimizes $\theta(x)$ over

$$\{x \in X \cap \mathbb{B}(x^*, \varepsilon^*) : g(x) \leq 0, h(x) = 0\}$$

if and only if it minimizes $\psi_N(x)$ over $X \cap \mathbb{B}(x^*, \varepsilon^*)$. (Note that such an x' is not necessarily a local minimizer of (1) because x' could lie on the boundary of $\mathbb{B}(x^*, \varepsilon^*)$.) Since S is compact and the open sets $\{\mathbb{B}(x^*, \varepsilon^*) : x^* \in S\}$ comprise an open covering of S, we can select a finite number of these sets, $\mathbb{B}(x_1^*, \varepsilon_1^*), ..., \mathbb{B}(x_k^*, \varepsilon_k^*)$, to cover S. Let $N \equiv \max\{N(x_1^*), ..., N(x_k^*)\}$.

We claim that with the above choice of N, the set S is contained in the set of local minimizers of (4) on X. In fact, any $x' \in S$ must belong to $\mathbb{B}(x_i^*, \varepsilon_i^*)$ for some $1 \leq i \leq k$. By the construction of the neighborhoods, x' and x_i^* lie in the same connected component in S. Thus $\theta(x') = \theta(x_i^*)$. So, both x' and x_i^* are global minimizers of $\theta(x)$ over the set

$$\{x \in \mathbb{B}(x_i^*, \varepsilon_i^*) \cap X : g(x) \leq 0, h(x) = 0\}.$$

Hence, x' is also a global minimizer of $\theta(x) + r(x)^{1/N(x_i^*)}$ over $X \cap \mathbb{B}(x_i^*, \varepsilon_i^*)$. By our choice of N, x' also minimizes $\theta(x) + r(x)^{1/N}$ over $X \cap \mathbb{B}(x_i^*, \varepsilon_i^*)$, thus showing that x' is a local minimizer of ψ_N on X.   **Q.E.D.**

In general, the function ψ_N may contain additional local minimizers on X other than those of (1). It remains to be seen if there is a simple condition under which the two sets of local minimizers are the same. Incidentally, the local minimizers on X of the second function in (4) also contain those of (1), and this containment is valid without any assumption on the latter.

2.2 Penalty Results for MPEC

In Subsection **1.3.2**, we have shown (Theorem **1.3.5**) that under the SBCQ, the inner VI of the MPEC (1.1.1) can be formulated in terms of its equivalent KKT conditions, which are a system of nonlinear equalities and inequalities. In this way, the MPEC (1.1.1) is reduced to the standard nonlinear program (1.3.8). The special feature of this NLP formulation of the MPEC is that it contains the complementarity conditions $\lambda_i g_i(x, y) = 0$ as constraints. In what follows, we use Theorem **2.1.2** to obtain a penalty equivalent of this KKT constrained MP in which the complementarity and other conditions are placed in the objective function through a penalty term. The resulting penalty equivalent of the MPEC (1.1.1) has as constraints only the set Z and some simple bounds on the multipliers λ; presumably, these constraints are much simpler than the original restriction $y \in \mathcal{S}(x)$.

2.2.1 Theorem. Let F be a continuous subanalytic function, Z be a compact subanalytic set, and f be Lipschitz on Z. Let the constraint map C be given by (1.3.3), where each g_i is continuous and subanalytic and for each $x \in X$, $g_i(x, \cdot)$ is convex. Assume that each $\nabla_y g_i$ exists and is continuous and subanalytic on an open set containing $\mathcal{F}$ (the feasible set of the MPEC). Assume further that $X \subseteq \mathrm{dom}(C)$, $\mathcal{F} \neq \emptyset$, and that the SBCQ holds on $\mathcal{F}$ for the multiplier map M. Let $c > 0$ be the constant as specified in Theorem **1.3.5**. Then there exist a scalar $\alpha^* > 0$ and an integer $N^* > 0$ such that for all scalars $\alpha \geq \alpha^*$ and $N \geq N^*$, any vector (x^*, y^*) solves (1.1.1) if and only if for some $\lambda^* \in \Re^\ell$, the triple (x^*, y^*, λ^*) solves the following (one-level) constrained minimization problem in the variables (x, y, λ):

$$
\begin{aligned}
\text{minimize} \quad & f(x,y) + \alpha r(x,y,\lambda)^{1/N} \\
\text{subject to} \quad & (x,y,\lambda) \in Z \times (\mathbb{B}(0,c) \cap \Re_+^\ell),
\end{aligned}
\tag{1}
$$

where

$$
r(x,y,\lambda) \equiv \left\| F(x,y) + \sum_{i=1}^{\ell} \lambda_i \nabla_y g_i(x,y) \right\| + \sum_{i=1}^{\ell} ((g_i(x,y))_+ + \lambda_i |g_i(x,y)|),
$$

and $\mathbb{B}(0,c)$ is the closed ball in $\Re^\ell$ with center at the origin and radius c.

Proof. By Theorem **1.3.5**, the MPEC (1.1.1) is equivalent to the following minimization problem in the variables (x,y,λ):

$$
\begin{aligned}
\text{minimize} \quad & f(x,y) \\
\text{subject to} \quad & (x,y,\lambda) \in Z \times (\mathbb{B}(0,c) \cap \Re_+^\ell), \\
& F(x,y) + \sum_{i=1}^{\ell} \lambda_i \nabla_y g_i(x,y) = 0, \\
& g_i(x,y) \leq 0, \quad \lambda_i g_i(x,y) = 0.
\end{aligned}
\tag{2}
$$

Theorem **2.1.2** can now be applied to the latter problem and the desired conclusion follows.   **Q.E.D.**

2.2.2 Remark. It would be desirable if the ball $\mathbb{B}(0, c)$ could be removed from the problem (1) and Theorem **2.2.1** (as well as some subsequent results of the same type) would remain valid. However, this seems difficult for the following reason. By the SBCQ, there exists a $c_* > 0$ such that (2) is equivalent to the MPEC (1.1.1) for all $c \geq c_*$. By the exact penalty result, Theorem **2.1.2**, for each $c \geq c_*$ there exist $\alpha(c)$ and $N^*(c)$ such that (1) is equivalent to (2) for all $\alpha \geq \alpha^*(c)$ and $N \geq N^*(c)$. Suppose now (x^*, y^*, λ^*) solves the problem (1) without the ball $\mathbb{B}(0, c)$ for some $\alpha \geq \alpha^*(c)$ and $N \geq N^*(c)$, for some $c \geq c_*$. If $\|\lambda^*\| > c$ and one of the penalty parameters $\alpha^*(\|\lambda^*\|)$ and $N^*(\|\lambda^*\|)$ exceeds the corresponding α and N, we cannot apply the penalty result (Theorem **2.1.2**) to conclude that (x^*, y^*) solves the MPEC (1.1.1).

Instead of assuming that Z is compact, we can derive a similar penalty equivalent for (1.1.1) under the weaker assumption that $Z \cap \mathrm{Gr}(C)$ is compact. In this case, the penalty equivalent is:

$$
\begin{aligned}
&\text{minimize} \quad f(x, y) + \alpha \tilde{r}(x, y, \lambda)^{1/N} \\
&\text{subject to} \quad (x, y, \lambda) \in (Z \cap \mathrm{Gr}(C)) \times (\mathbb{B}(0, c) \cap \mathfrak{R}_+^\ell),
\end{aligned}
\tag{3}
$$

where

$$
\tilde{r}(x, y, \lambda) = \left\| F(x, y) + \sum_{i=1}^{\ell} \lambda_i \nabla_y g_i(x, y) \right\| + \sum_{i=1}^{\ell} \lambda_i |g_i(x, y)|.
$$

The difference between the latter problem and (1) is rather obvious. In (1), the constraints involve only the set Z and the bounds on the multipliers, but the objective function contains the violation of the constraints $g_i(x, y) \leq 0$. In (3), the latter constraints remain as constraints but the residual function $\tilde{r}$ no longer includes their violation.

Similarly to Corollary **2.1.4**, a local version of Theorem **2.2.1** (and its modification as just described) can be stated for a local minimum of the problem (1.1.1). The local results do not add much value to the discussion except for the following details. One, the boundedness assumption on either Z or $Z \cap \mathrm{Gr}(C)$ is no longer needed; and two, although a local minimum of (1.1.1) must be a local minimum of its penalty equivalent, the converse need not hold unless the latter minimum is already feasible to (1.1.1). We illustrate this point using an LCP constrained MP as an example.

2.2.3 Example. Let $Z = \{(x,y) : x + 3y \geq 3, \ x + y \geq 1\}$. Consider the following LCP constrained MP:

$$\text{minimize} \quad f(x,y) \equiv 0$$

$$\text{subject to} \quad (x,y) \in Z,$$

$$y \geq 0, \ x^2 + x + y \geq 0,$$

$$y(x^2 + x + y) = 0.$$

Clearly, the inner LCP is strongly monotone. As a result, for each x, it has a unique solution $y(x) = \max(0, -x^2 - x)$. Thus, the feasible region of this MPEC is $\mathcal{F} = \{(x,y) : x \geq 3, \ y = 0\}$. Since $f(x,y)$ is identically zero, any point in $\mathcal{F}$ is a globally optimal solution. Let us consider the penalized program

$$\text{minimize} \quad r(x,y) \equiv y(x^2 + x + y)$$

$$\text{subject to} \quad x + 3y \geq 3,$$

$$x + y \geq 1, \quad (4)$$

$$x^2 + x + y \geq 0, \quad y \geq 0.$$

We claim that $(x^*, y^*) \equiv (0, 1)$ is an isolated local minimizer of this penalized program. Clearly, $(0,1)$ is feasible for (4), and the active constraints at this point are

$$g_1(x,y) \equiv x + 3y - 3 \geq 0 \quad \text{and} \quad g_2(x,y) \equiv x + y - 1 \geq 0.$$

Notice that

$$\nabla r(0,1) = \begin{pmatrix} 1 \\ 2 \end{pmatrix}, \quad \nabla g_1(0,1) = \begin{pmatrix} 1 \\ 3 \end{pmatrix}, \quad \nabla g_2(0,1) = \begin{pmatrix} 1 \\ 1 \end{pmatrix};$$

therefore

$$\nabla r(0,1) - \tfrac{1}{2}\nabla g_1(0,1) - \tfrac{1}{2}\nabla g_2(0,1) = 0.$$

Moreover, we have

$$\nabla^2 r(0,1) = \begin{bmatrix} 2 & 1 \\ 1 & 2 \end{bmatrix},$$

which is positive definite. Thus the standard first- and second-order sufficient conditions for optimality are satisfied by (x^*, y^*) for (4); it follows that (x^*, y^*) is indeed an isolated local minimizer of the penalized program. However, notice that $r(0,1) = 1 > 0$, so $(0,1)$ is not even feasible, not to say a local minimizer, for the original LCP constrained MP. In fact, it is clear that the vector $(0,1)$ remains an isolated local minimizer for the penalized program even if $r(x,y)$ is raised to any power $1/N$.

Like the general situation in Theorem **2.1.2**, the constants α^* and N^* in Theorem **2.2.1** are in general very difficult to obtain; in addition, so is the quantity c. In the remainder of this chapter, we consider some special classes of MPEC for which the integer N^* can be taken to be either 2 or 1 (these are the smallest possible values for N^*). In the case where $N^* = 1$, some further necessary conditions for optimality can be established; see Subsection **2.4.3** for a preliminary discussion of these conditions.

2.3 Improved Error Bounds

In Section **2.2**, we have used Lojasiewicz' error bound (Theorem **2.1.1**) to derive various exact penalty functions for MPEC. Unfortunately, these penalty functions all involve a theoretically guaranteed, but practically unknown, exponent $1/N$, thus significantly diminishing their practical values. To find the value of N, we need to determine the exponent in Lojasiewicz' error bound result for subanalytic systems of inequalities. This motivates the development of new error bounds with explicit error exponents.

In this section, we establish a number of improved error bounds for some special systems of inequalities. These include quadratic systems satisfying a nonnegativity assumption, parametric affine variational inequalities, and parametric nonlinear complementarity problems satisfying certain uniform P property. The new error bounds derived herein all have exponents γ equal to $\frac{1}{2}$ or 1. They will be used later in Section **2.4** to derive some improved exact penalty functions for MPEC.

A synopsis of error bound results

Due to the fundamental importance of error bounds in mathematical programming in general and for the study of MPEC in particular, we digress from our discussion of MPEC and give a detailed treatment of the theory of error bounds.

Many mathematical programming problems can be formulated as a problem of finding a feasible point satisfying a finite set of equalities and inequalities. For example, the nonlinear complementarity problem is to find a feasible point x satisfying the inequalities $F(x) \geq 0, x \geq 0$, and the equality $x^T F(x) = 0$, where $F : \Re^n \to \Re^n$ is a given function. For another example, each linear programming problem is, in the primal-dual formulation, equivalent to the problem of finding a feasible point satisfying a given set of linear inequalities. As a third example, every differentiable nonlinear program, through its KKT optimality conditions, can be formulated as the problem of finding a feasible point satisfying a set of equalities and inequalities.

Suppose S is the solution set of an optimization problem and assume S can be described by a finite set of equalities and inequalities:

$$S \equiv \{x \in \Re^n : f(x) \leq 0, \ g(x) = 0\}, \qquad (1)$$

where f and g are vector-valued functions. To solve the optimization problem, we usually employ an iterative algorithm which invariably generates a sequence of points approaching the solution set S. Thus, an important issue in all such algorithms is to decide how close a given iterate is to the solution set S.

More generally, given an arbitrary set S of the form (1), we can ask the question: how can we compute the distance from an arbitrary point $x \in \Re^n$ to S? This question itself leads to a projection problem: given $x \in \Re^n$, solve

$$\text{minimize} \quad \|y - x\|$$

$$\text{subject to} \quad y \in S.$$

The answer to the question is quite difficult in general. A simpler problem is to decide if x is a feasible solution by direct substitution. More precisely, $x \in S$ if and only if x satisfies the equality and inequality constraints. This leads to the concept of a residual function $r(x)$ for the set S, which is by definition a nonnegative function defined on the domain of the functions f and g and attains zero if and only if $x \in S$. Obviously, we wish to have a residual function $r(x)$ that is very easy to compute. One natural choice of a residual function $r(x)$ for the set S given by (1) is the sum of the functional values of the constraints evaluated at x; that is,

$$r(x) \equiv \|f^+(x)\| + \|g(x)\|, \qquad (2)$$

where

$$f^+(x) \equiv \max(0, f(x)).$$

Note that we need to use the nonnegative part of $f(x)$ in this residual function because of the corresponding inequality constraint in S.

With a residual function $r(x)$ for S, the problem of deciding if $x \in S$ becomes that of verifying if $r(x) = 0$. In particular, $x \notin S$ when $r(x) > 0$. We can now ask the following question: given a test set $T \subseteq \Re^n$, can we use $r(x)$ to bound the distance from $x \in T$ to S, i.e., do there exist positive constants γ, γ' and τ, τ' such that for all $x \in T$,

$$\tau' \, r(x)^{\gamma'} \leq \text{dist}(x, S) \leq \tau \, r(x)^{\gamma}. \tag{3}$$

Inequalities such as the above are called *error bounds* for the set S. In essence, Lojasiewicz' result, Theorem **2.1.1**, asserts that if f and g are continuous subanalytic (vector-valued) functions and T is a compact subanalytic set, then (3) holds with the residual function given by (2).

For the most part, the theory of error bounds deals only with the right-hand inequality in (3); the reason is that if the residual function $r(x)$ is Hölder continuous on the set T with a fixed modulus and positive exponent (which is clearly the case for $r(x)$ given by (2) with f and g having the same property), then the left-hand inequality trivially holds. In general, the difficulty of the analysis lies in the demonstration of the right-hand bound.

An error bound of the type (3) has interesting implications on the solutions of the system

$$f(x) \leq 0, \quad g(x) = 0$$

when the functions f and g are perturbed; among other things, we may obtain quantitative bounds on the change of the solutions of the perturbed system in terms of the magnitude of the perturbation. The following result makes this statement precise.

2.3.1 Proposition. Consider two sets defined by:

$$S_1 \equiv \{x \in R^n : f^1(x) \leq 0, \ g^1(x) = 0\},$$
$$S_2 \equiv \{x \in R^n : f^2(x) \leq 0, \ g^2(x) = 0\},$$

and assume that both S_1 and S_2 are nonempty. Let r_1 and r_2 be, respectively, the residual functions for S_1 and S_2 given by (2). Suppose there exist positive constants τ and γ such that

$$\text{dist}(x, S_i) \leq \tau\, r_i(x)^\gamma, \quad \text{for } i = 1, 2,$$

for all x in a set T that contains S_1 and S_2. Let

$$\varepsilon \equiv \sup_{x \in T} \left(\|f_1(x) - f_2(x)\| + \|g_1(x) - g_2(x)\| \right).$$

Then for $i, j = 1, 2$,

$$S_i \subseteq S_j + \tau\varepsilon^\gamma \mathbb{B}(0, 1). \tag{4}$$

Proof. It suffices to establish the inclusion (4) for $i = 1$ and $j = 2$. Let $x \in S_1$. We have

$$\text{dist}(x, S_2) \leq \tau\, r_2(x)^\gamma.$$

Moreover, since $f^1(x) \leq 0$ and $g^1(x) = 0$, we deduce

$$
\begin{aligned}
r_2(x) &= \|\max(0, f^2(x))\| + \|g^2(x)\| \\
&\leq \|\max(0, f^2(x) - f^1(x))\| + \|g^2(x) - g^1(x)\| \\
&\leq \|f^2(x) - f^1(x)\| + \|g^2(x) - g^1(x)\| \leq \varepsilon.
\end{aligned}
$$

Thus the inclusion (4) holds for $i = 1$ and $j = 2$. **Q.E.D.**

The study of error bounds in the context of mathematical programming has a rather long history. The first and also the best known error bound result was obtained in 1952 by A.J. Hoffman [116] who proved that (3) holds globally (i.e., with $T = \Re^n$) with $\gamma_1 = \gamma_2 = 1$ for all linear systems of equations and inequalities (i.e., with f and g both affine). Subsequently, Hoffman's error bound has been refined [184, 153] and extended in several directions [246, 155, 167, 171, 281], with the refinements coming mostly in the form of sharper estimates of the multiplicative constant τ, and the extensions being to more general systems (possibly nonconvex, nonlinear, and nonsmooth).

Along a different line of research and apparently unaware of Hoffman's result, Lojasiewicz obtained his inequality in 1959 as a positive response to a hypothesis used by M.L. Schwartz in the distribution theory of functions. Lojasiewicz' result was originally developed for analytic functions;

he first announced his result in [163] and later extended it to semianalytic sets in the paper [164]. Lojasiewicz' result includes the special case of a polynomial equation studied by Hörmander [117] who was concerned with the fundamental solutions of certain partial differential equations with constant coefficients. Lojasiewicz' result was further refined and extended to subanalytic functions by Hironaka [112, 113] and Bierstone and Milman [26].

Error bounds have found many important applications in mathematical programming. In particular, they have been used as termination criteria for iterative descent algorithms [190, 217, 229, 90]. They have also been used in sensitivity analysis [183, 195, 244] of linear programming and linear complementarity problems. Error bounds have been successfully applied to the convergence analysis of some iterative descent methods for solving certain degenerate problems. The methods studied include gradient projection algorithms [174], matrix splitting methods for linear complementarity problems [175, 155], coordinate descent methods [176], descent methods for convex quadratic splines [156], and primal-dual interior point algorithms for convex quadratic programs [169]. Previously, only subsequential convergence results were known for these algorithms, unless the problems to be solved satisfied a certain strong convexity assumption. Another area where error bounds play an important role is in the study of minimizing sequences and stationary sequences of optimization problems [12, 11, 10, 49]. From the last two sections, we have seen that error bounds are the cornerstone for obtaining exact penalty functions for constrained optimization problems, including MPECs. This last area of application is our main motivation for devoting this entire section to a detailed study of error bounds.

2.3.1 *Hoffman's error bound for linear systems*

Due to its importance, we start our discussion of error bound theory with Hoffman's 1952 result for a linear system consisting of a finite number of linear inequalities and equalities. The two most noteworthy characteristics of Hoffman's bound are: (i) it holds globally for all test vectors $x \in \Re^n$, and (ii) it holds with the residual itself as the upper bound for the distance function. Let

$$S = \{x \in \Re^n : Ax \leq a, \ Bx = b\}$$

be a nonempty polyhedral set, where A and B are some matrices of appropriate dimensions, and a and b are some vectors of matching dimensions. The precise statement of Hoffman's bound is given below.

2.3.2 Theorem. There exists a scalar $\tau > 0$ such that

$$\text{dist}(x, S) \leq \tau \left(\| [Ax - a]^+ \| + \| Bx - b \| \right), \qquad \forall x \in \Re^n. \qquad (5)$$

The constant τ plays a major role in a certain "condition number" of the polyhedral set S; in particular, τ contains quantitative information about the change in the solution set S when the right-hand side vectors a and b are perturbed. This is especially clear when S is a subspace (i.e., the inequalities $Ax \leq a$ are vacuous); in this case τ is seen to be related to the inverse of the determinant of some nonsingular submatrix of B.

Hoffman's error bound has many important applications in mathematical programming. For one thing, it can be used to establish error bounds for such mathematical programs as affine variational inequalities, linear complementarity problems, linear programs, and certain linearly constrained strictly convex programs. For example, in the case of a linear program, it can be shown that the solution set is a Lipschitz continuous (multi)function of the right-hand side vectors of the linear constraints; see Li [153], Mangasarian and Shiau [195]. More precisely, let $X(a, b)$ denote the nonempty solution set of the following linear program:

$$\begin{aligned} \text{minimize} \quad & c^T x \\ \text{subject to} \quad & Ax \leq a, \ Bx = b. \end{aligned} \qquad (6)$$

There exists then a positive constant τ such that

$$\text{dist}(x, X(a, b)) \leq \tau \left(\| b - b' \| + \| a - a' \| \right), \qquad \forall x \in X(a', b'), \ \forall a', b'.$$

Related to this Lipschitz continuity property is the so-called *weak sharp minima* property of linear programs, which is also a consequence of Theorem **2.3.2**. Specifically, it was shown in [192] that for the linear program (6), there exists some scalar $\lambda > 0$ such that

$$c^T x \geq c^* + \lambda \, \text{dist}(x, X(a, b)), \qquad \forall x \text{ satisfying } Ax \leq a, \ Bx = b, \qquad (7)$$

where c^* denotes the minimum value of this program. Part of the significance of the inequality (7) is that for the linear program (6), the deviation of the objective value from the optimal value, i.e., the difference $c^T x - c^*$, is therefore a legitimate residual measure for the optimal solution set of the program with reference to its feasible vectors. Incidentally, the concept of weak sharp minima was formally defined in [70] and its relationship to penalty functions was explored therein. This concept turns out to have many important algorithmic implications and has a much closer tie to error bounds than stated here. The reference [71] gives an extensive treatment of this topic in the context of an affine variational inequality and lists many related references.

Hoffman's error bound can be extended to integer linear systems. In particular, Cook, Gerard, Schrijver and Tardös [53] considered the following solution set of an integer linear system:

$$S_{\text{in}} \equiv \{x \,:\, Ax \leq a, \ Bx = b, \ x \text{ integer}\},$$

and proved that

$$\text{dist}(x, S_{\text{in}}) \leq \tau \left(\|[Ax - a]^+\| + \|Bx - b\| \right), \qquad \forall x \text{ integer},$$

where τ is some positive constant. Similar to the case of a linear program, this bound can be used to establish the Lipschitz continuity of the solutions of the integer linear program when the right-hand vectors of its linear constraints are perturbed.

Apart from these applications, there have been several attempts to obtain tight estimates of the multiplicative constant τ in Hoffman's error bound (5). For example, Mangasarian [181] proved that the constant τ can be bounded above by

$$\sup \left\{ \|(u, v)\| \,:\, \|A^T u + B^T v\| = 1, \ u \geq 0, \text{ the columns of } [A^T \ B^T] \right.$$

$$\text{corresponding to nonzero components of } (u, v) \qquad (8)$$

$$\left. \text{are linearly independent}\right\}.$$

Furthermore, Mangasarian gave an example to show that the above estimate of τ is actually tight. In addition to this estimate, there have been several other similar estimates of τ; see, for example, Bergthaller and Singer [22], Güler, Hoffman, and Rothblum [102], and Li [153]. Finally, we mention the work of Luo and Tseng [177] which studies the "stability" of the constant τ in (5) as the data (A, B, a, b) all undergo small changes.

2.3.2 *Error bounds for AVI and LCP*

Recalling the notation and discussion of the AVI (q, M, X) where X is a polyhedron in $\Re^n$, $q \in \Re^n$, and $M \in \Re^{n \times}$, we apply Hoffman's error bound to derive error bound results for AVIs and LCPs. The following error bound result is due to Luo and Tseng [174].

2.3.3 Theorem. Assume that $\text{SOL}(q, M, X)$ is nonempty. Then there exist constants $\tau > 0$ and $\delta > 0$ such that

$$\text{dist}(x, \text{SOL}(q, M, X)) \leq \tau \, \|x - \Pi_X(x - Mx - q)\| \qquad (9)$$

for all $x \in \Re^n$ with $\|x - \Pi_X(x - Mx - q)\| \leq \delta$.

Theorem **2.3.3** was derived using the fact that $\text{SOL}(q, M, X)$ is the union of finitely many polyhedra to which Hoffman's error bound can be applied. An alternative way to establish this result is to invoke a fundamental result of Robinson [249] which states that every polyhedral multifunction is locally upper Lipschitz continuous. (A polyhedral multifunction is a set-valued mapping whose graph is the union of finitely many polyhedra; we refer to this reference for the definition of locally upper Lipschitz continuity of a multifunction.)

In essence, Theorem **2.3.3** asserts that the error bound (9) holds for all vectors x with a sufficiently small residual, that is, the bound holds locally near the solution set S. Based on this interpretation, we call (9) a *local* error bound for the AVI (q, M, X). The following example which originates from [195, Example 2.10] shows that (9) cannot hold globally for all $x \in X$.

2.3.4 Example. Consider the LCP (q, M) with

$$M = \begin{bmatrix} 0 & 1 \\ 1 & 1 \end{bmatrix}, \quad q = \begin{pmatrix} -1 \\ -2 \end{pmatrix}.$$

It is easily seen that $\text{SOL}(q, M) = \{(1, 1), (0, 2)\}$. Consider the sequence of vectors $x(t) = (t, 1)$, where $t \in [0, \infty)$ is a parameter. Then, as $t \to \infty$, we have $\text{dist}(x(t), \text{SOL}(q, M)) \to \infty$. Since

$$x(t) - \Pi_{\Re^2_+}(x(t) - Mx(t) - q) = \min(x(t), q + Mx(t)),$$

we see that $\|x(t) - \Pi_{\Re^2_+}(x(t) - Mx(t) - q)\|$ remains bounded as $t \to \infty$. This shows that the error bound (9) cannot hold globally in this case.

It is possible to globalize a local error bound by introducing an extra multiplicative factor $(1 + \|x\|)$ into the right-hand side [170, 186]. To see this, we partition $\Re^n$ into two regions depending on whether

$$\|x - \Pi_X(x - Mx - q)\| \leq \delta \quad \text{or} \quad \|x - \Pi_X(x - Mx - q)\| > \delta.$$

In the first region, the local error bound applies, whereas in the second region, $\text{dist}(x, \text{SOL}(q, M, X))$ is bounded by a constant multiple of $(1+\|x\|)$. Combining these two cases yields the following global error bound

$$\text{dist}(x, \text{SOL}(q, M, X)) \leq \tau' \, (1 + \|x\|) \, \|x - \Pi_X(x - Mx - q)\|, \qquad \forall x \in \Re^n,$$

where τ' is some suitable constant.

In the case of the LCP (q, M), it is possible to characterize the class of matrices M for which the error bound (9) holds globally for any given vector q. The following interesting result is due to Mangasarian and Ren [194] and Luo and Tseng [168].

2.3.5 Theorem. Let $M \in \Re^n$ be given. The following two statements are equivalent.

(a) For all vectors $q \in \Re^n$, there exists $\tau > 0$ such that the error bound (9) holds globally for all $x \in \Re^n$.

(b) The homogeneous LCP $(0, M)$ has zero as its unique solution.

A matrix M satisfying condition (b) in the above theorem is called an R_0 matrix. It is known [54] that the class of R_0 matrices includes positive definite matrices, strictly copositive matrices, P matrices (the matrices whose principal minors are positive), and nondegenerate matrices (the matrices whose principal minors are nonzero). Notice that the condition "M being an R_0 matrix" is necessary only if we require the error bound (9) to hold globally for every choice of q. A similar characterization in terms of (q, M) for which the error bound (9) holds globally for this given q is obtained in [168]. This result extends the earlier work of [176] which in addition assumes that M is "positive semidefinite plus". (A positive semidefinite plus matrix M is a positive semidefinite matrix which satisfies the implication $x^T M x = 0 \Rightarrow M x = 0$. It has been shown in [173] a matrix M is positive semidefinite plus if and only if $M = E^T A E$ for some arbitrary matrix E and some positive definite matrix A.)

In the remainder of this subsection we will consider error bounds for parametric AVIs. These problems are of interest because they appear as the inner problems of an MPAEC (i.e., an AVI constrained MP). As shown in Section **2.4**, the error bounds obtained below will be very useful in deriving exact penalty functions for such an MPEC.

Consider the parametric affine variational inequality defined as the problem of finding, for each given $x \in \Re^n$, a vector $y^* \in C(x)$ satisfying

$$(y - y^*)^T F(x, y^*) \geq 0, \quad \forall y \in C(x),$$

where

$$C(x) \equiv \{y : Dx + Ey + b \leq 0\}$$

and

$$F(x, y) \equiv Px + Qy + q,$$

for some given matrices $D \in \Re^{\ell \times n}, E \in \Re^{\ell \times m}, P \in \Re^{m \times n}, Q \in \Re^{m \times m}$, and vectors $b \in \Re^\ell, q \in \Re^m$. For each x, the equivalent KKT formulation of this AVI is given by

$$
\begin{aligned}
Px + Qy + q + E^T \lambda = 0, \quad \lambda \geq 0, \\
Dx + Ey + b \leq 0, \quad \lambda^T(Dx + Ey + b) = 0.
\end{aligned}
\tag{10}
$$

Let $\Omega \subseteq \Re^{n+m+\ell}$ denote the set of all triples (x, y, λ) satisfying the above KKT system; let W denote the superset of Ω that consists of all triples (x, y, λ) satisfying all the conditions in (10) except the last complementarity condition $\lambda^T(Dx + Ey + b) = 0$. The sets Ω and W are, respectively, the solution set and feasible region of the system (10).

2.3.6 Theorem. Suppose $\Omega \neq \emptyset$. There exist scalars $\tau > 0$ and $\varepsilon > 0$ such that

$$\mathrm{dist}((x, y, \lambda), \Omega) \leq \tau \sum_{i=1}^{\ell} \min(\lambda_i, -(Dx + Ey + b)_i),$$

for all $(x, y, \lambda) \in W$ with $\sum_{i=1}^{\ell} \min(\lambda_i, -(Dx + Ey + b)_i) \leq \varepsilon$.

Proof. We first claim that there exists some $\varepsilon > 0$ such that for any $(x, y, \lambda) \in W$ with $\sum_{i=1}^{\ell} \min(\lambda_i, -(Dx + Ey + b)_i) \leq \varepsilon$, the following system has a solution:

$$(Du + Ev + b)_{\mathcal{I}} = 0, \quad \mu_{\mathcal{J}} = 0, \quad (u, v, \mu) \in W, \tag{11}$$

where

$$\mathcal{I} \equiv \{i : -(Dx + Ey + b)_i \leq \lambda_i\} \tag{12}$$

and $\mathcal{J}$ is the complement of $\mathcal{I}$ in $\{1, \ldots, \ell\}$. Suppose otherwise; then there exists a sequence $\{(x^k, y^k, \lambda^k)\} \subseteq W$ and a constant index set $\mathcal{I}$, with complement $\mathcal{J}$, such that

$$\sum_{i=1}^{\ell} \min(\lambda_i^k, -(Dx^k + Ey^k + b)_i) \to 0,$$

$$-(Dx^k + Ey^k + b)_{\mathcal{I}} \leq \lambda_{\mathcal{I}}^k \quad \text{and} \quad -(Dx^k + Ey^k + b)_{\mathcal{I}} \geq \lambda_{\mathcal{J}}^k$$

for all k. Since W is polyhedral, it follows from Hoffman's inequality (Theorem **2.3.2**) that there exists a new sequence in W which has the same properties as $\{(x^k, y^k, \lambda^k)\}$ and is bounded. For notational convenience, let us still use $\{(x^k, y^k, \lambda^k)\}$ to denote this new bounded sequence. Now any accumulation point of $\{(x^k, y^k, \lambda^k)\}$ can easily be seen to be a solution of the system (11), a contradiction.

Next, we observe that any solution of such a system (11) must be an element of Ω. Moreover, for any index subset $\mathcal{I} \subseteq \{1, \ldots, \ell\}$ such that (11) is consistent, Hoffman's error bound (2.3.5) implies that there exists a scalar $\tau_{\mathcal{I}} > 0$ such that for all $(x, y, \lambda) \in W$,

$$\text{dist}((x, y, \lambda), S_{\mathcal{I}}) \leq \tau_{\mathcal{I}} \left(\|(Dx + Ey + b)_{\mathcal{I}}\| + \|\lambda_{\mathcal{J}}\| \right),$$

where $S_{\mathcal{I}}$ denotes the solution set of (11). Let $\tau \equiv \max_{\mathcal{I}} \tau_{\mathcal{I}}$. Take any $(x, y, \lambda) \in W$ with $\sum_{i=1}^{\ell} \min(\lambda_i, -(Dx + Ey + b)_i) \leq \varepsilon$ and let $\mathcal{I}$ be given by (12). Then the system (11) is consistent. Thus we have

$$\|(x, y, \lambda) - (\bar{u}, \bar{v}, \bar{\mu})\| \leq \tau' \left(\|(Dx + Ey + b)_{\mathcal{I}}\| + \|\lambda_{\mathcal{J}}\| \right),$$

for some solution $(\bar{u}, \bar{v}, \bar{\mu}) \in S_{\mathcal{I}}$, which must necessarily belong to Ω as we have noted before. Consequently, we obtain

$$\text{dist}((x, y, \lambda), \Omega) \leq \tau \sum_{i=1}^{\ell} \min(\lambda_i, -(Dx + Ey + b)_i), \tag{13}$$

provided that $(x, y, \lambda) \in W$ and satisfies

$$\sum_{i=1}^{\ell} \min(\lambda_i, -(Dx + Ey + b)_i) \leq \varepsilon.$$

Q.E.D.

2.3.7 Remark. Similar to Theorem **2.3.3**, the validity of (13) when the residual in the right-hand side is small is also a consequence of the upper Lipschitz continuity property of a polyhedral multifunction proved in Robinson [249]. Indeed, by defining the piecewise affine function

$$F(x, y, \lambda) \equiv \sum_{i=1}^{\ell} \min(\lambda_i, -(Dx + Ey + b)_i), \quad \text{for all } (x, y, \lambda) \in \Re^{n+m+\ell},$$

then applying Robinson's result to the polyhedral multifunction $F^{-1} \cap W$ at the origin and a simple manipulation, we can easily establish the desired local error bound (13).

The residual $\sum_{i=1}^{\ell} \min(\lambda_i, -(Dx + Ey + b)_i)$ in Theorem **2.3.6** is non-differentiable due to the presence of the "min" operator. This is an undesirable feature from the computational standpoint. In what follows we develop an error bound using the complementarity term $-\lambda^T(Dx + Ey + b)$ as the residual. This "product", or "complementarity gap", residual, as it is called, is a quadratic polynomial in the variables (x, y, λ) and is certainly differentiable. We first establish the following relation between these two residuals.

2.3.8 Lemma. Let W, Ω denote the feasible region and the solution set of (10) respectively. Suppose that each solution $(x, y, \lambda) \in \Omega$ is nondegenerate in the sense that $\lambda - (Dx + Ey + b) > 0$. Then for any $\rho > 0$, there exists a constant $\tau > 0$ such that for all $(x, y, \lambda) \in W \cap \mathbb{B}(0, \rho)$,

$$\sum_{i=1}^{\ell} \min(\lambda_i, -(Dx + Ey + b)_i) \leq -\tau \, \lambda^T(Dx + Ey + b). \quad (14)$$

Proof. The proof is by contradiction. Suppose that no such scalar τ exists. Then there exists a sequence $\{(x^k, y^k, \lambda^k)\} \subseteq (W \cap \mathbb{B}(0, \rho)) \setminus \Omega$ such that

$$\sum_{i=1}^{\ell} \min(\lambda_i^k, -(Dx^k + Ey^k + b)_i) > -k \, (\lambda^k)^T(Dx^k + Ey^k + b). \quad (15)$$

Since $\|(x^k, y^k, \lambda^k)\| \leq \rho$, we may assume without loss of generality that the sequence $\{(x^k, y^k, \lambda^k)\}$ converges to the limit $(\bar{x}, \bar{y}, \bar{\lambda}) \in W$. Thus

$$\bar{\lambda}^T(D\bar{x} + E\bar{y} + b) = 0,$$

which means that $(\bar{x}, \bar{y}, \bar{\lambda}) \in \Omega$. Thus, $\bar{\lambda} - (D\bar{x} + E\bar{x} + b) > 0$. Let

$$\mathcal{I} \equiv \{i \, : \, \bar{\lambda}_i > 0 = (D\bar{x} + E\bar{y} + b)_i\}$$

and

$$\mathcal{J} \equiv \{i \, : \, \bar{\lambda}_i = 0 > (D\bar{x} + E\bar{y} + b)_i\}.$$

By nondegeneracy, it follows that $\mathcal{I}$ and $\mathcal{J}$ partition the index set $\{1, \ldots, \ell\}$. Moreover, there exists an $\varepsilon > 0$, which depends on $(\bar{x}, \bar{y}, \bar{\lambda})$, such that for all k sufficiently large, we have

$$\min(\lambda_i^k, -(Dx^k + Ey^k + b)_i) = -(Dx^k + Ey^k + b)_i$$

and $\lambda_i^k \geq \varepsilon$ for all $i \in \mathcal{I}$, and

$$\min(\lambda_i^k, -(Dx^k + Ey^k + b)_i) = \lambda_i^k$$

and $-(Dx^k + Ey^k + b)_i \geq \varepsilon$ for all $i \in \mathcal{J}$. Consequently, (15) yields

$$\sum_{i=1}^{\ell} \min(\lambda_i^k, -(Dx^k + Ey^k + b)_i)$$

$$> k\,\varepsilon \sum_{i=1}^{\ell} \min(\lambda_i^k, -(Dx^k + Ey^k + b)_i) > 0.$$

Since this must hold for all k sufficiently large and ε is a fixed scalar, we obtain the contradiction that $1 > k\varepsilon$.   Q.E.D.

The nondegeneracy assumption in Lemma **2.3.8** is also known as a *strict complementarity* condition. Admittedly, the requirement that this condition hold for all $(x, y, \lambda) \in \Omega$ is rather strong. It is not clear to what extent this assumption can be relaxed. Nonetheless, assuming the nondegeneracy condition holds, we have the following error bound result.

2.3.9 Theorem. Assume the setting of Lemma **2.3.8** and Theorem **2.3.6**. Then, for each $\rho > 0$, there exists a scalar $\tau > 0$ such that

$$\mathrm{dist}((x, y, \lambda), \Omega) \leq -\tau \lambda^T (Dx + Ey + b)$$

for all $(x, y, \lambda) \in W \cap \mathbb{B}(0, \rho)$.

Proof. The desired inequality is a simple corollary of Lemma **2.3.8** and Theorem **2.3.6**. Specifically, let τ_1, ε be the constants given by Theorem **2.3.6** and let τ_2 be the constant given by Lemma **2.3.8**. Then the desired error bound holds with $\tau \geq \tau_1\tau_2$ for all $(x, y, \lambda) \in W \cap \mathbb{B}(0, \rho)$ such that $-\lambda^T(Dx + Ey + b) \leq \varepsilon/\tau_2$. If $(x, y, \lambda) \in W \cap \mathbb{B}(0, \rho)$ and $-\lambda^T(Dx + Ey + b) \geq \varepsilon/\tau_2$, then by the compactness of the ball $\mathbb{B}(0, \rho)$, the same inequality must still hold, possibly with a different constant τ' that is independent of $(x, y, \lambda) \in W \cap \mathbb{B}(0, \rho)$.   **Q.E.D.**

2.3.3 Error bounds for a quadratic system

In this subsection, we show that $\gamma = 1/2$ is a valid error bound exponent for a quadratic system satisfying a certain nonnegativity condition (but not necessarily convex). This result will be used subsequently to derive an exact penalty equivalent for the AVI constrained MP and to develop a penalty based interior point algorithm for solving this problem. The precise statement of the result is given below. This result and the subsequent Theorem **2.3.12** were first announced in [171] without proofs.

2.3.10 Theorem. Let $g_i : \Re^n \to \Re, i = 1, \ldots, \ell$ be quadratic functions which are nonnegative over a polyhedron P. If the set

$$S \equiv \{x \in P : g_i(x) = 0, \ i = 1, \ldots, \ell\}$$

is nonempty, then for any compact set $W \subset \Re^n$, there exists a constant $\tau > 0$ such that

$$\text{dist}(x, S) \leq \tau r(x)^{1/2}, \quad \text{for all } x \in P \cap W, \tag{16}$$

where

$$r(x) \equiv \sum_{i=1}^{\ell} |g_i(x)|, \quad x \in \Re^n$$

is the residual function.

The proof of Theorem **2.3.10** is rather involved. It makes use of an interesting property about the structure of the set of stationary points of a quadratic program. Let $f : \Re^n \to \Re$ be the quadratic function given by

$$f(x) \equiv q^T x + \tfrac{1}{2}x^T Q x, \quad x \in \Re^n \tag{17}$$

for some vector $q \in \Re$ and some symmetric matrix $Q \in \Re^{n \times n}$, and let P be a polyhedral set in $\Re^n$. Consider the following quadratic program

$$\begin{aligned} \text{minimize} \quad & f(x) \\ \text{subject to} \quad & x \in P. \end{aligned} \tag{18}$$

As with a standard optimization problem, solutions of the AVI (q, Q, P) are the *stationary* points of the quadratic program (18). The following result was established in [175]; see Lemma 3.1 therein.

2.3.11 Lemma. Let Q be a symmetric matrix, q be an arbitrary vector, and P be a polyhedron. Then $\mathrm{SOL}(q, Q, P)$ is a finite union of polyhedral sets. If $C_1, C_2, ..., C_t$ denote the connected components of $\mathrm{SOL}(q, Q, P)$, where t is some positive integer, then

$$\mathrm{SOL}(q, Q, P) = \bigcup_{i=1}^{t} C_i$$

and the following properties hold:

(a) each C_i is the union of a finite collection of polyhedral sets;

(b) the C_i's are properly separated from one another; in other words, $\mathrm{dist}(C_i, C_j) > 0$, for all $i \neq j$;

(c) the quadratic function f given by (17) is constant on each C_i.

It follows from part (c) of the above lemma that if $\mathcal{S}$ and $f_{\min}$ denote, respectively, the optimal solution set and optimum objective value of the quadratic problem (18), then $\mathcal{S}$ is equal to the union of those pieces C_i on which f attains the value $f_{\min}$. In particular, $\mathcal{S}$, like $\mathrm{SOL}(q, Q, P)$, is also equal to the finite union of polyhedra. Note that since no convexity is assumed on the quadratic function f, $\mathcal{S}$ is most likely a proper subset of $\mathrm{SOL}(q, Q, P)$.

Proof of Theorem 2.3.10. We will prove Theorem **2.3.10** in two steps. In the first step we prove (16) for a single quadratic function ($\ell = 1$); in the second step we argue that (16) holds for all $\ell > 1$ by reducing this case to the case of a single quadratic function.

Consider the case $\ell = 1$ and let the single quadratic function be denoted by

$$g(x) = \tfrac{1}{2}x^T Q x + q^T x + c;$$

let the polyhedral set P be denoted by

$$P = \{x : Ax \leq a\}.$$

We will consider two cases.

Case 1. $S \cap W = \emptyset$. In this case, the error bound (16) holds trivially for $x \in W \cap P$. In fact, $g(x)$ is positive everywhere in $W \cap P$. Since W is compact, so $g(x) \geq \mu > 0$, for all $x \in W \cap P$, where μ is a constant. On the other hand, we have for all $x \in W \cap P$,

$$\text{dist}(x, S) \leq \|x - \bar{x}\| \leq \|x\| + \|\bar{x}\| \leq \theta + \|\bar{x}\|,$$

where $\theta \equiv \sup_{x \in W} \|x\|$, which is finite, and $\bar{x}$ is an element of S (such an $\bar{x}$ exists since $S \neq \emptyset$). Then, it follows that

$$\text{dist}(x, S) \leq \frac{\theta + \|\bar{x}\|}{\sqrt{\mu}} \sqrt{g(x)}$$

$$= \frac{\theta + \|\bar{x}\|}{\sqrt{\mu}} \sqrt{r(x)}, \qquad \forall x \in W \cap P, \qquad (19)$$

which establishes the desired error bound for the case $S \cap W = \emptyset$.

Case 2. $S \cap W \neq \emptyset$. By scaling if necessary (i.e., dividing g by $\|Q\|$), we may assume without loss of generality that $\|Q\| \leq 1$. Consider the following quadratic program:

$$\begin{align} \text{minimize} \quad & g(x) \\ \text{subject to} \quad & x \in P. \end{align} \qquad (20)$$

Since $g(x)$ is nonnegative over the feasible region P, it follows that S is the set of optimal solutions of (20). Let $\hat{S}$ denote the set of stationary points of (20). Clearly, $S \subseteq \hat{S}$ and they are both composed of finitely many polyhedral sets. By the local error bound (9), there exist some positive constants δ and τ

$$\text{dist}(x, \hat{S}) \leq \tau \|x - \Pi_P(x - Qx - q)\|, \qquad (21)$$

for all $x \in P$ with $\|x - \Pi_P(x - Qx - q)\| \leq \delta$. Also, by Lemma **2.3.11**, we have a partition of $\hat{S} = S \cup \bar{S}$, where $\bar{S}$ is the union of a finite number of polyhedral sets and $\bar{S} \cap S = \emptyset$. In fact, $\bar{S}$ consists of those polyhedral pieces of $\hat{S}$ over which the value of g is strictly positive. (Recall that g takes the value of zero over S.) Therefore, for some positive constant $\varepsilon_1 > 0$, we have

$$g(x) \geq \varepsilon_1 > 0, \qquad \forall x \in \bar{S}$$

and
$$\text{dist}(S, \bar{S}) \equiv \min_{x \in S, y \in \bar{S}} \|x - y\| \equiv \rho_1 > 0.$$

We define
$$M(\varepsilon) \equiv \{x \in P \cap W : g(x) \leq \varepsilon\}.$$

The following list summarizes some simple properties of the sets $M(\varepsilon)$ for $\varepsilon \geq 0$.

1. $M(\varepsilon)$ is compact for each $\varepsilon \geq 0$ and $M(\varepsilon) \subseteq M(\varepsilon')$ for all $\varepsilon \leq \varepsilon'$.

2. $M(\varepsilon)$ contains $S \cap W$ for all $\varepsilon \geq 0$. $(M(0) = S \cap W.)$

3. There exists some ε_2 such that $\text{dist}(M(\varepsilon), \bar{S}) \geq \rho_1/2$, for all $\varepsilon \leq \varepsilon_2$.

4. There exists some $\varepsilon_3 > 0$ such that $\text{dist}(x, S) < \rho_1/2$, for all $x \in M(\varepsilon)$ and all $\varepsilon \leq \varepsilon_3$.

5. There exists some $\varepsilon_4 > 0$ such that $\|x - \Pi_P(x - Qx - q)\| \leq \delta$ for all $x \in M(\varepsilon_4)$.

The above can be argued easily using the definitions of $M(\varepsilon)$, $\bar{S}$ and the compactness of W.

Now, we let $\varepsilon^* \equiv \min(\varepsilon_1, \varepsilon_2, \varepsilon_3, \varepsilon_4)$. Then, $\varepsilon^* > 0$. Consider any $x \in P \cap W$. There are two cases.

Case 2.1. $x \notin M(\varepsilon^*)$. Thus, $g(x) \geq \varepsilon^*$. Similar to (19), we obtain

$$\begin{aligned}
\text{dist}(x, S) \leq \|x - x^*\| &\leq \|x\| + \|x^*\| \\
&\leq \frac{\theta + \|x^*\|}{\sqrt{\varepsilon^*}} \sqrt{g(x)} \\
&= \frac{\theta + \|x^*\|}{\sqrt{\varepsilon^*}} \sqrt{r(x)},
\end{aligned}$$

where x^* is an element of $S \cap W$ and θ is the maximum size of any vector in the compact set W. (Such an x^* exists since we have assumed that $S \cap W$ is nonempty.)

Case 2.2. $x \in M(\varepsilon^*)$. Then, $g(x) \leq \varepsilon^*$. Let $z \in \hat{S}$ be the Euclidean projection of x onto $\hat{S}$; thus

$$\text{dist}(x, \hat{S}) = \|x - z\| = \min(\text{dist}(x, S), \text{dist}(x, \bar{S})).$$

By properties (3) and (4), we have

$$\text{dist}(x, S) < \rho_1/2 \quad \text{and} \quad \text{dist}(x, \bar{S}) \geq \rho_1/2.$$

Consequently, it must hold that

$$\|x - z\| = \text{dist}(x, S) < \rho_1/2 \quad \text{and} \quad z \in S. \tag{22}$$

Since $\varepsilon^* \leq \varepsilon_4$, it follows that $x \in M(\varepsilon^*) \subseteq M(\varepsilon_4)$. Thus

$$\|x - \Pi_P(x - Qx - q)\| \leq \delta.$$

By (21), we obtain

$$\|x - z\| = \text{dist}(x, \hat{S}) \leq \tau \|x - \Pi_P(x - Qx - q)\|. \tag{23}$$

Let $y \equiv \Pi_P(x - Qx - q)$. Then y is the solution of the following quadratic program in the variable v:

$$\text{minimize} \quad \|v + Qx + q - x\|^2$$

$$\text{subject to} \quad Av \leq a.$$

Thus, y must satisfy the following optimality conditions

$$0 = y + Qx + q - x + A^T \lambda,$$
$$Ay - a \leq 0, \quad \lambda \geq 0, \quad \lambda^T(a - Ay) = 0, \tag{24}$$

where $\lambda \in \Re^m$ is the Lagrange multiplier vector associated with the constraint $Ay \leq a$. Consider the following:

$$\begin{aligned}
(x - y)^T(x - y) &= (x - y)^T(Qx + q + A^T \lambda) \\
&= (x - y)^T(A^T \lambda) + (x - y)^T(Qx + q) \\
&= (Ax - Ay)^T \lambda + (x - y)^T(Qx + q) \\
&= (Ax - a)^T \lambda + (x - y)^T(Qx + q) \\
&\leq (x - y)^T(Qx + q),
\end{aligned}$$

where the last two steps are due to (24) and $Ax \leq a$ (since $x \in P$). Using Taylor expansion, we obtain

$$(x - y)^T(Qx + q) = g(x) - g(y) + \frac{1}{2}(x - y)^T Q(x - y).$$

Substituting this in the preceding bound yields

$$\|x - y\|^2 \leq g(x) - g(y) + \tfrac{1}{2}(x - y)^T Q(x - y)$$
$$\leq g(x) + \tfrac{1}{2}(x - y)^T Q(x - y)$$
$$\leq g(x) + \tfrac{1}{2}\|Q\|\|x - y\|^2$$
$$\leq g(x) + \tfrac{1}{2}\|x - y\|^2,$$

where the second step is due to $g(y) \geq 0$ (since g is nonnegative over P), and the last step follows from the assumption that $\|Q\| \leq 1$. Rearranging the terms and simplifying gives

$$\|x - y\| \leq \sqrt{2g(x)}.$$

Combining this with (23) and noting that $y = \Pi_P(x - Qx - q)$ yields

$$\|x - z\| \leq \tau \|x - y\| \leq \sqrt{2}\tau \sqrt{g(x)}.$$

Since $z \in S$ (cf. (22)), it follows that

$$\mathrm{dist}(x, S) \leq \sqrt{2}\tau \sqrt{g(x)} = \sqrt{2}\tau \sqrt{r(x)}, \qquad \forall x \in P \cap W.$$

Now we let

$$\kappa' \equiv \max\left(\frac{\theta + \|x^*\|}{\sqrt{\varepsilon^*}}, \sqrt{2}\tau\right)$$

and then combine the bounds in **Case 2.1** and **Case 2.2** to obtain

$$\mathrm{dist}(x, S) \leq \kappa' \sqrt{r(x)}, \qquad \forall x \in P \cap W.$$

Finally, by combining **Case 1** and **Case 2**, we deduce

$$\mathrm{dist}(x, S) \leq \kappa \sqrt{r(x)}, \qquad \forall x \in P \cap W,$$

where

$$\kappa \equiv \max\left(\kappa', \frac{\theta + \|\bar{x}\|}{\sqrt{\mu}}\right).$$

This establishes (16) for the case $\ell = 1$ and $x \in W \cap P$.

If $\ell > 1$, then we let $g(x) = \sum_{i=1}^{\ell} g_i(x)$. Clearly, $g(x)$ is nonnegative over P. Moreover, any vector $x \in P$ satisfies $g(x) = 0$ if and only if $g_i(x) = 0$ for all i. We have

$$S = \{x \in P : g(x) = 0\}.$$

Since g is a single quadratic function nonnegative over P, by the above case, we obtain

$$\text{dist}(x, S) = \text{dist}(x, \bar{S}) \leq \kappa \sqrt{g(x)}, \qquad \forall x \in P \cap W,$$

thus completing the proof of Theorem **2.3.10**.   **Q.E.D.**

Theorem **2.3.10** applies only to the "feasible" vectors $x \in P \cap W$. It is possible to extend this result to include the "infeasible" vectors $x \in W \setminus P$. This is accomplished in the following theorem.

2.3.12 Theorem. Let $g_i : \Re^n \to \Re, i = 1, \ldots, \ell$ be quadratic functions which are nonnegative over the polyhedron P. If the set

$$S \equiv \{x \in P : g_i(x) = 0, \ i = 1, \ldots, \ell\}$$

is nonempty, then for any compact set $W \subset \Re^n$, there exists a constant $\tau > 0$ such that

$$\text{dist}(x, S) \leq \tau r(x)^{1/2}, \quad \text{for all } x \in W, \qquad (25)$$

where

$$r(x) \equiv \text{dist}(x, P) + \sum_{i=1}^{\ell} |g_i(x)|, \quad \forall x \in \Re^n$$

is the residual function.

Proof. We use Theorem **2.3.10** to prove the error bound (25) for an arbitrary vector $x \in W$. Naturally, an extra term, such as $\text{dist}(x, P)$, must be introduced to account for the amount of constraint violation associated with the polyhedral set P. The basic idea of the proof is first to project x onto P and then apply the error bound (25) at the projected point. For this purpose, let $W' \equiv \Pi_P(W)$ be the image of W under the Euclidean projector $\Pi_P(\cdot)$. Since this map is continuous and W is compact, W' is compact. Clearly, $W' \subseteq P$.

Let $g(x) \equiv \sum_{i=1}^{\ell} g_i(x)$. Then $g(x)$ is a quadratic function which can be written as

$$g(x) = \tfrac{1}{2} x^T Q x + q^T x + c$$

for some symmetric $Q \in \Re^{n \times n}, q \in \Re^n$, and $c \in \Re$. Clearly, $g(x)$ is nonnegative over the polyhedral set P, and $S = \{x \in P : g(x) = 0\}$.

Fix an arbitrary $x \in W$. Let $z \in W'$ be the orthogonal projection of x onto P. Thus, $\|x - z\| = \mathrm{dist}(x, P)$. By the error bound (16) applied to the set W', and in particular to the vector z, we deduce the existence of a scalar $\tau > 0$ which is independent of z and x such that

$$\mathrm{dist}(x, S) \leq \|x - z\| + \mathrm{dist}(z, S) \leq \mathrm{dist}(x, P) + \tau \sqrt{g(z)}, \qquad (26)$$

where the first inequality follows from the triangle inequality. We next bound the term $\sqrt{g(z)}$ by $\sqrt{|g(x)|}$. Let $u = z - x$. By Taylor expansion, we obtain

$$g(z) = g(x) + (Qx + q)^T u + \tfrac{1}{2} u^T Q u.$$

Therefore, we have

$$\sqrt{g(z)} \leq \sqrt{|g(x)|} + \sqrt{|(Qx + q)^T u|} + \sqrt{\tfrac{1}{2}|u^T Q u|}$$

$$\leq \sqrt{|g(x)|} + (\|Qx\|^{1/2} + \|q\|^{1/2})\|u\|^{1/2} + \frac{\|Q\|^{1/2}}{\sqrt{2}}\|u\|.$$

Substituting this bound into (26) and simplifying, we obtain

$$\mathrm{dist}(x, S)$$

$$\leq \|u\| + \tau \left(\sqrt{|g(x)|} + (\|Qx\|^{1/2} + \|q\|^{1/2})\|u\|^{1/2} + \frac{\|Q\|^{1/2}}{\sqrt{2}}\|u\| \right)$$

$$\leq \kappa' \left(\|u\|^{1/2} + \sqrt{|g(x)|} \right),$$

where the last step follows from choosing a suitably large constant κ' and noting that both x and u are bounded (since W and W' are compact) so each term in the preceding step can be bounded by either $\|u\|^{1/2}$ or $\sqrt{|g(x)|}$. Now, noting that $\|u\| = \mathrm{dist}(x, P)$ in the above expression, we immediately obtain the desired error bound (25) for all $x \in W$. **Q.E.D.**

By Hoffman's error bound (2.3.5), the term $\mathrm{dist}(x, P)$ in the residual function $r(x)$ of (25) can in turn be replaced by the residual $\|[Ax - a]^+\|$. In so doing, the constant τ may need to be modified accordingly. We also point out that the nonnegativity assumption in Theorems **2.3.10** and **2.3.12** cannot be removed in general, as the next example shows.

2.3.13 Example. Consider the quadratic system in n-variables:

$$x_1 = x_2^2, \quad x_2 = x_3^2, \quad ...., \quad x_n = x_{n+1}^2, \quad x_1 = 0.$$

Clearly, the solution set $S = \{(0, ..., 0)\}$ which is a singleton. Consider the vector $x(\varepsilon) = (0, \varepsilon^{2^{n-1}}, ..., \varepsilon^2, \varepsilon)$ where ε is an arbitrary positive parameter. Clearly, $\text{dist}(x(\varepsilon), S) \geq \varepsilon$. Nevertheless, the residual $r(x(\varepsilon))$ is seen to be ε^{2^n}. Therefore the error bound (25) cannot hold for the above quadratic system with an exponent γ greater than $1/2^n$. The reason why the exponent $1/2$ fails to hold for this system is that it does not satisfy the nonnegativity assumption.

The next example shows that the error bound (25) cannot hold globally in general.

2.3.14 Example. Consider the polynomial system in $\Re^2$

$$x_1 x_2 = 0, \qquad x_1 \geq 0, \qquad x_2 \geq 1.$$

Clearly, the solution set is given by $S = \{(0, x_2) : x_2 \geq 1\}$. Consider the sequence of vectors parametrized by $\{(t, 0)\}$ with $t \geq 0$. Then the residual function evaluated at $(t, 0)$ is equal to 1, for all $t \geq 0$. On the other hand, $\text{dist}((t, 0), S) = \sqrt{t^2 + 1}$, which is unbounded as t tends to ∞. Thus, the residual function cannot be used to bound the distance to the solution set S for all vectors in $\Re^2$. The restriction to a compact set is necessary in general.

When the defining quadratic polynomials are convex (i.e., each Q_i is positive semidefinite), it is possible to obtain an exponent of 1, provided that the inequality system satisfies the Slater condition. This is a result due to Luo and Luo [167].

2.3.4 Error bounds for NCP

Suppose $F : \Re^{n+m} \to \Re^m$ is a continuous subanalytic mapping and $Z \subset \Re^{n+m}$ is a compact subanalytic set. Let X be the projection of Z onto $\Re^n$; i.e.,

$$X = \{x \in \Re^n : (x, y) \in Z \text{ for some } y\}.$$

Consider the following parametric NCP with y as the primary variable and $x \in X$ as the parameter:

$$y \geq 0, \quad F(x, y) \geq 0, \quad y^T F(x, y) = 0. \tag{27}$$

Notice that when $C(x) = \Re_+^m$ (the nonnegative orthant), the inner VI of the MPEC (1.1.1) can be cast exactly in the above form. Let S denote the solution set of the above parametric NCP; that is

$$S = \left\{ (x,y) \in \Re^{n+m} : y^T F(x,y) = 0, \ F(x,y) \geq 0, \ y \geq 0 \right\}. \qquad (28)$$

Notice that if $n = 0$ (equivalently, if x is absent), then (27) becomes the standard NCP. By Lojasiewicz' error bound, Theorem **2.1.1**, there exist some positive constants τ and γ such that

$$\operatorname{dist}((x,y), S) \leq \tau \left(r(x,y) \right)^\gamma, \qquad \forall (x,y) \in Z,$$

where the residual function $r(x,y)$ is defined by

$$r(x,y) \equiv \|y^+\| + \|F^-(x,y)\| + |y^T F(x,y)|,$$

with

$$F^-(x,y) \equiv \max(0, -F(x,y)).$$

In fact, this is the error bound we have used in Section **2.2** to derive the exact penalty functions for MPEC.

It is possible to show that $\gamma = 1$ is a valid exponent when the mapping F possesses certain uniform properties. This is the main topic of discussion in the rest of this subsection. Specifically, let us make the following two assumptions:

(UNI-P) the mapping $F(x, \cdot)$ is uniform P with the same modulus for all $x \in X$; in other words, there exists a positive constant c such that

$$\max_{1 \leq i \leq m} (y - y')_i (F(x,y) - F(x,y'))_i \geq c \, \|y - y'\|^2, \qquad (29)$$

for all x in X and all y and y' in $\Re_+^m$;

(LIP) the mapping F is Lipschitz continuous on $X \times \Re_+^m$; in other words, there exists a positive constant $\gamma > 0$ such that

$$\|F(x,y) - F(x',y')\| \leq \gamma \left(\|x - x'\| + \|y - y'\| \right), \qquad (30)$$

for all x and x' in X and all y and y' in $\Re_+^m$.

Note that the same constant c in the UNI-P property applies to all vectors $x \in X$; this requirement is rather restrictive. Nevertheless, it seems essential for the kind of global error bound that we are trying to obtain herein. One situation in which this property holds is when $F(x, y)$ is separable in x and y, say $F(x, y) \equiv f(x) + g(y)$, and the function g is uniform P. Then clearly F has the UNI-P property.

Given that $F(x, \cdot)$ is a uniform P function, the NCP (27) has a unique solution for each fixed $x \in X$. Let us denote this solution by $y(x)$. The following result summarizes several important properties of this solution as a function of $x \in X$.

2.3.15 Lemma. Let $X \subset \Re^n$ be compact. Suppose that F has the UNI-P and LIP properties, then the following statements hold.

(a) For any scalar $\delta > 0$, the level set

$$L_\delta \equiv \left\{ (x, y) \in X \times \Re_+^m \ : \ y^T F(x, y) \le \delta, \ F(x, y) \ge 0 \right\}$$

is compact.

(b) The solution function $y(x)$ is Lipschitz continuous on X.

(c) There exists a constant $\tau > 0$ such that for all $(x, y) \in X \times \Re_+^m$,

$$\| y - y(x) \| \le \tau \, \| \min(y, F(x, y)) \|.$$

(d) For any scalar $\delta > 0$, there exists a scalar $\varepsilon > 0$ such that for all $(x, y) \in X \times \Re_+^m$ with either $\| \min(y, F(x, y)) \| \ge \delta$ or $y^T F(x, y) \ge \delta$,

$$\| y - y(x) \| \ge \varepsilon.$$

Proof. Fix any $\delta > 0$ and $\bar{x} \in X$. Let $\bar{y} \equiv y(\bar{x})$. Consider any pair $(x, y) \in L_\delta$. By the UNI-P property of F, there exists an index i such that

$$c \| y - \bar{y} \|^2$$

$$\le (y - \bar{y})_i (F(\bar{x}, y) - F(\bar{x}, \bar{y}))_i$$

$$= -(y - \bar{y})_i F_i(\bar{x}, \bar{y}) + (y - \bar{y})_i (F(\bar{x}, y) - F(x, y))_i + (y_i - \bar{y}_i) F_i(x, y)$$

$$\le \delta + \gamma \| y - \bar{y} \| \, \| x - \bar{x} \|,$$

where the second inequality follows from the LIP property of F and the given conditions: $y_i F_i(\bar{x}, \bar{y}) \ge 0$, $\bar{y}_i F_i(x, y) \ge 0$, and $\bar{y}_i F_i(\bar{x}, \bar{y}) = 0$. The above is a quadratic inequality in $\| y - \bar{y} \|$. Since $\| x - \bar{x} \|$ is bounded (by the compactness of X), this inequality implies that $\|y - \bar{y}\|$ is bounded. Thus, the level set L_δ is bounded. Since F is continuous and X is closed, it follows that L_δ is compact.

To prove part (b), let $x^1, x^2 \in X$ be arbitrary. Let $y^i \equiv y(x^i)$ for $i = 1, 2$. By the same reasoning as above, we have, for some index i

$$c \|y^1 - y^2\|^2$$

$$\leq (y^1 - y^2)_i (F(x^1, y^1) - F(x^1, y^2))_i$$

$$\leq -(y^1 - y^2)_i (F(x^1, y^2) - F(x^2, y^2))_i$$

$$\leq \gamma \|y^1 - y^2\| \, \|x^1 - x^2\|,$$

which yields

$$\|y^1 - y^2\| \leq \frac{\gamma}{c} \|x^1 - x^2\|,$$

establishing the desired Lipschitz continuity of the solution function $y(x)$ on X.

To prove part (c), let $(x, y) \in X \times \Re^m_+$ be given. We write

$$z \equiv \min(y, F(x, y)).$$

Then we have

$$u \equiv y - z \geq 0, \quad v \equiv F(x, y) - z \geq 0, \quad u^T v = 0.$$

By the uniform P property of F, there exists an index i such that

$$(y - y(x))_i (F(x, y) - F(x, y(x)))_i \geq c \|y - y(x)\|^2.$$

For this index i, we have

$$0 \geq (u - y(x))_i (v - F(x, y(x)))_i$$

$$= (y - y(x))_i (F(x, y) - F(x, y(x)))_i - (y - y(x))_i z_i$$

$$\quad - z_i (F(x, y) - F(x, y(x)))_i + z_i^2,$$

which implies by the LIP property of F

$$c \|y - y(x)\|^2 \leq (1 + \gamma) \|y - y(x)\| \, \|z\|.$$

Hence,

$$\|y - y(x)\| \leq \frac{1 + \gamma}{c} \| \min(y, F(x, y)) \|.$$

For part (d), we note that the continuity of $y(x)$ on X implies

$$\rho \equiv \sup_{x \in X} \|y(x)\| < \infty.$$

The scalar function

$$\theta(x, y) \equiv \|y - y(x)\|$$

is continuous and positive on the compact set

$$\{(x, y) \in X \times (\mathbb{B}(0, 2\rho) \cap \Re_+^m) : y^T F(x, y) \geq \delta,$$
$$\| \min(y, F(x, y))\| \geq \delta\}; \tag{31}$$

thus θ attains its minimum on this set (provided that the set is nonempty) and the minimum value is positive. Consequently, for some positive scalar $\varepsilon_1 > 0$, we have

$$\|y - y(x)\| \geq \varepsilon_1$$

for all (x, y) in the set (31), if this set is nonempty. Let $\varepsilon_1 \equiv \infty$ if (31) is empty. For $(x, y) \in X \times \Re_+^m$ with $\|y\| > 2\rho$, we clearly have $\|y - y(x)\| \geq \rho$. Consequently part (d) follows with $\varepsilon \equiv \min(\varepsilon_1, \rho)$.   **Q.E.D.**

Part (c) of the above lemma implies that $r(x, y) \equiv \| \min(y, F(x, y))\|$ is a legitimate residual for the parametric NCP (27) with reference to the test set $X \times \Re_+^m$, under the assumptions UNI-P and LIP on F. For the standard NCP (i.e., $n = 0$ and the parameter x is absent) this global error bound is somewhat well known although it has never been documented exactly as such; see, e.g., [217, 202]. In essence, the proof of Lemma **2.3.15**(c) is an extension of a standard argument used for the non-parametrized LCP and for a non-parametrized strongly monotone variational inequality.

Another consequence of Lemma **2.3.15** is that for any scalar $\delta > 0$, the set

$$\{(x, y) \in X \times \Re_+^m : \| \min(y, F(x, y))\| \leq \delta\}$$

is compact. Indeed this follows easily from parts (b) and (c) of the lemma. This observation is the analog of part (a) where the set L_δ involves the product residual.

As in Lemma **2.3.8**, we can show that the min residual is bounded above by the product residual under a further nondegeneracy assumption.

2.3.16 Lemma. Let $X \subset \Re^n$ be compact. Suppose that F has the UNI-P and LIP properties and that for every $x \in X$, the solution $y(x)$ of the NCP (27) is nondegenerate; i.e., $y(x) + F(x, y(x)) > 0$. Then for any scalar $\delta > 0$, there exists a constant $\tau > 0$ such that for all $(x, y) \in X \times \Re_+^m$ with $F(x, y) \geq 0$ and $\| \min(y, F(x, y)) \| \leq \delta$,

$$\| \min(y, F(x, y)) \| \leq \tau \, y^T F(x, y).$$

Proof. The proof is parallel to that of Lemma **2.3.8** and makes use of the observation made above. Let $\delta > 0$ be given. Assume for the sake of contradiction that there exists a sequence $\{(x^k, y^k)\} \subset X \times \Re_+^m$ such that for each k, $F(x^k, y^k) \geq 0$, and

$$\delta \geq \| \min(y^k, F(x^k, y^k)) \| > k(y^k)^T F(x^k, y^k) > 0.$$

It follows that this sequence is bounded. Without loss of generality we may assume that $\{(x^k, y^k)\}$ converges to a limit (x^∞, y^∞) which must satisfy: $x^\infty \in X$, and

$$y^\infty \geq 0, \quad F(x^\infty, y^\infty) \geq 0, \quad (y^\infty)^T F(x^\infty, y^\infty) = 0.$$

By the uniqueness of solution of the NCP (27), it follows that $y^\infty = y(x^\infty)$ and $y^\infty + F(x^\infty, y^\infty) > 0$. Let

$$\mathcal{I} \equiv \{i : (y^\infty)_i > 0 = F_i(x^\infty, y^\infty)\}$$
$$\mathcal{J} \equiv \{i : (y^\infty)_i = 0 < F_i(x^\infty, y^\infty)\}.$$

By nondegeneracy, $\mathcal{I}$ and $\mathcal{J}$ partition the index set $\{1, \dots, m\}$. By a continuity argument similar to that used at the end of the proof of Lemma **2.3.8**, we can derive a contradiction. The details are omitted.   **Q.E.D.**

Using the above lemmas, we can now establish the main result of this subsection which gives a global error bound of order 1 for a parametric NCP satisfying the UNI-P and LIP properties.

2.3.17 Theorem. Under the assumptions of Lemma **2.3.16**, there exists a scalar $\tau > 0$ such that

$$\text{dist}((x, y), S) \leq \tau y^T F(x, y), \tag{32}$$

for all $(x, y) \in X \times \Re_+^m$ with $F(x, y) \geq 0$, where S is given by (28).

Proof. Let $\delta > 0$ be an arbitrary scalar. Let $\varepsilon > 0$ be the scalar associated with this δ as asserted by Lemma **2.3.15**(d). Consider any $(x, y) \in X \times \Re^m_+$ with $F(x, y) \geq 0$. There are two cases.

Case 1. $\| \min(y, F(x, y)) \| \geq \delta$. By part (d) of Lemma **2.3.15**, we have $\| y - y(x) \| \geq \varepsilon$. To establish the desired error bound, we use the UNI-P property of F to obtain, for some index i,

$$
\begin{aligned}
c \| y - y(x) \|^2 &\leq (y - y(x))_i (F(x, y) - F(x, y(x)))_i \\
&= y_i F_i(x, y) - y_i(x) F_i(x, y) - y_i F_i(x, y(x)) \\
&\leq y^T F(x, y),
\end{aligned}
$$

where the last step follows from the nonnegativity of the terms involved. Now the desired error bound follows immediately:

$$
\begin{aligned}
\mathrm{dist}((x, y), S) &\leq \| y - y(x) \| \\
&\leq \frac{1}{\varepsilon} \| y - y(x) \|^2 \leq \frac{1}{c\varepsilon} y^T F(x, y).
\end{aligned}
$$

This completes the proof for **Case 1**.

Case 2. $\| \min(y, F(x, y)) \| \leq \delta$. In this case, it suffices to combine Lemma **2.3.15**(c) and Lemma **2.3.16**.   **Q.E.D.**

Unlike the min residual, a desirable feature of the error bound (32) is that the residual $y^T F(x, y)$ is a differentiable function, provided that F is smooth. In Section **2.4.2** we use this (uniform) error bound to derive a differentiable exact penalty function for a mathematical program with parametric NCP constraints.

Theorem **2.3.17** can be compared to Theorem **2.3.9**. One could consider the former as a restricted, nonlinear generalization of the latter. In fact, both results use the complementarity term to bound the distance function; their assumptions are also similar: both assume X is compact and both require nondegeneracy conditions. Nevertheless, there are also several differences. For one thing, Theorem **2.3.17** requires a very strong P property on the function F, whereas Theorem **2.3.9** applies to all parametric affine variational inequalities without assuming any strong monotonicity or P property. Of course, one implication of the uniform P assumption in Theorem **2.3.17** is the uniqueness of solution to the parametric NCP

(27); there was no such uniqueness assumption or implication in the other theorem.

A result similar to Theorem **2.3.17** was obtained by Marcotte and Zhu [201] for parametric strongly monotone variational inequality problems defined on a polyhedral set. In principle, we believe that the results established herein could be generalized to this case. The assumptions used in the cited reference are somewhat different; it requires, among other things, that $\dot{F}$ be continuously differentiable and that the Jacobian ∇F be uniformly Lipschitz continuous (our analysis does not require F to be differentiable), and that the problem be (uniformly) geometrically stable. The latter notion is related to the nondegeneracy condition.

2.4 Improved Penalty Results for MPEC

In this section, we apply the error bound results developed in the previous section to establish some improved exact penalty functions for MPEC. Unlike the exact penalty function of Theorem **2.2.1**, these new exact penalty functions all have explicit exponents (equal to 1 or 1/2), thus making them good candidates for use in numerical computation.

2.4.1 *AVI constrained mathematical program*

We consider a sharpening of Theorem **2.2.1** in the case where the inner VI is an affine variational inequality defined on a polyhedron of special type. More specifically, suppose that

$$g(x,y) = Dx + Ey + b, \quad \text{for all } (x,y) \in \Re^{n+m} \tag{1}$$

for some given matrices $D \in \Re^{\ell \times n}$ and $E \in \Re^{\ell \times m}$ and vector $b \in \Re^{\ell}$. Also assume for the sake of discussion that

$$F(x,y) = Px + Qy + q, \quad \text{for all } (x,y) \in \Re^{n+m} \tag{2}$$

for some given matrices $P \in \Re^{m \times n}$ and $Q \in \Re^{m \times m}$ and vector $q \in \Re^{m}$.

The following is the MPAEC to be studied in this subsection:

minimize $f(x, y)$

subject to $(x, y) \in Z,$

$$Dx + Ey + b \leq 0, \tag{3}$$

$$(y' - y)^T (Px + Qy + q) \geq 0,$$

for all y' such that $Dx + Ey' + b \leq 0.$

We also assume throughout this section that Z is polyhedral. Within this setting, our goal is first to obtain an exact penalty function for (3) with an exponent of $1/2$ (i.e., $N^* = 2$) in the penalty equivalents, (2.2.1) and (2.2.3), then we give a sufficient condition for the exponent to be equal to 1 in the exact penalty function for MPEC. We also obtain an alternative exact penalty function for MPAEC which is based on the "min" function.

Under the above specification of the function g, the CRCQ holds for the inner VIs; thus so does SBCQ. Theorem **1.3.5** therefore yields a constant $c > 0$ which bounds some multiplier $\lambda \in M(x, y)$ for all (x, y) feasible to (3). Now the problem (2.2.2), which is equivalent to (3), has the simplified form

minimize $f(x, y)$

subject to $(x, y, \lambda) \in Z \times (\mathbb{B}(0, c) \cap \Re_+^\ell),$

$$Px + Qy + q + E^T \lambda = 0, \tag{4}$$

$$Dx + Ey + b \leq 0, \quad \lambda^T (Dx + Ey + b) = 0.$$

In the next result, we consider the penalization of the complementarity condition $\lambda^T (Dx + Ey + b) = 0$ in (4). The resulting penalty equivalent is a linearly constrained nonlinear program with a square root term in the objective function. In principle, under the assumption that Z is compact, we could also penalize the linear constraints $Px + Qy + q + E^T \lambda = 0$ and $Dx + Ey + b \leq 0$. That is, the latter constraints could also be transferred to the objective function via a penalty term; but we refrain from doing so to avoid repeating similar arguments. Note, however, that we cannot penalize all the constraints in (4) because we need some compact set W in order to apply Theorem **2.3.12**. (The proof of the following result will elucidate this point.) In other words, we do not have an equivalent unconstrained minimization formulation for (3) even under the affine assumptions made herein. The following is the promised exact penalty function formulation for the MPAEC.

2.4.1 Theorem. Suppose that (3) is feasible, the set

$$Z \cap \mathrm{Gr}(C) = \{(x,y) \in Z : Dx + Ey + b \le 0\}$$

is a compact polyhedron, and f is Lipschitz on $Z \cap \mathrm{Gr}(C)$. Then there exist positive scalars α^* and c such that for all scalars $\alpha \ge \alpha^*$, any vector (x^*, y^*) solves (3) if and only if for some $\lambda^* \in \Re^\ell$, the triple (x^*, y^*, λ^*) solves the following minimization problem in the variables (x, y, λ):

$$
\begin{aligned}
\text{minimize} \quad & f(x,y) + \alpha\sqrt{-\lambda^T(Dx + Ey + b)} \\
\text{subject to} \quad & (x,y) \in Z, \quad \lambda \ge 0, \quad \|\lambda\| \le c, \\
& Px + Qy + q + E^T\lambda = 0, \\
& Dx + Ey + b \le 0.
\end{aligned}
\tag{5}
$$

Proof. Let $c > 0$ be the constant as specified by Theorem **1.3.5** specialized to the polyhedral map C in the present context. Let $W \subseteq \Re^{n+m+\ell}$ denote the feasible region of the problem (5). Then W is a compact subset of $\Re^{n+m+\ell}$ and the quadratic function $-\lambda^T(Dx+Ey+b)$ is clearly nonnegative on W. Thus, by Theorem **2.3.12**, there exists a constant $\rho > 0$ such that

$$\rho \, \mathrm{dist}((x,y,\lambda), \Omega) \le \sqrt{-\lambda^T(Dx + Ey + b)}, \quad \text{for all } (x,y,\lambda) \in W,$$

where $\Omega \subseteq \Re^{n+m+\ell}$ is the feasible region of the problem (4). The existence of the constant α^* with the desired property now follows exactly as in the proof of Theorem **2.1.2**.   **Q.E.D.**

A bilevel linear program is a further special case of the MPEC where the inner problem (1.1.3) is a linear program, the outer objective function f is also linear, and the set Z is a polyhedron. (This corresponds to the case where in (2), the matrices P and Q are both zero.) For the bilevel linear program, exact penalty functions have been obtained in [5, 23, 284]. In [5], it was shown that the complementarity term itself (without the square root), $-\lambda^T(Dx + Ey + b)$, can be used in the penalty function in (5). In

[23], it was shown that the residual function $\sum_{i=1}^{\ell} \min(\lambda_i, -(Dx+Ey+b)_i)$ is another exact penalty; nevertheless, this reference required the strict complementarity condition to hold for each primal-dual optimal solution to the lower-level linear program. In Theorem **2.4.3**, we remove this strict complementarity assumption.

In what follows, we give a simple example to illustrate that it is generally not possible to remove the square root in the penalty term in (5). The interesting feature of this example is that the inner problem is a very simple convex quadratic program. Thus, as long as the inner problem is not a linear program, there is a good chance that there will not be a differentiable exact penalty equivalent for the MPEC, unless there is some rather strong assumption satisfied by the inner program; see Theorem **2.4.4**.

2.4.2 Example. Consider the following bilevel program in $\Re^2$:

$$\begin{aligned} &\text{minimize} \quad x - y \\ &\text{subject to} \quad (x,y) \geq 0, \\ &\text{and} \quad\quad y \in \operatorname{argmin}\{ \tfrac{1}{2}y^2 \ : \ x+y \geq 0 \}. \end{aligned}$$

Clearly, this problem has a unique solution $(0,0)$. It is also easy to verify that for all α sufficiently large, $(0,0,0)$ is the unique solution of the following penalized program in the variables (x, y, λ):

$$\begin{aligned} &\text{minimize} \quad x - y + \alpha\sqrt{\lambda(x+y)} \\ &\text{subject to} \quad (x,y,\lambda) \geq 0, \\ &\quad\quad\quad\quad x + y \geq 0, \quad y - \lambda = 0. \end{aligned}$$

We have thus verified the conclusion of Theorem **2.4.1** for this bilevel program (even though the feasible region is not compact). Now consider the following problem which is obtained from the last program by removing the square root in the objective function:

$$\begin{aligned} &\text{minimize} \quad x - y + \alpha\lambda(x+y) \\ &\text{subject to} \quad (x,y,\lambda) \geq 0, \\ &\quad\quad\quad\quad x + y \geq 0, \quad y - \lambda = 0. \end{aligned}$$

This is clearly equivalent to

$$\text{minimize} \quad x - y + \alpha y(x + y)$$
$$\text{subject to} \quad (x, y) \geq 0,$$
$$x + y \geq 0.$$

It can be seen that for any $\alpha > 0$, the optimal value of the latter quadratic program is negative and $(0, 0)$ is not an optimal solution. In fact, consider $x = 0$; the objective function is negative when y is small (but positive). This shows that in general the square-root operator cannot be removed.

The reason why, in the above example, the complementarity term alone (without the square root) cannot be an exact penalty function is twofold: neither the strict complementarity nor a "regularity" condition holds for the lower-level problem at the optimal solution $(0, 0)$. In Section **4.4**, we treat the MPEC satisfying certain regularity assumptions. In what follows, we give an alternative exact penalty function for the AVI constrained MP. This alternative function does not involve the square root of the residual function of the complementarity condition.

2.4.3 Theorem. In the setting of Theorem **2.4.1**, there exist positive scalars α^* and c such that for all scalars $\alpha \geq \alpha^*$, any vector (x^*, y^*) solves (3) if and only if for some $\lambda^* \in \Re^\ell$, the triple (x^*, y^*, λ^*) solves the following minimization problem in the variables (x, y, λ):

$$\text{minimize} \quad f(x, y) + \alpha \sum_{i=1}^{\ell} \min(\lambda_i, -(Dx + Ey + b)_i)$$
$$\text{subject to} \quad (x, y) \in Z, \quad \lambda \geq 0, \quad \|\lambda\| \leq c,$$
$$Px + Qy + q + E^T \lambda = 0,$$
$$Dx + Ey + b \leq 0.$$

(6)

Proof. As in the proof of Theorem **2.4.1**, let W and Ω be the feasible regions of the problems (6) and (4) respectively. It suffices to show that there exists a scalar τ such that

$$\text{dist}((x, y, \lambda), \Omega) \leq \tau \sum_{i=1}^{\ell} \min(\lambda_i, -(Dx + Ey + b)_i),$$

for all $(x, y, \lambda) \in W$. By Theorem **2.3.6**, the above inequality holds for all $(x, y, \lambda) \in W$ with $\sum_{i=1}^{\ell} \min(\lambda_i, -(Dx + Ey + b)_i) \leq \varepsilon$, where ε is a positive constant. If $(x, y, \lambda) \in W$ and $\sum_{i=1}^{\ell} \min(\lambda_i, -(Dx + Ey + b)_i) > \varepsilon$, then the same inequality must still hold, possibly with a different constant τ' dependent on ε but independent of $(x, y, \lambda) \in W$, by the compactness of the W.   **Q.E.D.**

In contrast to (5) which is an exact penalty equivalent of order $1/2$ of the AVI constrained MP (4), the problem (6) is a (generally nondifferentiable) exact penalty equivalent of order 1 of the same problem. Note that the square root function in (5) is not even Lipschitz in a neighborhood of zero, hence in general the penalty term in (5) is not Lipschitz near feasible points of (4); in contrast, the "min" function in (6) is globally Lipschitz. Moreover, being a piecewise linear operator, the min function possesses some useful differentiability properties [219] that can be used to derive optimality conditions for the program (4); see Subsection **2.4.3** for a preliminary discussion of these conditions and Section **3.2** for the detailed development of these conditions.

In what follows, we show that if each feasible solution of the problem (4) satisfies the strict complementarity condition, i.e., if $\lambda - (Dx + Ey + b) > 0$ for all $(x, y, \lambda) \in \Omega$, then the square root of the penalized complementarity term in (5) can be removed. The proof of this result is a straightforward application of Theorem **2.3.9** and is therefore omitted.

2.4.4 Theorem. In the setting of Theorem **2.4.1**, if the assumptions of Lemma **2.3.8** hold, then there exist positive constants α^* and c such that for all scalars $\alpha \geq \alpha^*$, any vector (x^*, y^*) solves (3) if and only if for some $\lambda^* \in \Re^{\ell}$, the triple (x^*, y^*, λ^*) solves the following minimization problem in the variables (x, y, λ):

$$
\begin{aligned}
\text{minimize} \quad & f(x, y) - \alpha \lambda^T (Dx + Ey + b) \\
\text{subject to} \quad & (x, y) \in Z, \quad \lambda \geq 0, \quad \|\lambda\| \leq c, \\
& Px + Qy + q + E^T \lambda = 0, \\
& Dx + Ey + b \leq 0.
\end{aligned}
\tag{7}
$$

Between the two penalty equivalents (6) and (7), the former has a nondifferentiable objective function (of a special type), whereas the latter has a smooth objective function. Nevertheless, the former Theorem **2.4.3** does

not require the kind of strict complementarity assumption on the inner AVI
as does the latter Theorem **2.4.4**.

In contrast to the global results, Theorems **2.4.1**, **2.4.3**, and **2.4.4**,
where each of the exact penalty equivalents is a linearly constrained optimization problem, it is possible to obtain an unconstrained penalty equivalent for a local minimizer of (4). To derive the latter penalty equivalent,
we write

$$Z \equiv \{(x, y) \in \Re^{n+m} : Gx + Hy + a \le 0\} \tag{8}$$

for some matrices G and H and some vector a of appropriate dimensions.
Since the proof of this local result is similar to that of Corollary **2.1.4**, we
simply state the result without proof.

2.4.5 Corollary. Let Z, g, and F be given, respectively, by (8), (1), and
(2). Let $(x^*, y^*) \in \mathcal{F}$. Suppose that f is Lipschitz continuous in a neighborhood of (x^*, y^*). Then there exists a positive constant $\alpha^* > 0$ such that
for all $\alpha \ge \alpha^*$, (x^*, y^*) is a local minimum of (3) if and only if for some
$\lambda^* \in \Re^\ell$, the triple (x^*, y^*, λ^*) is an unconstrained local minimum of the
function $f(x, y) + \alpha r(x, y, \lambda)$, where

$$r(x, y, \lambda) \equiv \|[Gx + Hy + a]^+\| + \|\lambda^-\| + \|Px + Qy + q + E^T\lambda\| +$$
$$\|[Dx + Ey + b]^+\| + \sqrt{|\lambda^T(Dx + Ey + b)|}.$$

2.4.6 Remark. The remark following Corollary **2.1.4** regarding the assumption $(x^*, y^*) \in \mathcal{F}$ also applies to Corollary **2.4.5**.

We end this subsection with two further remarks. First, Theorems **2.4.3**
and **2.4.4** could be used to generate first-order optimality conditions for
the AVI constrained MP. Subsection **2.4.3** discusses the application of the
former theorem in this regard. Second, at the end of Subsection **3.3.1**, a
pointwise version of the strict complementarity hypothesis of Lemma **2.3.8**
is examined more fully in the context of the general MPEC (1.1.1). Here
we simply acknowledge that this hypothesis is very strong.

2.4.2 NCP constrained mathematical program

In Subsection **2.4.1** we have sharpened Theorem **2.2.1** in the case where
the inner VI is affine. In this subsection we consider another sharpening of
this theorem for the case where the inner VI is a parametric NCP which

satisfies the assumptions of Lemma **2.3.16**. For ease of reference, we state the problem to be considered as follows:

$$\text{minimize} \quad f(x,y)$$
$$\text{subject to} \quad (x,y) \in Z, \tag{9}$$
$$y \geq 0, \quad F(x,y) \geq 0, \quad y^T F(x,y) = 0.$$

As before, we let $\mathcal{F}$ be the feasible region of this problem.

2.4.7 Theorem. Let $Z \in \Re^{n+m}$ be a compact set and X be the projection of Z onto $\Re^n$. Let f be Lipschitz on Z. Suppose that F satisfies the UNI-P and LIP properties as stated in Subsection **2.3.4** and that $y + F(x,y) > 0$ for all $(x,y) \in \mathcal{F}$ which is assumed nonempty. Then there exists a scalar $\alpha^* > 0$ such that for all scalars $\alpha \geq \alpha^*$ any vector (x^*, y^*) solves (9) if and only if it solves

$$\text{minimize} \quad f(x,y) + \alpha y^T F(x,y)$$
$$\text{subject to} \quad (x,y) \in Z, \tag{10}$$
$$y \geq 0, \quad F(x,y) \geq 0.$$

Proof. Let S be given by (2.3.28). The set X is compact since Z is compact. Therefore we can invoke Theorem **2.3.17** to obtain

$$\text{dist}((x,y), S) \leq \tau y^T F(x,y),$$

for all $(x,y) \in X \times \Re^m_+$ with $F(x,y) \geq 0$, where τ is a positive constant. By the proof of Theorem **2.1.2**, the desired conclusion follows readily. **Q.E.D.**

The main improvement of the exact penalty function formulation (10) over the formulation in Theorem **2.2.1** is that the former is a differentiable program (assuming F is smooth). This improvement is made possible by the error bound of Theorem **2.3.17** under the UNI-P, LIP, and nondegeneracy assumptions. Admittedly these assumptions are rather restrictive and it is not clear whether or how they can be relaxed.

2.4.3 *Optimality conditions: preliminary discussion*

Based on the improved exact penalty results established in the previous subsections and an elementary optimality theory from standard nonlinear programming, we can derive some first-order optimality conditions for a local minimizer of an MPEC under the assumptions made therein. In what follows, we take a preliminary stroll into this vast subject of optimality conditions for MPEC and postpone the full treatment until the next chapter.

Consider the MPAEC (2.4.3) with a polyhedral Z. Assume that (x^*, y^*) is a local minimum of this problem. Write $z^* \equiv (x^*, y^*)$. Then there exist a scalar $\alpha > 0$ and a vector λ^* such that the triple (x^*, y^*, λ^*) is a local minimizer of the following linearly constrained nonlinear program in the variables (x, y, λ):

$$
\begin{aligned}
\text{minimize} \quad & f(x, y) + \alpha \sum_{i=1}^{\ell} \min(\lambda_i, -(Dx + Ey + b)_i) \\
\text{subject to} \quad & (x, y, \lambda) \in Z \times \Re_+^\ell, \\
& Px + Qy + q + E^T \lambda = 0, \\
& Dx + Ey + b \le 0.
\end{aligned}
\tag{11}
$$

We note that $\min(\lambda^*, -(Dx^* + Ey^* + b)) = 0$ because (x^*, y^*) is feasible to (2.4.3). The objective function of the program (11) is not Fréchet differentiable, but is directionally differentiable, provided that f is so; see Definition **4.2.1** for a review of these differentiability concepts. The following lemma gives an explicit formula for the directional derivative of the min function of two arguments.

2.4.8 Lemma. Let $\theta(u, v) \equiv \min(u, v)$ for $(u, v) \in \Re^2$. For any two pairs (u, v) and (du, dv) in $\Re^2$, it holds that

$$
\theta'((u, v), (du, dv)) = \begin{cases}
du & \text{if } u < v \\
\min(du, dv) & \text{if } u = v \\
dv & \text{if } v < u.
\end{cases}
\tag{12}
$$

Proof. This is fairly easy.   **Q.E.D.**

Assume that the function f is continuously differentiable in a neighborhood of (x^*, y^*); let $W \subset \Re^{n+m+\ell}$ denote the feasible region of the program (11). Motivated by the formula (12), we define three index sets of which

$\mathcal{J}$ is the set of degenerate indices:

$$\mathcal{I} \equiv \{i : 0 = \lambda_i^* < -(Dx^* + Ey^* + b)_i\},$$
$$\mathcal{J} \equiv \{i : \lambda_i^* = 0 = -(Dx^* + Ey^* + b)_i\},$$
$$\mathcal{K} \equiv \{i : \lambda_i^* > -(Dx^* + Ey^* + b)_i = 0\}.$$

Since z^* is a local minimizer of (11) and W is a polyhedron, by (12) for the directional derivative of the min function, it follows that

$$(dx, dy, d\lambda) \in (W - (x^*, y^*, \lambda^*)) \implies$$
$$\nabla_x f(z^*)^T dx + \nabla_y f(z^*)^T dy + \alpha \left(\sum_{i \in \mathcal{I}} d\lambda_i + \right.$$
$$\left. \sum_{i \in \mathcal{J}} \min(d\lambda_i, -(Ddx + Edy)_i) - \sum_{i \in \mathcal{K}} (Ddx + Edy)_i \right) \geq 0,$$

where

$$W - (x^*, y^*, \lambda^*) \equiv \{(x, y, \lambda) - (x^*, y^*, \lambda^*) : (x, y, \lambda) \in W\}.$$

Consequently for any pair of index sets $(\mathcal{J}_1, \mathcal{J}_2)$ partitioning $\mathcal{J}$, we have

$$(dx, dy, d\lambda) \in (W - (x^*, y^*, \lambda^*)) \implies$$
$$\nabla_x f(z^*)^T dx + \nabla_y f(z^*)^T dy + \qquad\qquad (13)$$
$$\alpha \left(\sum_{i \in \mathcal{I} \cup \mathcal{J}_1} d\lambda_i - \sum_{i \in \mathcal{K} \cup \mathcal{J}_2} (Ddx + Edy)_i \right) \geq 0.$$

In summary we have shown the following result which gives a preliminary set of first-order necessary conditions for (x^*, y^*) to be a local minimizer of (3).

2.4.9 Proposition. Let $z^* \equiv (x^*, y^*)$ be a feasible vector to (3) and λ^* be such that $(z^*, \lambda^*) \in W$. If z^* is a local minimizer of (3), then (13) holds for any pair of index sets $(\mathcal{J}_1, \mathcal{J}_2)$ partitioning $\mathcal{J}$.

The converse of the above result is actually correct, provided that f is pseudoconvex in its arguments. We will postpone the proof of this converse until later; see Proposition **4.1.4**. The implications (13) for various pairs $(\mathcal{J}_1, \mathcal{J}_2)$ are of the primal type; further discussion of these implications, including their equivalent primal-dual formulations, occurs in the next two chapters; see in particular Theorem **4.1.2**.

In a similar fashion, we can use Theorem **2.4.7** to derive a result similar to the above proposition for the NCP constrained MP (9). This will entail the introduction of constraint qualifications for the penalized nonlinear program (10). Since the assumptions in Theorem **2.4.7** are already fairly restrictive in the first place, we do not devote more discussion to this approach for obtaining optimality conditions for (9).

We end this chapter by mentioning the paper [193] which contains some further exact penalty function results for mathematical programs with LCP constraints and concave objective functions which include the case of an LPEC.

3

First-Order Optimality Conditions

This is the first of three chapters in which we derive some necessary optimality conditions for the MPEC (1.1.1). This chapter is concerned with the fundamental first-order conditions; Chapter **4** deals with the verification of hypotheses required for these first-order conditions; and Chapter **5** is concerned with the second-order conditions. Although we have seen that the MPEC (1.1.1) can be formulated as a standard nonlinear program via the KKT formulation of the inner VIs (see, e.g., Theorem **1.3.5**), for reasons to be given in this chapter a straightforward application of the optimality conditions in classical nonlinear programming is inappropriate.

One approach to deriving optimality conditions of an MPEC is to use a differentiable exact penalty equivalent of the problem, such as (2.4.7) or (2.4.10), and apply classical results to such a formulation. A drawback of this approach is that strong assumptions are required for the MPEC to be equivalent to a differentiable constrained optimization problem; see Theorem **2.4.7**. Even in the case of the MPAEC (Theorem **2.4.4**), the equivalent formulation requires the restrictive nondegeneracy assumption. For the MPAEC, one could use the less restrictive Theorem **2.4.3**; although the min function is nondifferentiable, it is simple enough to make the formulation (2.4.6) a viable candidate for this purpose; see Subsection **2.4.3**. Proposition **2.4.9** provides a preliminary set of necessary conditions for a local minimum of the MPAEC.

In this chapter, we abandon the exact penalty function approach for the derivation of optimality conditions of MPEC. Instead, we resort to first principles to understand the stationarity conditions and the tangent cone of the feasible region of MPEC. The chapter contains four sections. Section **3.1** introduces some basic concepts associated with stationarity and gives examples to illustrate that a standard nonlinear programming treatment of MPEC is inappropriate. Section **3.2** examines in detail the tangent cone of an MPEC at a feasible point; the results therein lead to the definition of appropriate constraint qualifications for the MPEC under which stationarity can be characterized. The chapter concludes with some further discussion of the connection between the MPEC (1.1.1) and its KKT formulation (1.3.8).

3.1 Elementary Stationarity Concepts

Stationarity conditions for optimization problems, also called first-order necessary (optimality) conditions, are among the most basic tools used for existence results, algorithms, and computation. These conditions are important because of the fact that each local minimizer must be a stationary point. The classical stationarity result for a smooth function is that unconstrained extrema of the function are zeros or roots of its gradient.

To ease the referencing, we repeat the statement of the MPEC (1.1.1):

$$
\begin{aligned}
\text{minimize} \quad & f(x,y) \\
\text{subject to} \quad & (x,y) \in Z \subseteq \Re^{n+m}, \text{ and} \\
& y \in \mathcal{S}(x) \equiv \text{SOL}(F(x,\cdot), C(x)),
\end{aligned}
\tag{1}
$$

and for

$$
C(x) \equiv \{ y \in \Re^m : g(x,y) \leq 0 \},
\tag{2}
$$

the formulation of the constraint $y \in \mathcal{S}(x)$ in terms of its KKT conditions:

$$\text{minimize} \quad f(x,y)$$

$$\text{subject to} \quad (x,y) \in Z,$$

$$F(x,y) + \sum_{i=1}^{\ell} \lambda_i \nabla_y g_i(x,y) = 0, \tag{3}$$

$$\lambda \geq 0, \quad g(x,y) \leq 0, \quad \lambda^T g(x,y) = 0.$$

Let $\mathcal{F} \subseteq \Re^{n+m}$ and $\mathcal{F}^{\text{KKT}} \subseteq \Re^{n+m+\ell}$ denote the feasible region of (1) and (3) respectively. Under the SBCQ on $\mathcal{F}$ and the convexity assumption of $g(x,\cdot)$ for $x \in X$, we have $(x,y) \in \mathcal{F}$ if and only if there exists $\lambda \in M(x,y)$ such that $(x,y,\lambda) \in \mathcal{F}^{\text{KKT}}$.

Formally, a stationary point of (1) is a vector $\bar{z} \equiv (\bar{x}, \bar{y})$ in $\mathcal{F}$ such that

$$\nabla_x f(\bar{z})^T dx + \nabla_y f(\bar{z})^T dy \geq 0, \quad \text{for all } (dx, dy) \in \mathcal{T}(\bar{z}; \mathcal{F}),$$

where the tangent cone $\mathcal{T}(\bar{z}; \mathcal{F})$ is defined in the usual way (see Subsection **1.3.1**). As a reminder, for any set $S \subset \Re^N$ containing $\bar{z}$, $\mathcal{T}(\bar{z}; S)$ is the set of all vectors $dx \in \Re^N$ for which there exist a sequence of positive scalars $\{\tau_k\}$ converging to zero and a sequence of vectors $\{x^k\} \subset S$ converging to $\bar{x}$ such that $dx = \lim(x^k - \bar{x})/\tau_k$.

It is an easy exercise to show that if $\bar{z}$ is a local minimizer of (1), then it is a stationary point of the MPEC. This stationarity condition is presented in the "primal" form, that is, each vector in the tangent cone of the feasible set at a local minimizer is a direction of non-descent for the gradient of the objective function.

In traditional nonlinear programming, such a stationarity condition is made computationally practical by a constraint qualification (CQ) which rules out pathological behavior of the constraint functions and ensures that the tangent cone is polyhedral. Under such a CQ the primal stationarity condition can be equivalently phrased in "primal-dual" form, that is, there exist Lagrange or KKT multipliers, one per constraint, which are directly linked to the dual of the tangent cone.

In MPEC, unfortunately, the tangent cone of $\mathcal{F}$ at a feasible point is generally the union of polyhedral convex cones; therefore the standard nonlinear programming approach is not immediately applicable. Nevertheless, we are able to characterize this tangent cone in a practical way by assuming a reasonable "MPEC CQ" (see Corollary **3.3.1**) which paves the way to a set of practical first-order optimality conditions. More specifically, under the latter CQ we give a primal-dual formulation of stationarity by explicitly decomposing the tangent cone into finitely many polyhedral cones. Sufficient conditions for the satisfaction of this MPEC CQ are discussed in detail in the next chapter.

In general, without referring to the inner VIs (in particular, without requiring the SBCQ and/or the convexity of $g(x, \cdot)$), we could define the stationarity of a triple $(x, y, \lambda) \in \mathcal{F}^{\mathrm{KKT}}$ for the problem (3), itself considered a constrained optimization problem. Section **3.4** is devoted to a detailed discussion of the connection between the two problems (1) and (3) as far as their local minima and stationary points are concerned (assuming the SBCQ and convexity).

Background

To motivate the following subsections, we return to the familiar nonlinear program (2.1.1):

$$\begin{aligned} \text{minimize} \quad & \theta(x) \\ \text{subject to} \quad & x \in X \\ & g(x) \leq 0, \quad h(x) = 0, \end{aligned} \tag{4}$$

where we also require differentiability of $\theta : \Re^n \to \Re$, $g : \Re^n \to \Re^m$, and $h : \Re^n \to \Re^q$, and convexity and closedness of X. We recall the notation $\mathcal{F}^{\mathrm{NLP}}$ which denotes the feasible region of this nonlinear program.

Let $\bar{x}$ be feasible to this NLP. If $\bar{x}$ is a local minimizer of (4), then it must satisfy the well-known stationarity condition

$$\nabla\theta(\bar{x})^T d \geq 0, \quad \text{for all } d \in \mathcal{T}(\bar{x}; \mathcal{F}^{\mathrm{NLP}}).$$

In general, due to the abstract definition of the cone $\mathcal{T}(\bar{x}; \mathcal{F}^{\mathrm{NLP}})$, the stationarity condition as stated does not yield useful information about $\bar{x}$. Indeed, a practical stationarity condition must rely on a computable representation of the tangent cone. To this end it is a common practice to consider the linearized cone:

$$\mathcal{L}(\bar{x}; \mathcal{F}^{\mathrm{NLP}})$$
$$= \{dx \in \mathcal{T}(\bar{x}; X) : \nabla g_i(\bar{x})^T dx \leq 0 \text{ if } g_i(\bar{x}) = 0, \ \nabla h(\bar{x})dx = 0\}.$$

It is easy to see that $\mathcal{T}(\bar{z}; \mathcal{F}^{\mathrm{NLP}})$ is always contained in $\mathcal{L}(\bar{z}; \mathcal{F}^{\mathrm{NLP}})$. The most basic CQ in NLP is apparently that $\mathcal{T}(\bar{x}; \mathcal{F}^{\mathrm{NLP}})$ contains $\mathcal{L}(\bar{x}; \mathcal{F}^{\mathrm{NLP}})$, that is, $\mathcal{T}(\bar{x}; \mathcal{F}^{\mathrm{NLP}}) = \mathcal{L}(\bar{x}; \mathcal{F}^{\mathrm{NLP}})$; see Guignard [101]. Under this CQ, we obtain the following equivalent form of the stationarity condition:

$$\nabla\theta(\bar{x})^T d \geq 0, \quad \text{for all } d \in \mathcal{L}(\bar{x}; \mathcal{F}^{\mathrm{NLP}}). \tag{5}$$

The basic CQ holds in many situations [73], for instance, for linearly constrained nonlinear programs. In fact, it can be argued that without the basic CQ, the representation of $\mathcal{F}^{\mathrm{NLP}}$ using X, g, and h is numerically ill-posed. If $\mathcal{T}(\bar{x}; X)$ is polyhedral, then so is the tangent cone $\mathcal{T}(\bar{x}; \mathcal{F}^{\mathrm{NLP}})$ if the basic CQ holds; the polyhedrality property of $\mathcal{T}(\bar{x}; \mathcal{F}^{\mathrm{NLP}})$ is a reasonable assumption in the practice of NLP.

We present some simple MPECs to illustrate that naively applying these results from NLP is inappropriate. Indeed the naive approach was the source of errors in early bilevel programming research; see the discussion at the end of Section **1.1**.

3.1.1 Example. Consider the MPEC (1) when $(x, y) \in \Re^2$, $F(x, y) = y$ and $g(x, y) = -y$. Rewrite this problem using the KKT formulation (3); this results in the following standard NLP:

$$\begin{aligned} \text{minimize} \quad & f(x, y) \\ \text{subject to} \quad & y - \lambda = 0, \quad y\lambda = 0, \quad y, \lambda \geq 0. \end{aligned} \tag{6}$$

Denote the feasible region by $\mathcal{F}^{\mathrm{NLP}}$; clearly $\mathcal{F}^{\mathrm{NLP}} = \Re \times \{(0,0)\}$. Thus $\mathcal{T}((x, y, \lambda); \mathcal{F}^{\mathrm{NLP}}) = \Re \times \{(0,0)\}$ at any feasible point $(x, y, \lambda) = (x, 0, 0)$. It is easy to see that $\mathcal{L}((x, y, \lambda); \mathcal{F}^{\mathrm{NLP}}) = \Re \times \{(dy, dy) : dy \in \Re_+\}$. So even though $\mathcal{T}((x, y, \lambda); \mathcal{F}^{\mathrm{NLP}})$ is polyhedral, it is far from being equal to $\mathcal{L}((x, y, \lambda); \mathcal{F}^{\mathrm{NLP}})$. Thus the basic CQ for the NLP is violated. Moreover, with a simple objective function such as $f(x, y) = -y$, it is easily seen that every feasible point of (6) is stationary, but (5) fails to hold (with $\bar{x} = (x, y, \lambda)$ and $\theta(x, y, \lambda) = -y$).

3.1.2 Example. Consider an MPEC in $\Re^3$:

$$\begin{aligned} \text{minimize} \quad & x^2 + y_1^2 + y_2^2 \\ \text{subject to} \quad & x^2 + x + y_1 \geq 0, \quad y_1 \geq 0, \quad y_1(x^2 + x + y_1) = 0, \\ & x^3 + y_2 \geq 0, \quad y_2 \geq 0, \quad y_2(x^3 + y_2) = 0. \end{aligned} \tag{7}$$

This problem is a realization of (1) with $Z = \Re^3$,

$$F(x, y_1, y_2) = (x^2 + x + y_1, x^3 + y_2), \quad \text{and} \quad g(x, y_1, y_2) = -(y_1, y_2),$$

and recasting the inner VI $(F(x, \cdot), C(x))$ as the complementarity system in (7). The unique optimal solution of this MPEC is

$$\bar{z} = (\bar{x}, \bar{y}_1, \bar{y}_2) = (0, 0, 0).$$

Clearly

$$\mathcal{S}(x) = \left\{ \left(\max(0, -x^2 - x), \max(0, -x^3) \right) \right\};$$

thus the feasible region $\mathcal{F}$ of (7) is equal to the union of the following three sets:

$$\{(x, 0, 0) : x \geq 0\}, \quad \{(x, 0, -x^3) : x \leq -1\},$$

and

$$\{(x, -x^2 - x, -x^3) : x \in [-1, 0]\}.$$

It is not difficult to verify that

$$\mathcal{T}(\bar{z}; \mathcal{F}) = \{(a, 0, 0) : a \geq 0\} \cup \{(-a, a, 0) : a \geq 0\},$$

which is the union of two half-rays; and

$$\mathcal{L}(\bar{z}; \mathcal{F}^{\text{NLP}}) = \{(a, b, c) : a + b \geq 0, \ (b, c) \geq 0\}.$$

Clearly, $\mathcal{T}(\bar{z}; \mathcal{F})$ is *nonconvex* and its convex hull is a proper subset of $\mathcal{L}(\bar{z}; \mathcal{F}^{\text{NLP}})$. Immediately we see that the basic CQ for NLP is violated.

We hope to convince the reader in the sequel that we can recognize and appropriately deal with the (possible) nonconvexity of $\mathcal{T}(\bar{z}; \mathcal{F})$ and the failure of the basic CQ in NLP, in order to produce first-order optimality conditions for the MPEC that are theoretically and computationally satisfactory.

3.2 The Tangent Cone

Throughout this section, we assume that f and F are continuously differentiable in an open set U containing the feasible region $\mathcal{F}$ of (3.1.1) and that for each $x \in X$, $C(x)$ is given by (3.1.2) with each g_i being twice continuously differentiable in U and $g_i(x, \cdot)$ being convex.

Given the definition of stationarity above, our first order of business is to obtain a good understanding of the (possibly) nonconvex cone $\mathcal{T}((\bar{x}, \bar{y}); \mathcal{F})$. Like the analysis of a standard NLP, i.e., (3.1.4) with $X = \Re^n$, our initial attempt is to show that the cone $\mathcal{T}((\bar{x}, \bar{y}); \mathcal{F})$ is contained in a certain "linearized cone" of the region $\mathcal{F}$. Unlike the case of a standard NLP, however, the latter cone will not be polyhedral in the present context. In what follows, we show how to construct this cone.

Throughout the rest of this section, the pair $(\bar{x}, \bar{y}) \in \mathcal{F}$ is fixed. We write $\bar{z} \equiv (\bar{x}, \bar{y})$. We recall the index set of active constraints at $\bar{z}$; that is,

$$\mathcal{I}(\bar{z}) = \{i : g_i(\bar{z}) = 0\}.$$

In Subsection **1.3.2**, we have introduced the SBCQ and discussed the role it plays in the study of the MPEC; see Theorems **1.3.4**, **1.3.5**, and **2.2.1**. There, the SBCQ was required to hold at all vectors in the feasible region $\mathcal{F}$ of (3.1.1). In what follows, we shall need this CQ to hold at the particular vector $\bar{z} \in \mathcal{F}$. For ease of reference, we rephrase the SBCQ at $\bar{z}$ as follows:

(SBCQ) for any sequence $\{(x^k, y^k)\} \subseteq \mathcal{F}$ converging to $\bar{z}$, there exists for each k a multiplier vector $\lambda^k \in M(x^k, y^k)$ and $\{\lambda^k\}$ is bounded.

This assumption implies that $M(x, y) \neq \emptyset$ for all $(x, y) \in \mathcal{F}$ sufficiently close to $\bar{z}$.

For any multiplier $\lambda \in M(\bar{z})$, we define the *lifted critical cone* at $(\bar{x}, \bar{y}, \lambda)$ as the following polyhedral cone in $\Re^{n+m}$:

$$\mathcal{K}(\bar{z}, \lambda) \equiv \{(dx, dy) \in \Re^{n+m} :$$

$$dx^T \nabla_x g_i(\bar{z}) + dy^T \nabla_y g_i(\bar{z}) \leq 0, \text{ for } i \in \mathcal{I}(\bar{z}) \text{ with } \lambda_i = 0, \qquad (1)$$

$$dx^T \nabla_x g_i(\bar{z}) + dy^T \nabla_y g_i(\bar{z}) = 0, \text{ for } i \text{ such that } \lambda_i > 0\}.$$

If $\nabla_x g_i(\bar{z}) = 0$ for all $i \in \mathcal{I}(\bar{z})$, then the projection of this cone onto $\Re^m$ reduces to the "critical cone" of the VI $(F(\bar{x}, \cdot), C(\bar{x}))$ at the solution $\bar{y}$ [220]; i.e., the projection $\mathcal{K}(\bar{z}, \lambda)$ onto $\Re^m$ is equal to the intersection of the linearized cone of $C(\bar{x})$ at $\bar{y}$ with the orthogonal complement of $F(\bar{x}, \bar{y})$. For any vector $dx \in \Re^n$, let $\mathcal{K}(\bar{z}, \lambda; dx)$ denote the following polyhedral set in $\Re^m$, which we call the *directional critical set* along the direction dx:

$$\mathcal{K}(\bar{z}, \lambda; dx) \equiv \{dy \in \Re^m : (dx, dy) \in \mathcal{K}(\bar{z}, \lambda)\}. \qquad (2)$$

Note that $\mathcal{K}(\bar{z}, \lambda; dx)$ is a (possibly empty) polyhedral set; it is a "section" of the cone $\mathcal{K}(\bar{z}, \lambda)$. Moreover $\mathcal{K}(\bar{z}, \lambda; 0)$ is precisely the critical cone of the VI $(F(\bar{x}, \cdot), C(\bar{x}))$ at $\bar{y}$ mentioned above. At the end of this section, we give alternative descriptions of $\mathcal{K}(\bar{z}, \lambda)$ and $\mathcal{K}(\bar{z}, \lambda; dx)$.

As defined in (1.3.10), the Lagrangean function of the VI $(F(x, \cdot), C(x))$ is given by

$$L(z, \lambda) \equiv F(z) + \sum_{i=1}^{\ell} \lambda_i \nabla_y g_i(z), \quad \text{for } (z, \lambda) \in \Re^{n+m+\ell}. \tag{3}$$

3.2.1 Lemma. If $dz = (dx, dy) \in \mathcal{T}(\bar{z}; \mathcal{F})$ and the SBCQ holds at $\bar{z} \in \mathcal{F}$, then $dz \in \mathcal{T}(\bar{z}; Z)$ and there exists a $\bar{\lambda} \in M(\bar{z})$ such that dy solves the

$$\text{AVI } \left(\nabla_x L(\bar{z}, \bar{\lambda}) dx, \nabla_y L(\bar{z}, \bar{\lambda}), \mathcal{K}(\bar{z}, \bar{\lambda}; dx) \right). \tag{4}$$

Proof. By definition, there exist a sequence of vectors $\{z^k\} \subseteq \mathcal{F}$ converging to $\bar{z}$, where $z^k \equiv (x^k, y^k)$ for each k, and a sequence of positive scalars $\{\tau_k\}$ converging to 0 such that $dz = \lim_{k \to \infty} \tau_k^{-1}(z^k - \bar{z})$. Clearly, $dz \in \mathcal{T}(\bar{z}; Z)$; it remains to verify the claim about dy. Since the SBCQ holds at $\bar{z}$, we may assume without loss of generality, by working with an appropriate subsequence if necessary, that for each k, a λ^k exists which satisfies:

$$F(z^k) + \sum_{i=1}^{\ell} \lambda_i^k \nabla_y g_i(z^k) = 0,$$
$$\lambda_i^k \geq 0, \quad g_i(z^k) \leq 0, \quad \lambda_i^k g_i(z^k) = 0, \tag{5}$$

and the sequence $\{\lambda^k\}$ converges to a limit $\bar{\lambda}$. Clearly, $\bar{\lambda} \in M(\bar{z})$. For each k, we have

$$0 \geq g_i(z^k) = g_i(\bar{z}) + \nabla g_i(\bar{z})^T(z^k - \bar{z}) + o(\|z^k - \bar{z}\|),$$

which implies, by the fact that $g_i(\bar{z}) = 0$ for $i \in \mathcal{I}(\bar{z})$ and upon dividing by τ_k and taking the limit $k \to \infty$,

$$0 \geq \nabla g_i(\bar{z})^T dz, \quad \text{for all } i \in \mathcal{I}(\bar{z});$$

moreover, equality must hold for an index i such that $\bar{\lambda}_i > 0$. Thus, $dy \in \mathcal{K}(\bar{z}, \bar{\lambda}; dx)$. For each k, we may write

$$F(z^k) = F(\bar{z}) + \nabla F(\bar{z})(z^k - \bar{z}) + o(\|z^k - \bar{z}\|),$$

$$\nabla_y g_i(z^k) = \nabla_y g_i(\bar{z}) + \nabla \left(\nabla_y g_i(\bar{z}) \right) (z^k - \bar{z}) + o(\|z^k - \bar{z}\|).$$

Substituting these expressions into the first equation of (5) and using the fact that $\bar{\lambda} \in M(\bar{z})$, we obtain

$$
\begin{aligned}
0 \;=\; & \nabla_z L(\bar{z}, \lambda^k)(z^k - \bar{z}) + \sum_{i:\bar{\lambda}_i > 0} (\lambda_i^k - \bar{\lambda}_i) \nabla_y g_i(\bar{z}) \\
& + \sum_{i:\bar{\lambda}_i = 0} \lambda_i^k \nabla_y g_i(\bar{z}) + o\left(\|z^k - \bar{z}\|\right).
\end{aligned}
\tag{6}
$$

Now let $dy' \in \mathcal{K}(\bar{z}, \bar{\lambda}; dx)$ and consider the inner product of $dy - dy'$ with the sum of terms at the right side in (6). By definition of $\mathcal{K}(\bar{z}, \bar{\lambda}; dx)$, for all i such that $\bar{\lambda}_i > 0$ we have

$$
0 \;=\; dx^T \nabla_x g_i(\bar{z}) + dy^T \nabla_y g_i(\bar{z}) \;=\; dx^T \nabla_x g_i(\bar{z}) + (dy')^T \nabla_y g_i(\bar{z}),
$$

thus $(dy - dy')^T \sum_{i:\bar{\lambda}_i > 0} \nabla_y g_i(\bar{z})$ is zero. For i such that $\bar{\lambda}_i = 0$ and

$$
dx^T \nabla_x g_i(\bar{z}) + dy^T \nabla_y g_i(\bar{z}) = 0,
$$

we have

$$
\lambda_i^k (dy' - dy)^T \nabla_y g_i(\bar{z}) \leq 0.
$$

For $i \in \mathcal{I}(\bar{z})$ with $\bar{\lambda}_i = 0$ and

$$
dx^T \nabla_x g_i(\bar{z}) + dy^T \nabla_y g_i(\bar{z}) < 0,
$$

it follows that for all k sufficiently large, we must have $g_i(z^k) < 0$ which implies $\lambda_i^k = 0$ by complementarity. Consequently, it follows that if dy' belongs to $\mathcal{K}(\bar{z}, \bar{\lambda}; dx)$, then for all k sufficiently large,

$$
(dy' - dy)^T \left[\nabla_z L(\bar{z}, \lambda^k)(z^k - \bar{z}) + o\left(\|z^k - \bar{z}\|\right) \right] \geq 0,
$$

which yields, upon dividing by τ_k and letting $k \to \infty$,

$$
(dy' - dy)^T \nabla_z L(\bar{z}, \bar{\lambda}) dz \geq 0,
$$

establishing the desired property of the vector dy.   **Q.E.D.**

We make some remarks about the AVI (4) in two special cases. If $F(x, y)$ is the (partial) gradient map of a real-valued function of y, that is, if

$$
F(x, y) \equiv \nabla_y \theta(x, y), \quad \text{for } (x, y) \in \Re^{n+m},
$$

for some twice continuously differentiable function $\theta : \Re^{n+m} \to \Re$, then the matrix

$$\nabla_y L(\bar{z}, \bar{\lambda}) = \nabla_{yy}^2 \theta(\bar{x}, \bar{y}) + \sum_{i=1}^{\ell} \bar{\lambda}_i \nabla_{yy}^2 g_i(\bar{x}, \bar{y})$$

is symmetric, and the AVI (4) is equivalent to the stationarity problem of the quadratic program in the variable $dy \in \Re^m$:

minimize $dy^T \nabla_x L(\bar{z}, \bar{\lambda}) dx + \frac{1}{2} dy^T \nabla_y L(\bar{z}, \bar{\lambda}) dy$

subject to $dy \in \mathcal{K}(\bar{z}, \bar{\lambda}; dx)$.

The other special case is the MP with NCP constraints. This is the situation where

$$g(x, y) \equiv -y, \quad \text{for all } (x, y) \in \Re^{n+m}.$$

In this case, $M(\bar{z})$ is a singleton and equal to $\{F(\bar{z})\}$. We introduce three index sets associated with a solution $\bar{z} \equiv (\bar{x}, \bar{y})$ of the NCP:

$$y \geq 0, \quad F(x, y) \geq 0, \quad y^T F(x, y) = 0;$$

these are

$$\alpha(\bar{z}) \equiv \{i : \bar{y}_i > 0 = F_i(\bar{z})\},$$

$$\beta(\bar{z}) \equiv \{i : \bar{y}_i = 0 = F_i(\bar{z})\}, \tag{7}$$

$$\gamma(\bar{z}) \equiv \{i : \bar{y}_i = 0 < F_i(\bar{z})\}.$$

Incidentally, the indices $i \in \beta(\bar{z})$ (and the corresponding variables y_i) are said to be *degenerate*. For any $dx \in \Re^n$, the directional critical set at $\bar{z}$ along dx is equal to the constant cone

$$\Re^{|\alpha|} \times \Re_+^{|\beta|} \times \{0\}^{|\gamma|}.$$

Therefore for a given dx, (4) is equivalent to the mixed LCP (dropping $\bar{z}$ from the notation of the index sets):

$$\nabla_x F_\alpha(\bar{z}) dx + \nabla_\alpha F_\alpha(\bar{z}) dy_\alpha + \nabla_\beta F_\alpha(\bar{z}) dy_\beta = 0$$

$$\nabla_x F_\beta(\bar{z}) dx + \nabla_\alpha F_\beta(\bar{z}) dy_\alpha + \nabla_\beta F_\beta(\bar{z}) dy_\beta \geq 0$$

$$dy_\alpha \text{ free}, \quad dy_\beta \geq 0, \quad dy_\gamma = 0, \tag{8}$$

$$(dy_\beta)^T [\nabla_x F_\beta(\bar{z}) dx + \nabla_\beta F_\alpha(\bar{z}) dy_\alpha + \nabla_\beta F_\beta(\bar{z}) dy_\beta] = 0,$$

where, for any two index subsets S and T of $\{1, \ldots, m\}$, $\nabla_S F_T$ refers to the $|T| \times |S|$ partial Jacobian matrix of the subfunction F_T with respect to the variables y_S.

In general, the problem (4) is obtained by "linearizing" the parametric VI $(F(x, \cdot), C(x))$ at $\bar{z}$; in particular, the VI becomes an AVI and the NCP becomes a mixed LCP.

Motivated by Lemma **3.2.1**, we define, for each $\lambda \in M(\bar{z})$, the set-valued map $\mathcal{LS}_{(\bar{z}, \lambda)} : \Re^n \to \Re^m$ with $\mathcal{LS}_{(\bar{z}, \lambda)}(dx)$ being the solution set of the AVI $(\nabla_x L(\bar{z}, \lambda)dx, \nabla_y L(\bar{z}, \lambda), \mathcal{K}(\bar{z}, \lambda; dx))$; i.e., in the notation introduced in Subsection **2.3.2**,

$$\mathcal{LS}_{(\bar{z}, \lambda)}(dx) \equiv \mathrm{SOL}\left(\nabla_x L(\bar{z}, \lambda)dx, \nabla_y L(\bar{z}, \lambda), \mathcal{K}(\bar{z}, \lambda; dx)\right).$$

Lemma **3.2.1** then says that

$$\mathcal{T}(\bar{z}; \mathcal{F}) \subseteq \mathcal{L}(\bar{z}; \mathcal{F}) \equiv \mathcal{T}(\bar{z}; Z) \bigcap \left(\bigcup_{\lambda \in M(\bar{z})} \mathrm{Gr}(\mathcal{LS}_{(\bar{z}, \lambda)}) \right). \tag{9}$$

We note that the set $\mathcal{L}(\bar{z}; \mathcal{F})$ is a (possibly) nonconvex closed cone. In general, the cone

$$\bigcup_{\lambda \in M(\bar{z})} \mathrm{Gr}(\mathcal{LS}_{(\bar{z}, \lambda)}) \tag{10}$$

is difficult to manipulate; it can be thought of as the "linearized cone" of the complicated constraints $(x, y) \in \mathrm{Gr}(\mathcal{S})$, in (3.1.1). For a given $\lambda \in M(\bar{z})$, by the polyhedrality of the set $\mathcal{K}(\bar{z}, \lambda; dx)$ and the linearity of the function $\nabla_z L(\bar{z}, \lambda)$, it is not difficult to show that $\mathrm{Gr}(\mathcal{LS}_{(\bar{z}, \lambda)})$ is the union of a finite number of polyhedral cones in $\Re^{n+m}$.

When there is a unique multiplier, the structure of the cone $\mathcal{L}(\bar{z}; \mathcal{F})$ is relatively simple. For example, in the NCP constrained MP,

$$\mathcal{L}(\bar{z}; \mathcal{F}) = \mathcal{T}(\bar{z}; Z) \cap \{(dx, dy) \in \Re^{n+m} : (dx, dy) \text{ satisfy } (8)\}. \tag{11}$$

Subsequently, the existence of a unique multiplier in the general case is discussed further.

If a CQ stronger than the SBCQ holds at $\bar{z}$, then we may reduce the complexity of the definition of the cone $\mathcal{L}(\bar{z}; \mathcal{F})$ by considering the union in (10) over those multipliers $\lambda \in M(\bar{z})$ that are also extreme points in $M(\bar{z})$. More specifically, suppose that the CRCQ holds at $\bar{z} \in \mathcal{F}$ [120]; i.e., suppose that there exists a neighborhood $V \subseteq \Re^{n+m}$ of $\bar{z}$ such that for all index sets $K \subseteq \mathcal{I}(\bar{z})$, the family of gradient vectors,

$$\{\nabla_y g_i(x,y) : i \in K\},$$

has the same rank (which depends on K) for all vectors $(x,y) \in V$. By letting $M^e(\bar{z})$ be the set of extreme points of $M(\bar{z})$ and

$$\mathcal{L}^e(\bar{z};\mathcal{F}) \equiv \mathcal{T}(\bar{z};Z) \bigcap \left(\bigcup_{\lambda \in M^e(\bar{z})} \mathrm{Gr}(\mathcal{LS}_{(\bar{z},\lambda)}) \right),$$

then a slight refinement of the proof of Lemma **3.2.1** shows that

$$\mathcal{T}(\bar{z};\mathcal{F}) \subseteq \mathcal{L}^e(\bar{z};\mathcal{F}); \tag{12}$$

moreover, throughout Corollary **3.3.1**, we may replace the cone $\mathcal{L}(\bar{z};\mathcal{F})$ by its subcone $\mathcal{L}^e(\bar{z};\mathcal{F})$. Indeed, we may choose each λ^k satisfying (5) to be an extreme point of $M(z^k)$; then by the CRCQ, the limit $\bar{\lambda}$ must be an extreme point of $M(\bar{z})$, hence an element of $M^e(\bar{z})$. One major difference between the two cones $\mathcal{L}(\bar{z};\mathcal{F})$ and $\mathcal{L}^e(\bar{z};\mathcal{F})$ is that the former cone is obtained by taking the union over a possibly infinite set of multipliers λ, whereas the latter involves only a finite number of multipliers (because $M^e(\bar{z})$ is a finite set). Since for each $\lambda \in M(\bar{z})$, $\mathrm{Gr}(\mathcal{LS}_{(\bar{z},\lambda)})$ is the union of a finite number of polyhedral cones, it follows that the cone $\mathcal{L}^e(\bar{z};\mathcal{F})$ is also the union of a finite number of polyhedral cones, provided that $\mathcal{T}(\bar{z};Z)$ is polyhedral.

We summarize the above discussion in the result below. Incidentally, the CRCQ has a very important role to play in the next chapter.

3.2.2 Proposition. If CRCQ holds at $\bar{z} \in \mathcal{F}$, then the inclusion (12) holds. If in addition $\mathcal{T}(\bar{z};Z)$ is polyhedral, then $\mathcal{L}^e(\bar{z};\mathcal{F})$ is the union of a finite number of polyhedral cones.

Another simplified situation about the cone $\mathcal{L}(\bar{z};\mathcal{F})$ that is worth noting occurs when the *strict Mangasarian-Fromovitz constraint qualification* holds at a given triple $(\bar{x},\bar{y},\bar{\lambda})$, where $\bar{\lambda} \in M(\bar{z})$. This CQ states that

(SMFCQ) the gradients

$$\{\nabla_y g_i(\bar{x},\bar{y}) : \text{ for } i \text{ such that } \bar{\lambda}_i > 0\}$$

are linearly independent, and there exists a vector $\bar{v} \in \Re^m$ such that

$$\nabla_y g_i(\bar{x}, \bar{y})^T \bar{v} = 0, \quad \text{for all } i \in \mathcal{I}(\bar{x}, \bar{y}) \text{ such that } \bar{\lambda}_i > 0,$$

$$\nabla_y g_i(\bar{x}, \bar{y})^T \bar{v} < 0, \quad \text{for all } i \in \mathcal{I}(\bar{x}, \bar{y}) \text{ such that } \bar{\lambda}_i = 0.$$

It is not difficult to show [148] that the SMFCQ holds at $(\bar{x}, \bar{y}, \bar{\lambda})$ if and only if $M(\bar{z}) = \{\bar{\lambda}\}$; this is further equivalent to the following implication holding:

$$\left. \begin{array}{c} \sum_{i \in \mathcal{I}(\bar{z})} \delta_i \nabla_y g_i(\bar{x}, \bar{y}) = 0 \\ \delta_i \geq 0, \ \forall i \in \mathcal{I}(\bar{x}, \bar{y}) \text{ such that } \bar{\lambda}_i = 0 \end{array} \right\} \implies \delta_i = 0, \ \forall i \in \mathcal{I}(\bar{z}). \quad (13)$$

Clearly, SMFCQ implies MFCQ, but not conversely. Under the SMFCQ, we have

$$\mathcal{L}(\bar{z}; \mathcal{F}) = \mathcal{T}(\bar{z}; Z) \bigcap \text{Gr}(\mathcal{LS}_{(\bar{z}, \bar{\lambda})}).$$

We close this section with further discussion of the directional critical set $\mathcal{K}(\bar{z}, \lambda; dx)$. For this purpose, let us introduce, for any given vector $dx \in \Re^n$, the following linear program:

$$\text{maximize} \quad \sum_{i=1}^{\ell} \lambda_i dx^T \nabla_x g_i(\bar{z})$$

$$\text{subject to} \quad \lambda \in M(\bar{z}).$$

Let $M^c(\bar{z}; dx)$ denote the (possibly empty) optimal solution set of this program; the superscript c denotes the "critical" multipliers corresponding to the direction dx. Substituting the explicit polyhedral representation of $M(\bar{z})$, we can restate this linear program equivalently as:

$$\text{maximize} \quad \sum_{i \in \mathcal{I}(\bar{z})} \lambda_i dx^T \nabla_x g_i(\bar{z})$$

$$\text{subject to} \quad F(\bar{z}) + \sum_{i \in \mathcal{I}(\bar{z})} \lambda_i \nabla_y g_i(\bar{z}) = 0$$

$$\lambda_i \geq 0, \quad \text{for } i \in \mathcal{I}(\bar{z}),$$

which has as variables only those λ_i for $i \in \mathcal{I}(\bar{z})$. (Note: $\lambda_i = 0$ for all $\lambda \in M(\bar{z})$ and $i \notin \mathcal{I}(\bar{z})$.) The dual of the latter program is: with dx given, find $dy \in \Re^m$,

$$\text{minimize} \quad dy^T F(\bar{z})$$

$$\text{subject to} \quad dx^T \nabla_x g_i(\bar{z}) + dy^T \nabla_y g_i(\bar{z}) \leq 0, \quad \text{for } i \in \mathcal{I}(\bar{z}).$$

(14)

In terms of this pair of linear programs, we have the following result for the directional critical set; see also Dempe [60].

3.2.3 Proposition. Let $\lambda \in M(\bar{z})$. Then $\mathcal{K}(\bar{z}, \lambda; dx) \neq \emptyset$ if and only if $\lambda \in M^c(\bar{z}; dx)$. Moreover, $\mathcal{K}(\bar{z}, \lambda; dx)$ is equal to a constant set for all $\lambda \in M^c(\bar{z}; dx)$; indeed, for any such λ, $\mathcal{K}(\bar{z}, \lambda; dx)$ coincides with the set of optimal solutions of the dual program (14).

Proof. Suppose $\mathcal{K}(\bar{z}, \lambda; dx) \neq \emptyset$. Let $dy \in \mathcal{K}(\bar{z}, \lambda; dx)$; then dy is feasible to the dual program (14). Moreover, it is easy to verify that for all $i \in \mathcal{I}(\bar{z})$,

$$\lambda_i \left[dx^T \nabla_x g_i(\bar{z}) + dy^T \nabla_y g_i(\bar{z}) \right] = 0. \tag{15}$$

Hence $\lambda \in M^c(\bar{z}; dx)$. Conversely, if $\lambda \in M^c(\bar{z}; dx)$, then the dual program (14) has an optimal solution dy satisfying the complementary slackness condition (15) for all $i \in \mathcal{I}(\bar{z})$. The latter condition easily implies that $dy \in \mathcal{K}(\bar{x}, \lambda; dx)$. This establishes the first assertion of the proposition. The characterization of $\mathcal{K}(\bar{z}, \lambda; dx)$ for any $\lambda \in M^c(\bar{z}; dx)$ also follows easily from this proof.   **Q.E.D.**

3.3 Stationarity Under the Full CQ for MPEC

Motivated by the discussion in the last section, we say that the *basic CQ* holds for the MPEC (3.1.1) at the pair $\bar{z} \equiv (\bar{x}, \bar{y}) \in \mathcal{F}$ if there exists a nonempty subset $M' \subseteq M(\bar{z})$ such that

$$\mathcal{T}(\bar{z}; \mathcal{F}) = \mathcal{L}'(\bar{z}; \mathcal{F}), \tag{1}$$

where

$$\mathcal{L}'(\bar{z}; \mathcal{F}) \equiv \mathcal{T}(\bar{z}; Z) \bigcap \left(\bigcup_{\lambda \in M'} \mathrm{Gr}(\mathcal{LS}_{(\bar{z}, \lambda)}) \right).$$

The basic CQ holds in MPAEC, the analog of linearly constrained nonlinear programming; see Section 4.1 for details. If (1) holds with $M' = M^e(\bar{z})$, $(M' = M(\bar{z}))$, then we say that the *extreme (full) CQ* holds for the MPEC (3.1.1) at $\bar{z}$. Clearly, if $M(\bar{z})$ is a singleton, then all these CQs for the MPEC are the same. In general, the following implications are easily seen to be valid:

full CQ $\implies$ extreme CQ,   provided that CRCQ holds;

extreme CQ $\implies$ basic CQ.

Our focus in this monograph is on the full CQ and the extreme CQ. In particular, in later sections we present sufficient conditions that will ensure the validity of either one of these two CQs for MPEC at a feasible point. It seems difficult, in general, to derive sufficient conditions that will ensure the basic CQ to hold, without requiring both the full and extreme CQ to hold; however this does not mean that if the basic CQ for the MPEC holds, then either the full or extreme CQ must hold; see Example **3.4.2**.

Under the CRCQ (SBCQ) for the inner VI at $\bar{z}$, the extreme (full) CQ for MPEC at $\bar{z}$ holds if and only if the reverse inclusion in (3.2.12) ((3.2.9) respectively) holds. In general, if (1) holds, then $\mathcal{L}'(\bar{z}; \mathcal{F})$ is a closed set (regardless of whether M' is finite or not); this is because the tangent cone $\mathcal{T}(\bar{z}; \mathcal{F})$ is always closed. In particular, if the full CQ holds, then $\mathcal{L}(\bar{z}; \mathcal{F})$ must be closed. This observation allows us to obtain the following corollary of Lemma **3.2.1** which gives various necessary and sufficient conditions for a feasible vector $\bar{z} \in \mathcal{F}$ to be a stationary point of (3.1.1).

3.3.1 Corollary. Suppose the SBCQ holds at $\bar{z} \in \mathcal{F}$ and the full CQ holds for the MPEC at $\bar{z}$; that is,

$$\mathcal{L}(\bar{z}; \mathcal{F}) = \mathcal{T}(\bar{z}; \mathcal{F}). \qquad (2)$$

The following statements are then equivalent:

(a) $\bar{z}$ is a stationary point of (3.1.1);

(b) the following implication holds:

$$dz \in \mathcal{L}(\bar{z}; \mathcal{F}) \implies \nabla f(\bar{z})^T dz \geq 0;$$

i.e., $\nabla f(\bar{z}) \in \mathcal{L}(\bar{z}; \mathcal{F})^*$;

(c) $\bar{d}z \equiv 0$ is an optimal solution of the following minimization problem:

$$\text{minimize} \quad \nabla f(\bar{z})^T dz$$
$$\text{subject to} \quad dz \in \mathcal{L}(\bar{z}; \mathcal{F});$$

(d) for any scalar $\alpha \geq \|\nabla f(\bar{z})\|$ and any $dz \in \Re^{n+m}$,

$$\nabla f(\bar{z})^T dz + \alpha \, \text{dist}(dz, \mathcal{L}(\bar{z}; \mathcal{F})) \geq 0.$$

Proof. That (a) $\Leftrightarrow$ (b) is an immediate consequence of the definition of a stationary point of (3.1.1) and the condition (1). Moreover, (b) $\Leftrightarrow$ (c) is obvious; so is (d) $\Rightarrow$ (c). Finally, (c) $\Rightarrow$ (d) follows either by quoting Proposition 2.4.3 in [51] or by a simple proof; in what follows, we give a direct argument for this implication. Let α and dz be as given. Since $\mathcal{L}(\bar{z}; \mathcal{F})$ is closed, there exists $dz' \in \mathcal{L}(\bar{z}; \mathcal{F})$ such that $\|dz - dz'\| = \mathrm{dist}(dz, \mathcal{L}(\bar{z}; \mathcal{F}))$. We have

$$\nabla f(\bar{z})^T dz + \alpha \, \mathrm{dist}(dz, \mathcal{L}(\bar{z}; \mathcal{F}))$$
$$\geq \nabla f(\bar{z})^T dz' + (\alpha - \|\nabla f(\bar{z})\|) \, \|dz - dz'\| \geq 0,$$

where the last inequality holds by (c) and the choice of dz'.   **Q.E.D.**

If, in addition to the hypothesis of Corollary **3.3.1**, the CRCQ and the extreme CQ holds at $\bar{z}$, then the statements (a)–(d) of Corollary **3.3.1** are equivalent with $\mathcal{L}^e(\bar{z}; \mathcal{F})$ replacing $\mathcal{L}(\bar{z}; \mathcal{F})$.

The cone $\mathcal{L}(\bar{z}; \mathcal{F})$ is generally simpler to manipulate than $\mathcal{T}(\bar{z}; \mathcal{F})$. We shall show in a later subsection that if the lower-level problem is an AVI, then both the SBCQ and the full CQ (2) hold at $\bar{z} \in \mathcal{F}$.

3.3.2 Example. Consider the MPEC (3.1.7) in Example **3.1.2**, and the solution $\bar{z} = (0, 0, 0)$. Since $C(x)$ is just the nonnegative orthant in $\Re^2$ for all x, the CRCQ holds at $\bar{z}$. Moreover, it is not hard, using $Z = \Re^3$ and (3.2.11), to verify that the cone $\mathcal{L}(\bar{z}; \mathcal{F})$ consists of all triples $(dx, dy_1, 0)$ in $\Re^3$ such that

$$dy_1 \geq 0, \quad dx + dy_1 \geq 0, \quad dy_1(dx + dy_1) = 0.$$

This cone $\mathcal{L}(\bar{z}; \mathcal{F})$ coincides with the tangent cone $\mathcal{T}(\bar{z}; \mathcal{F})$ as given before. Thus the assumptions of Corollary **3.3.1** hold for this MPEC.

The following example illustrates an instance of an NCP constrained MP where the full CQ fails to hold.

3.3.3 Example. Consider the MPEC with 1 upper-level and 1 lower-level variable and with parametric NCP constraints: $Z \equiv \Re_+ \times \Re$,

$$F(x, y) \equiv y^2 + x^2, \quad C(x) \equiv \Re_+.$$

In this case, we have $\mathcal{F} = \Re_+ \times \{0\}$. Consider the point $\bar{z} \equiv (0, 0) \in \mathcal{F}$. We have $\mathcal{T}(\bar{z}; \mathcal{F}) = \mathcal{F}$; nevertheless, from (3.2.11), $\mathcal{L}(\bar{z}; \mathcal{F}) = \Re_+^2$ which is clearly a proper superset of $\mathcal{T}(\bar{z}; \mathcal{F})$.

3.3.1 *Primal-dual characterization of stationarity*

Conditions (b), (c), and (d) in Corollary **3.3.1** are "primal" characterizations of stationarity for the MPEC. Similar to the KKT theory of nonlinear programming, we could derive some further characterizations of stationarity by "dualizing" these conditions, or equivalently, by introducing multipliers associated with the constraints of the MPEC. The resulting "primal-dual" stationarity conditions form the analog of the KKT conditions for the MPEC; they must be satisfied by any local minimizer of MPEC and therefore are the first-order necessary optimality conditions for MPEC. Due to the (possibly) nonconvex nature of the cone $\mathcal{L}(\bar{z};\mathcal{F})$, these optimality conditions are somewhat more complicated than their counterparts (namely, the KKT conditions) for a standard nonlinear program.

For simplicity of notation, we consider only the case where the set Z is polyhedral and is given by

$$Z = \{(x,y) \in \Re^{n+m} : Gx + Hy + a \leq 0\} \tag{3}$$

with $G \in \Re^{s \times n}, H \in \Re^{s \times m}$, and $a \in \Re^s$. The generalization to a set Z defined by (nonlinear) differentiable inequalities and equations satisfying some standard constraint qualifications in nonlinear programming is straightforward and will not be discussed. In essence, as long as the tangent cone $\mathcal{T}(\bar{z};Z)$ is polyhedral, the following analysis can easily be extended.

Let $\lambda \in M(\bar{z})$ and $dz \in \text{Gr}(\mathcal{LS}_{(\bar{z},\lambda)})$ be given. Write

$$\mathcal{I}_0(\bar{z},\lambda) \equiv \{i \in \mathcal{I}(\bar{z}) : \lambda_i = 0\}, \quad \text{and} \quad \mathcal{I}_+(\bar{z},\lambda) \equiv \{i \in \mathcal{I}(\bar{z}) : \lambda_i > 0\}.$$

Also write $dz \equiv (dx, dy)$. The indices in $\mathcal{I}_0(\bar{z},\lambda)$ are called *degenerate*. By the definition of $\mathcal{LS}_{(\bar{z},\lambda)}$, it follows that dy is a solution of the AVI $(\nabla_x L(\bar{z},\lambda)dx, \nabla_y L(\bar{z},\lambda), \mathcal{K}(\bar{z},\lambda; dx))$. Hence there exist multipliers $\{d\lambda_i : i \in \mathcal{I}(\bar{z})\}$ with $d\lambda_i \geq 0$ for $i \in \mathcal{I}_0(\bar{z},\lambda)$ such that

$$\nabla_x L(\bar{z},\lambda)dx + \nabla_y L(\bar{z},\lambda)dy + \sum_{i \in \mathcal{I}(\bar{z})} d\lambda_i \nabla_y g_i(\bar{z}) = 0,$$
$$d\lambda_i \left(dx^T \nabla_x g_i(\bar{z}) + dy^T \nabla_y g_i(\bar{z})\right) = 0, \quad \text{for all } i \in \mathcal{I}_0(\bar{z},\lambda). \tag{4}$$

(This is just the mixed LCP formulation of the AVI in question.) Furthermore, by partitioning the set $\mathcal{I}_0(\bar{z},\lambda)$ into two disjoint subsets corresponding to the active and inactive constraints in

$$dx^T \nabla_x g_i(\bar{z}) + dy^T \nabla_y g_i(\bar{z}) \leq 0,$$

the system (4) can be used to characterize vectors in the set $\mathrm{Gr}(\mathcal{LS}_{(\bar{z},\lambda)})$.
Indeed, for each subset α of $\mathcal{I}_0(\bar{z}, \lambda)$, let

$$\mathcal{K}_\alpha(\bar{z}, \lambda) \equiv$$
$$\{(dx, dy) \in \mathcal{K}(\bar{z}, \lambda) : dx^T \nabla_x g_i(\bar{z}) + dy^T \nabla_y g_i(\bar{z}) = 0 \text{ for } i \in \alpha\}.$$

Alternatively, by partitioning $\mathcal{I}(\bar{z}, \lambda)$ into two sets,

$$\bar{\alpha} \equiv \mathcal{I}_0(\bar{z}, \lambda) \setminus \alpha \quad \text{and} \quad \tilde{\alpha} \equiv \mathcal{I}_+(\bar{z}, \lambda) \cup \alpha, \tag{5}$$

we see that

$$\mathcal{K}_\alpha(\bar{z}, \lambda) \equiv \{(dx, dy) \in \Re^{n+m} \ :$$
$$dx^T \nabla_x g_i(\bar{z}) + dy^T \nabla_y g_i(\bar{z}) \leq 0, \text{ for } i \in \bar{\alpha}$$
$$dx^T \nabla_x g_i(\bar{z}) + dy^T \nabla_y g_i(\bar{z}) = 0, \text{ for } i \in \tilde{\alpha}\}.$$

A vector dz belongs to $\mathrm{Gr}(\mathcal{LS}_{(\bar{z},\lambda)})$ if and only if there exists a subset α
of $\mathcal{I}_0(\bar{z}, \lambda)$ such that $dz \in \mathcal{K}_\alpha(\bar{z}, \lambda)$ and for some multipliers $\{d\lambda_i : i \in \tilde{\alpha}\}$
with $d\lambda_i \geq 0$ for all $i \in \alpha$,

$$\nabla_x L(\bar{z}, \lambda)dx + \nabla_y L(\bar{z}, \lambda)dy + \sum_{i \in \tilde{\alpha}} d\lambda_i \nabla_y g_i(\bar{z}) = 0.$$

Consequently, it follows that the implication

$$dz \in \mathcal{T}(\bar{z}; Z) \cap \mathrm{Gr}(\mathcal{LS}_{(\bar{z},\lambda)}) \implies \nabla f(\bar{z})^T dz \geq 0 \tag{6}$$

holds if and only if for all index sets $\alpha \subseteq \mathcal{I}_0(\bar{z}, \lambda)$ with $\bar{\alpha}$ and $\tilde{\alpha}$ defined in
(5), the following implication in the variables dz and $\{d\lambda_i\}_{i \in \tilde{\alpha}}$ holds:

$$\left.\begin{array}{r} dz \in \mathcal{T}(\bar{z}; Z) \\[4pt] d\lambda_i \geq 0, \text{ for } i \in \alpha \\[4pt] \nabla_z L(\bar{z}, \lambda)dz + \sum_{i \in \tilde{\alpha}} d\lambda_i \nabla_y g_i(\bar{z}) = 0 \\[4pt] dz^T \nabla g_i(\bar{z}) \leq 0, \text{ for } i \in \bar{\alpha} \\[4pt] dz^T \nabla g_i(\bar{z}) = 0, \text{ for } i \in \tilde{\alpha} \end{array}\right\} \implies \nabla f(\bar{z})^T dz \geq 0.$$

By a theorem of the alternatives, it therefore follows that (6) holds if and only if there exist multipliers $\zeta \in \Re^s$, $\{\eta_i : i \in \mathcal{I}(\bar{z})\}$ with $\eta_i \geq 0$ for all $i \in \bar{\alpha}$, and $\pi \in \Re^m$ such that

$$\nabla_x f(\bar{z}) + G^T \zeta + \sum_{i \in \mathcal{I}(\bar{z})} \eta_i \nabla_x g_i(\bar{z}) = \nabla_x L(\bar{z}, \lambda)^T \pi$$

$$\nabla_y f(\bar{z}) + H^T \zeta + \sum_{i \in \mathcal{I}(\bar{z})} \eta_i \nabla_y g_i(\bar{z}) = \nabla_y L(\bar{z}, \lambda)^T \pi$$

$$\pi^T \nabla_y g_i(\bar{z}) \leq 0, \quad \text{for } i \in \alpha$$

$$\pi^T \nabla_y g_i(\bar{z}) = 0, \quad \text{for } i \in \mathcal{I}_+(\bar{z}, \lambda)$$

$$\zeta \geq 0, \quad \zeta^T(G\bar{x} + H\bar{y} + a) = 0.$$

We conclude that the implication (6) holds if and only if for all index sets $\alpha \subseteq \mathcal{I}_0(\bar{z}, \lambda)$ there exist multipliers $\zeta \in \Re^s$, $\eta \in \Re^\ell$ and $\pi \in \Re^m$ such that

$$\nabla_x f(\bar{z}) + G^T \zeta + \nabla_x g(\bar{z})^T \eta = \nabla_x L(\bar{z}, \lambda)^T \pi$$

$$\nabla_y f(\bar{z}) + H^T \zeta + \nabla_y g(\bar{z})^T \eta = \nabla_y L(\bar{z}, \lambda)^T \pi$$

$$\pi^T \nabla_y g_i(\bar{z}) \leq 0, \quad \text{for } i \in \alpha$$

$$\pi^T \nabla_y g_i(\bar{z}) = 0, \quad \text{for } i \in \mathcal{I}_+(\bar{z}, \lambda) \qquad (7)$$

$$\eta_i \geq 0, \quad \text{for all } i \in \bar{\alpha}$$

$$\eta_i = 0, \quad \text{for all } i \notin \mathcal{I}(\bar{z})$$

$$\zeta \geq 0, \quad \zeta^T(G\bar{x} + H\bar{y} + a) = 0.$$

We may now summarize the above discussion and give a further characterization of a stationary point of MPEC. There is no need for further proof of the following result.

3.3.4 Theorem. Under the assumptions of Corollary **3.3.1**, $\bar{z}$ is a stationary point of (3.1.1) with Z given by (3) if and only if for each vector $\lambda \in M(\bar{z})$ and each pair of index sets α and $\bar{\alpha}$ partitioning $\mathcal{I}_0(\bar{z}, \lambda)$, there exist multipliers $\zeta \in \Re^s$, $\eta \in \Re^\ell$ and $\pi \in \Re^m$ such that (7) holds.

3.3.5 Remark. If in place of the SBCQ and the full CQ, the CRCQ and the extreme CQ hold at $\bar{z}$, then the conclusion Theorem **3.3.4** remains valid if $M(\bar{z})$ is replaced by $M^e(\bar{z})$. The significance of this observation is that the resulting optimality condition consists of *finitely* many linear inequality systems.

The stationarity conditions (7) for various sets $\alpha \subseteq \mathcal{I}_0(\bar{z}, \lambda)$ may appear somewhat unnatural at first observation. In what follows we relate these conditions to the standard KKT conditions of a certain family of nonlinear programs equivalent to the MPEC. Indeed, the MPEC (3.1.1) can be formulated as (3.1.3) where the lower VI $(F(x, \cdot), C(x))$ is cast in terms of its KKT conditions. For ease of explanation, we restate the formulation (3.1.3) as follows:

$$
\begin{aligned}
\text{minimize} \quad & f(x, y) \\
\text{subject to} \quad & Gx + Hy + a \leq 0, \\
& L(x, y, \lambda) = 0, \\
& g(x, y) \leq 0, \quad \lambda \geq 0, \quad \lambda^T g(x, y) = 0,
\end{aligned}
\tag{8}
$$

where we have substituted the inequality representation (3.3.3) for the set Z and use the VI Lagrangean function $L(x, y, \lambda)$ given in (3.2.3) to write the Lagrangean equation of the VI $(F(x, \cdot), C(x))$. Define the MPEC Lagrangean function: for $(x, y, \lambda, \zeta, \pi, \eta) \in \Re^{n+m+\ell+s+m+\ell}$,

$$
\begin{aligned}
\mathcal{L}^{\text{MPEC}}&(x, y, \lambda, \zeta, \pi, \eta) \equiv \\
& f(x, y) + \zeta^T(Gx + Hy + a) - \pi^T L(x, y, \lambda) + \eta^T g(x, y).
\end{aligned}
\tag{9}
$$

The vectors ζ, π, and η can be thought of, respectively, as the *MPEC multipliers* for the constraints:

$$
Gx + Hy + a \leq 0, \quad L(x, y, \lambda) = 0, \quad g(x, y) \leq 0,
$$

in (8). Note that there are no (MPEC) multipliers for the nonnegativity constraint $\lambda \geq 0$ and the complementarity condition $\lambda^T g(x, y) = 0$ in the function $\mathcal{L}^{\text{MPEC}}$. The fact that no multiplier is needed for the latter complementarity constraint is a special feature of the MPEC that distinguishes it from a standard nonlinear program. Indeed, as made clear in the discussion below, the complementarity condition $\lambda^T g(x, y) = 0$ will be decomposed into a disjunction of finitely many systems of equalities and inequalities.

For each pair of index sets $\mathcal{J}_1$ and $\mathcal{J}_2$ partitioning $\{1, \ldots, \ell\}$, consider the nonlinear program in the variables (x, y, λ):

$$
\begin{aligned}
\text{minimize} \quad & f(x, y) \\
\text{subject to} \quad & Gx + Hy + a \leq 0, \\
& L(x, y, \lambda) = 0, \\
& \left.\begin{aligned} \lambda_i &= 0 \\ g_i(x, y) &\leq 0 \end{aligned}\right\} \quad \forall i \in \mathcal{J}_1, \\
& \left.\begin{aligned} \lambda_i &\geq 0 \\ g_i(x, y) &= 0 \end{aligned}\right\} \quad \forall i \in \mathcal{J}_2.
\end{aligned}
\tag{10}
$$

Clearly, (x, y, λ) is feasible to (8) if and only if there exist two such index sets $\mathcal{J}_1$ and $\mathcal{J}_2$ such that (x, y, λ) is feasible to (10). Moreover, for each $\lambda \in M(\bar{z})$ and each $\alpha \subseteq \mathcal{I}_0(\bar{z}, \lambda)$, the triple $(\bar{x}, \bar{y}, \lambda)$ is feasible to (10) for

$$
\mathcal{J}_1 \equiv \bar{\alpha} \cup \{i : g_i(\bar{z}) < 0\}, \quad \text{and} \quad \mathcal{J}_2 \equiv \tilde{\alpha},
\tag{11}
$$

where $\bar{\alpha}$ and $\tilde{\alpha}$ are defined in (5). For each pair $(\mathcal{J}_1, \mathcal{J}_2)$ as stated, the KKT system of (10) at $(\bar{x}, \bar{y}, \lambda)$ is

$$
\begin{aligned}
\nabla_x \mathcal{L}^{\text{MPEC}}(\bar{x}, \bar{y}, \lambda, \zeta, \pi, \eta) &= 0, \\
\nabla_y \mathcal{L}^{\text{MPEC}}(\bar{x}, \bar{y}, \lambda, \zeta, \pi, \eta) &= 0, \\
\nabla_{\lambda_{\mathcal{J}_2}} \mathcal{L}^{\text{MPEC}}(\bar{x}, \bar{y}, \lambda, \zeta, \pi, \eta) &\geq 0, \quad \lambda_{\mathcal{J}_2} \geq 0, \\
\lambda_{\mathcal{J}_2}^T \nabla_{\lambda_{\mathcal{J}_2}} \mathcal{L}^{\text{MPEC}}(\bar{x}, \bar{y}, \lambda, \zeta, \pi, \eta) &= 0, \\
\eta_{\mathcal{J}_1} \geq 0, \quad g_{\mathcal{J}_1}(\bar{z}) \leq 0, \quad \eta_{\mathcal{J}_1}^T g_{\mathcal{J}_1}(\bar{z}) &= 0, \\
\zeta \geq 0, \quad G\bar{x} + H\bar{y} + a \leq 0, \quad \zeta^T(G\bar{x} + H\bar{y} + a) &= 0.
\end{aligned}
\tag{12}
$$

It is easy to see that this system is exactly the same as (7) under the above definition of $\mathcal{J}_1$ and $\mathcal{J}_2$ in terms of α.

In summary, the conditions (7), which are equivalent to stationarity of the MPEC (8) if the basic CQ holds, are the KKT conditions of the various nonlinear programs (10) at $(\bar{x}, \bar{y}, \lambda)$ for all multipliers $\lambda \in M(\bar{z})$ of the lower-level VI $(F(\bar{x}, \cdot), C(\bar{x}))$ and various index sets $\alpha \subseteq \mathcal{I}_0(\bar{z}, \lambda)$. Notice that if $M(\bar{z})$ is not a singleton, then, by convexity of $M(\bar{z})$, there will be a continuum of multipliers $\lambda \in M(\bar{z})$. Consequently, infinitely many systems (7) will result from the nonuniqueness of multipliers $\lambda \in M(\bar{z})$. Nevertheless, there are only finitely many index subsets α in all, regardless of the cardinality of $M(\bar{z})$. Even if $M(\bar{z})$ is a singleton, say $\{\bar{\lambda}\}$, the number of systems (7) is equal to 2^c, where $c \equiv |\mathcal{I}_0(\bar{z}, \bar{\lambda})|$ is a measure of degeneracy of the multiplier $\bar{\lambda}$. Thus in general, the cardinality of $M(\bar{z})$ and the degeneracy of each $\lambda \in M(\bar{z})$ combined determine the complexity of the first-order necessary optimality conditions for the MPEC. The question of whether there is a small number of "effective" systems which will be equivalent to these necessary conditions deserves further investigation.

The above discussion makes it clear that the stationarity conditions of the MPEC (8) cannot be obtained by considering this problem as a single smooth nonlinear program. Instead, the correct way to derive these conditions is to view the MPEC, locally near $\bar{z}$, as a family of nonlinear programs of the type (10) for different pairs of index sets $(\mathcal{J}_1, \mathcal{J}_2)$ as defined in (11).

In the rest of this section, we consider two simplifications of the stationarity conditions (7). One simplification arises in the case of an NCP constrained MP. Specifically, consider the problem:

$$
\begin{aligned}
\text{minimize} \quad & f(x,y) \\
\text{subject to} \quad & (x,y) \in Z \\
& y \geq 0, \quad F(x,y) \geq 0, \quad y^T F(x,y) = 0.
\end{aligned}
\tag{13}
$$

We recall the special notation (3.2.7): $\alpha(\bar{z})$, $\beta(\bar{z})$, and $\gamma(\bar{z})$, for the lower NCP at a feasible solution $\bar{z} \in \mathcal{F}$. Analogous to Theorem **3.3.4** is the following result for the NCP constrained MP (13).

3.3.6 Theorem. Let $f : \Re^{n+m} \to \Re$ and $F : \Re^{n+m} \to \Re^m$ be continuously differentiable functions. Let Z be given by (3). If the full CQ (2) holds for the MPEC (13) at the vector $\bar{z} \in \mathcal{F}$, then $\bar{z}$ is a stationary point of (13) if and only if for each subset β_1 of the degenerate index set $\beta(\bar{z})$ defined in

(3.2.7), there exist multipliers $\zeta \in \Re^s$ and $\pi \in \Re^m$ such that

$$\nabla_x f(\bar{z}) + G^T \zeta - \nabla_x F(\bar{z})^T \pi = 0,$$

$$\frac{\partial f(\bar{z})}{\partial y_i} + (H_i)^T \zeta - \sum_{j=1}^m \pi_j \frac{\partial F_j(\bar{z})}{\partial y_i} = 0, \quad i \in \alpha(\bar{z}),$$

$$\frac{\partial f(\bar{z})}{\partial y_i} + (H_i)^T \zeta - \sum_{j=1}^m \pi_j \frac{\partial F_j(\bar{z})}{\partial y_i} \geq 0, \quad i \in \beta(\bar{z}) \setminus \beta_1, \qquad (14)$$

$$\pi_i \geq 0, \quad i \in \beta_1,$$

$$\pi_i = 0, \quad i \in \gamma(\bar{z}),$$

$$\zeta \geq 0, \quad \zeta^T (G\bar{x} + H\bar{y} + a) = 0,$$

where H_i denotes the i-th column of H.

The above theorem reaffirms the complexity of the necessary optimality conditions for MPEC even in the case where the multiplier set $M(\bar{z})$ is a singleton, such as the NCP constrained MP. Indeed, the number of sets β_1 in Theorem **3.3.6** is $2^{|\beta(\bar{z})|}$.

We illustrate Theorem **3.3.6** with an MPEC discussed previously.

3.3.7 Example. Consider the MPEC in $\Re^3$:

$$\text{minimize} \quad x^2 + x + 2y_1 + y_2^2$$

$$\text{subject to} \quad x^2 + x + y_1 \geq 0, \quad y_1 \geq 0, \quad y_1(x^2 + x + y_1) = 0, \qquad (15)$$

$$x^3 + y_2 \geq 0, \quad y_2 \geq 0, \quad y_2(x^3 + y_2) = 0,$$

which has the same constraints as in Example **3.1.2** but with a slightly different objective function. The vector $\bar{z} \equiv (0, 0, 0)$ is a globally optimal solution. See Example **3.3.2** for $\mathcal{L}(\bar{z}; \mathcal{F})$. We have $\beta(\bar{z}) = \{1, 2\}$ and $\alpha(\bar{z}) = \gamma(\bar{z}) = \emptyset$. We illustrate the satisfaction of the system (14) for subsets of $\beta(\bar{z})$.

For $\beta_1 = \emptyset$, the system (14) is

$$1 - \pi_1 = 0, \quad 2 - \pi_1 \geq 0, \quad 0 - \pi_2 \geq 0,$$

which is clearly satisfied by $(\pi_1, \pi_2) = (1, 0)$, for instance. For $\beta_1 = \{1\}$, the system (14) is

$$1 - \pi_1 = 0, \quad 0 - \pi_2 \geq 0, \quad \pi_1 \geq 0;$$

it is clearly satisfied by the same (π_1, π_2). Similarly, the same conclusion holds for the remaining two choices of β_1.

A second simplification of the stationarity conditions (7) arises in the case where the feasible solution $\bar{z}$ of the MPEC (3.1.1) is *strongly nondegenerate*. Specifically, we say that a feasible solution $\bar{z}$ of the MPEC (3.1.1) is strongly nondegenerate if $M(\bar{z}) \neq \emptyset$ and for every $\lambda \in M(\bar{z})$, $\lambda - g(\bar{x}, \bar{y}) > 0$. The strong nondegeneracy property was used in Lemmas **2.3.8** and **2.3.16** which were the basis for Theorems **2.4.4** and **2.4.7**; in turn, these theorems established, respectively, a differentiable exact penalty equivalent for the MPAEC and for an NCP constrained MP. In what follows, we give a characterization for the strong nondegeneracy property of a feasible solution of the MPEC (3.1.1). This characterization supports the feeling that the nondegeneracy assumption in Lemmas **2.3.8** and **2.3.16** is indeed very restrictive.

3.3.8 Proposition. Let $\bar{z} \in \mathcal{F}$. Then $\bar{z}$ is strongly nondegenerate if and only if the set of active gradients

$$\{ \nabla_y g_i(\bar{z}) \ : \ i \in \mathcal{I}(\bar{z}) \} \tag{16}$$

is linearly independent (hence the set $M(\bar{z})$ is a singleton, say $\{\bar{\lambda}\}$), and $\bar{\lambda} - g(\bar{x}, \bar{y}) > 0$.

Proof. Indeed, the "if" part is easy; for the "only if" part, suppose that the gradients (16) are linearly dependent. Then there exist scalars $c_i, i \in \mathcal{I}(\bar{z})$, not all zero, such that

$$\sum_{i \in \mathcal{I}(\bar{z})} c_i \, \nabla_y g_i(\bar{x}, \bar{y}) = 0.$$

Take any $\lambda \in M(\bar{z})$. By assumption, $\lambda_i > 0$ for all $i \in \mathcal{I}(\bar{z})$. By choosing a scalar ε appropriately, the scalars

$$\lambda_i(\varepsilon) \equiv \lambda_i - \varepsilon c_i, \quad \text{for all } i \in \mathcal{I}(\bar{z})$$

remain nonnegative and equal to zero for at least one i. For this choice of ε, the vector $\lambda(\varepsilon)$ (with $\lambda_i(\varepsilon) \equiv 0$ for $i \notin \mathcal{I}(\bar{z})$) remains an element of $M(\bar{z})$ but clearly violates the strict complementarity condition; hence a contradiction.   **Q.E.D.**

If $\bar{z}$ is strongly nondegenerate and $M(\bar{z}) = \{\bar{\lambda}\}$, the index set $\mathcal{I}_0(\bar{z}, \bar{\lambda})$ is empty and the cone $\mathcal{K}(\bar{z}, \bar{\lambda})$ is a linear subspace; hence for every $dx \in \Re^n$, the set $\mathcal{K}(\bar{z}, \bar{\lambda}; dx)$ is an affine subspace. As a consequence, each of the problems AVI $\left(\nabla_x L(\bar{z}, \bar{\lambda})dx, \nabla_y L(\bar{z}, \bar{\lambda}), \mathcal{K}(\bar{z}, \bar{\lambda}; dx)\right)$ is equivalent to a system of linear equations (instead of a mixed linear complementarity problem as in the general case). Moreover, the stationarity system (7) has a particularly simple form.

3.3.9 Proposition. Let Z be given by (3.3.3) and $\bar{z}$ be a strongly nondegenerate feasible solution of (1). Let $M(\bar{z}) = \{\bar{\lambda}\}$. If the full CQ (3.3.2) holds, then $\bar{z}$ is a stationary point of (1) if and only if there exist multipliers $\zeta \in \Re^s$, $\eta \in \Re^\ell$, and $\pi \in \Re^m$ such that

$$\nabla_x f(\bar{z}) + G^T \zeta + \nabla_x g(\bar{z})^T \eta = \nabla_x L(\bar{z}, \bar{\lambda})^T \pi,$$

$$\nabla_y f(\bar{z}) + H^T \zeta + \nabla_y g(\bar{z})^T \eta = \nabla_y L(\bar{z}, \bar{\lambda})^T \pi,$$

$$\pi^T \nabla_y g_i(\bar{z}) = 0, \quad i \in \mathcal{I}(\bar{z}),$$

$$\eta_i = 0, \quad i \notin \mathcal{I}(\bar{z}),$$

$$\zeta \geq 0, \quad \zeta^T(G\bar{x} + H\bar{y} + a) = 0.$$

Subsequently, we give a simple sufficient condition for the full CQ (3.3.2) to hold under the strong nondegeneracy assumption at $\bar{z}$; see Subsection **4.2.7**.

3.4 More About the KKT Formulation of MPEC

In the previous section, we have relied heavily on the KKT formulation (3.3.8) to obtain the stationarity conditions for the MPEC (3.1.1). In Theorem **1.3.5**, we have seen that, under an appropriate convexity assumption and the SBCQ on $\mathcal{F}$, these problems are equivalent as far as the global minima are concerned. The goal of this section is to clarify the connection between the local minima and stationary points of these problems.

3.4.1 Proposition. Let Z be an arbitrary subset of $\Re^{n+m}$ and X be the projection of Z onto $\Re^n$. Suppose that each $g_i(x, \cdot)$ is convex for each $x \in X$ and the SBCQ holds at $\bar{z} \in \mathcal{F}$. The pair $\bar{z} \equiv (\bar{x}, \bar{y})$ is a local minimizer of (3.1.1) if and only if for every $\bar{\lambda} \in M(\bar{z})$, the triple $(\bar{x}, \bar{y}, \bar{\lambda})$ is a local minimizer of

$$
\begin{aligned}
\text{minimize} \quad & f(x, y) \\
\text{subject to} \quad & (x, y) \in Z, \\
& L(x, y, \lambda) = 0, \\
& g(x, y) \leq 0, \quad \lambda \geq 0, \quad \lambda^T g(x, y) = 0.
\end{aligned}
\tag{1}
$$

If the CRCQ holds at $\bar{z}$, then the equivalence remains valid if $M(\bar{z})$ is replaced by $M^e(\bar{z})$.

Proof. Suppose $\bar{z} \equiv (\bar{x}, \bar{y})$ is a local minimizer of (3.1.1). There exists a neighborhood W of $\bar{z}$ such that for all $z \in W \cap \mathcal{F}$, $f(z) \geq f(\bar{z})$. Let $\bar{\lambda} \in M(\bar{z})$ be arbitrary. Consider the neighborhood $W \times \Re^\ell$ of the triple $(\bar{x}, \bar{y}, \bar{\lambda})$. If (x, y, λ) belongs to this neighborhood and is feasible to (1), then $(x, y) \in W \cap \mathcal{F}$, by the convexity assumption of $g_i(x, \cdot)$. It follows that $f(x, y) \geq f(\bar{x}, \bar{y})$, establishing that $(\bar{x}, \bar{y}, \bar{\lambda})$ is a local minimizer of (1). The CRCQ does not add anything to this part.

Conversely, suppose that for every $\bar{\lambda} \in M(\bar{z})$, the triple $(\bar{x}, \bar{y}, \bar{\lambda})$ is a local minimizer of (1). Assume for contradiction that there exists a sequence $\{z^k\} \subset \mathcal{F}$ converging to $\bar{z}$ and $f(z^k) < f(\bar{z})$ for all k. By the SBCQ, for large enough k, there exists $\lambda^k \in M(z^k)$ such that the sequence $\{\lambda^k\}$ converges to some $\lambda \in M(\bar{z})$. The hypothesis clearly guarantees $f(z^k) \geq f(\bar{z})$ for all k sufficiently large, a contradiction.

The assertion in the case of the CRCQ is obvious because we can take λ^k in the above proof to be in $M^e(z^k)$ for all k; by the CRCQ at $\bar{z}$, $\{\lambda^k\}$ will have a limit point which belongs to $M^e(\bar{z})$.   **Q.E.D.**

Theorem **1.3.5** asserts that if there exists a $\bar{\lambda} \in M(\bar{x}, \bar{y})$ such that the triple $(\bar{x}, \bar{y}, \bar{\lambda})$ is a global minimizer of (1), then the pair $(\bar{x}, \bar{y})$ is a global minimizer of (3.1.1). This is different from Proposition **3.4.1** which states that in order for $(\bar{x}, \bar{y})$ to be a local minimizer of (3.1.1), it is necessary and sufficient that $(\bar{x}, \bar{y}, \lambda)$ is a local minimizer of (1) for all $\lambda \in M(\bar{x}, \bar{y})$. The following example illustrates that in the case of local minima we cannot relax the "for all" to "there exists" in this equivalence. This example also illustrates further the various CQs for the MPEC defined in Section **3.3**.

3.4.2 Example. Consider the inner VI of an MPEC being defined by the following parametric nonlinear program with $y \in \Re^2$ as the primary variable and $x \in \Re^2$ as the parameter:

$$\text{minimize} \quad \tfrac{1}{2}(y - x)^T (y - x)$$
$$\text{subject to} \quad y \in C \equiv \{y \in \Re^2 : y^T y \leq 1, \, y_1 \leq 1\}. \tag{2}$$

This example is discussed in [240]. For each $x \in \Re^2$, the unique optimal solution is

$$y(x) \equiv \Pi_C(x),$$

which is a globally Lipschitz continuous function. Let $Z \equiv \Re^4$; we have $\mathcal{F} = \{(x, y(x)) \in \Re^4 : x \in \Re^2\}$. The equivalent KKT system of (2) is

$$y_1 - x_1 + 2\lambda_1 y_1 + \lambda_2 = 0,$$
$$y_2 - x_2 + 2\lambda_1 y_1 = 0,$$
$$y^T y \leq 1, \quad \lambda_1 \geq 0, \quad \lambda_1(1 - y^T y) = 0, \tag{3}$$
$$y_1 \leq 1, \quad \lambda_2 \geq 0, \quad \lambda_2(1 - y_1) = 0.$$

Consider the pair $(\bar{x}, \bar{y}) \equiv ((2,0),(1,0))$. It is not hard to see that $M(\bar{z})$ is the convex hull of $(1/2, 0)$ and $(0, 1)$. Thus $M^e(\bar{z}) = \{(1/2, 0), (0, 1)\}$ and the MFCQ holds at $(\bar{x}, \bar{y})$. However, CRCQ fails at this pair. When $x = (x_1, x_2)$ is near $\bar{x}$, it can be verified that

$$y_1(x) \equiv \frac{x_1}{\sqrt{x_1^2 + x_2^2}}, \quad y_2(x) \equiv \frac{x_2}{\sqrt{x_1^2 + x_2^2}};$$

moreover, if $x_2 \neq 0$, the only active constraint at $(x, y(x))$ is $y^T y \leq 1$; for such an x, we have $M(x, y(x)) = \{\lambda(x)\}$, where

$$\lambda(x) \equiv \left(\tfrac{1}{2} \left(\sqrt{x_1^2 + x_2^2} - 1 \right), 0 \right).$$

As such an x approaches $\bar{x}$, $\lambda(x)$ approaches $(1/2, 0)$. By direct verification, (or by Lemma **4.2.5**), we can establish

$$\mathcal{T}(\bar{x}; \mathcal{F}) = \{((dx_1, dx_2), (0, dx_2/2)) \in \Re^4 : (dx_1, dx_2) \in \Re^2\}.$$

For each $\lambda \in M(\bar{z})$ and $dx \in \Re^2$, it is easy to verify that

$$\mathcal{K}(\bar{x}, \lambda; dx) = \{0\} \times \Re$$

and the unique solution of the AVI $(\nabla_x L(\bar{z}, \lambda)dx, \nabla_y L(\bar{z}, \lambda), \mathcal{K}(\bar{z}, \lambda; dx))$, is $\left(0, \frac{dx_2}{1+2\lambda_1}\right)$. (Note: $\lambda_1 \in [0, 1]$ for all $\lambda \in M(\bar{z})$.) Hence

$$\text{Gr}(\mathcal{LS}_{(\bar{x}, \lambda)}) = \left\{ \left((dx_1, dx_2), \left(0, \frac{dx_2}{1+2\lambda_1}\right)\right) : (dx_1, dx_2) \in \Re^2 \right\}.$$

Consequently,

$$\mathcal{T}(\bar{x}; \mathcal{F}) = \text{Gr}(\mathcal{LS}_{(\bar{x}, \bar{\lambda})})$$

where $\bar{\lambda} = (1/2, 0) \in M^e(\bar{x})$; more importantly, $\mathcal{T}(\bar{x}; \mathcal{F})$ is a proper subset of $\mathcal{L}^e(\bar{x}; \mathcal{F})$. Thus the extreme and the full CQ both fail for this feasible set $\mathcal{F}$ at the given $\bar{z}$; yet the basic CQ is valid.

Let $f(x, y) = y_1$. We claim that $(\bar{x}, \bar{y}, \lambda^*)$ is a local minimizer of

$$\begin{array}{ll} \text{minimize} & y_1 \\ \text{subject to} & (3) \end{array} \tag{4}$$

for all $\lambda^* \in M(\bar{z})$ *except* $(1/2, 0)$. The reason is as follows. Suppose $(x, y(x), \lambda)$ with $\lambda \in M(x, y(x))$ is near $(\bar{x}, \bar{y}, \lambda^*)$ where $\lambda^* \neq (\frac{1}{2}, 0)$ and $\lambda^* \in M(\bar{z})$. Since λ is not near $1/2$, it follows from the above calculation that we must have $x_2 = 0$ and $y(x) = (1, 0)$. The fact that $((2, 0), (1, 0), (\frac{1}{2}, 0))$ is *not* a local minimizer of (4) is also clear because by letting x be arbitrarily close to $(2, 0)$ with $x_2 \neq 0$, we have $y_1(x) < 1$.

It can be seen that the vector $\bar{z}$ is not a local minimizer of the MPEC

$$\begin{array}{ll} \text{minimize} & y_1 \\ \text{subject to} & y(x) = \Pi_C(x) \end{array}$$

because any such minimizer must have $y_1 = -1$.

In summary, this example has illustrated two things: one, it is possible for both the extreme and the full CQ to fail for an MPEC and yet the basic CQ holds; two, the "for all" requirement in Proposition **3.4.1** is essential for the equivalence of local minima for the two problems (3.1.1) and (1).

The connection between stationarity of $\bar{z}$ for the MPEC (3.1.1) and stationarity of $(\bar{z}, \bar{\lambda})$ for the KKT formulation (1) also warrants discussion. Note that the feasible set of (1) is

$$\mathcal{F}^{\text{KKT}} \equiv (Z \times \Re^\ell) \cap \text{Gr}(M),$$

where the graph of the set-valued multiplier mapping $M(\cdot)$ is

$$\mathrm{Gr}(M) \equiv \{(z, \lambda) \in \Re^{n+m+\ell} : \lambda \in M(z)\}.$$

For each $dx \in \Re^n$ and $\lambda \in M(\bar{z})$, let $\mathcal{LS}_{(\bar{z},\lambda)}^{\mathrm{KKT}}(dx)$ denote the set of KKT pairs $(dy, d\lambda) \in \Re^m \times \Re^\ell$ of the AVI $(\nabla_x L(\bar{z}, \lambda) dx, \nabla_y L(\bar{z}, \lambda), \mathcal{K}(\bar{z}, \lambda; dx))$. Comparing with $\mathcal{LS}_{(\bar{z},\lambda)}$ defined in Section **3.2** and recalling the standard correspondence between solutions of an AVI and KKT points of the same AVI, we see that the canonical projection of $\mathcal{LS}_{(\bar{z},\lambda)}^{\mathrm{KKT}}(dx)$ onto $\Re^m$ is equal to $\mathcal{LS}_{(\bar{z},\lambda)}(dx)$. The SMFCQ and CRCQ play an important role in the next two results.

3.4.3 Proposition. Let Z be an arbitrary subset of $\Re^{n+m}$ and X be the projection of Z onto $\Re^n$. Suppose that each $g_i(x, \cdot)$ is convex for each $x \in X$; assume further that either the SMFCQ holds at $(\bar{z}, \bar{\lambda}) \in \mathcal{F}^{\mathrm{KKT}}$ or the CRCQ holds at $\bar{z} \in \mathcal{F}$. With

$$M' \equiv \begin{cases} \{\bar{\lambda}\} & \text{if the SMFCQ holds,} \\ M^e(\bar{z}) & \text{if the CRCQ holds,} \end{cases}$$

the following two statements hold.

(a) The cone $\mathcal{T}(\bar{z}; \mathcal{F})$ is the projection onto $\Re^{n+m}$ of

$$\bigcup_{\lambda \in M'} \mathcal{T}((\bar{z}, \lambda); \mathcal{F}^{\mathrm{KKT}}) \subseteq \Re^{m+n+\ell}.$$

(b) If the MPEC CQ holds:

$$\mathcal{T}(\bar{z}; \mathcal{F}) = \mathcal{L}'(\bar{z}; \mathcal{F}) \equiv \mathcal{T}(\bar{z}; Z) \bigcap \left(\bigcup_{\lambda \in M'} \mathrm{Gr}(\mathcal{LS}_{(\bar{z},\lambda)}) \right),$$

then

$$\bigcup_{\lambda \in M'} \mathcal{T}((\bar{z}, \lambda); \mathcal{F}^{\mathrm{KKT}}) = (\mathcal{T}(\bar{z}; Z) \times \Re^\ell) \bigcap \left(\bigcup_{\lambda \in M'} \mathrm{Gr}(\mathcal{LS}_{(\bar{z},\lambda)}^{\mathrm{KKT}}) \right).$$

Proof. We first show that $\mathcal{T}(\bar{z}; \mathcal{F})$ always contains the projection onto $\Re^{n+m}$ of $\mathcal{T}((\bar{z}, \lambda); \mathcal{F}^{\mathrm{KKT}})$ for any $\lambda \in M(\bar{z})$. Let $(dz, d\lambda)$ be the limit of $(z^k - \bar{z}, \lambda^k - \bar{\lambda})/\tau_k$ for some $\{(z^k, \lambda^k)\} \subset \mathcal{F}^{\mathrm{KKT}}$ converging to $(\bar{z}, \bar{\lambda})$ and $\tau_k \to 0+$, where $\bar{\lambda} \in M(\bar{z})$. Obviously $(z^k, \lambda^k) \to (\bar{z}, \bar{\lambda}) \in \mathrm{Gr}(M)$. By convexity of each function $g(x^k, \cdot)$, we deduce that each z^k lies in $\mathcal{F}$; hence $dz \in \mathcal{T}(\bar{z}; \mathcal{F})$.

For the reverse inclusion, we first assume that the SMFCQ holds at $(\bar{z}, \bar{\lambda}) \in \mathcal{F}^{\mathrm{KKT}}$. Following the proof of Lemma **3.2.1**, we see that in order to show that $\mathcal{T}(\bar{z}; \mathcal{F})$ is contained in the projection of $\mathcal{T}((\bar{z}, \bar{\lambda}); \mathcal{F}^{\mathrm{KKT}})$ onto $\Re^{n+m}$, it suffices to show that the sequence $\{\tau_k^{-1}(\lambda^k - \bar{\lambda})\}$ is bounded. Assume by way of contradiction that this sequence is unbounded; then without loss of generality, we may assume that the normalized sequence $\{(\lambda^k - \bar{\lambda})/\|\lambda^k - \bar{\lambda}\|\}$ converges to a nonzero vector $\bar{d}$ which must satisfy (i) $\bar{d}_i \geq 0$ for all i such that $\bar{\lambda}_i = 0$ and (ii) $\bar{d}_i = 0$ for all $i \notin \mathcal{I}(\bar{z})$. Dividing by $\|\lambda^k - \bar{\lambda}\|$ in equation (3.2.6) and passing to the limit $k \to \infty$, we deduce

$$\sum_{i \in \mathcal{I}(\bar{z})} \bar{d}_i \nabla_y g_i(\bar{z}) = 0.$$

This equation contradicts the SMFCQ at the tuple $(\bar{z}, \bar{\lambda})$, by (3.2.13). Hence part (a) is valid under the SMFCQ at $(\bar{z}, \bar{\lambda})$.

Suppose that the CRCQ holds at $\bar{z} \in \mathcal{F}$. As in the proof of Proposition **3.2.2**, we choose, for each k, the multiplier λ^k satisfying (3.2.5) to be an extreme point of $M(z^k)$; that is, $\lambda^k \in M^e(z^k)$. This implies that the gradients

$$\{\nabla_y g_i(z^k) : i \in \mathrm{supp}(\lambda^k)\}$$

are linearly independent, where $\mathrm{supp}(\lambda^k) \equiv \{i : \lambda_i^k > 0\}$ is the *support* of λ^k. We may assume, by taking an appropriate subsequence, that $\mathrm{supp}(\lambda^k)$ is equal to a constant set J for all k. Clearly J is a subset of $\mathcal{I}(\bar{z})$. By the CRCQ, the limiting gradients $\{\nabla_y g_i(\bar{z}) : i \in J\}$ are also linearly independent. Hence the sequence $\{\lambda^k\}$ converges to a limit $\bar{\lambda} \in M^e(\bar{z})$. The equation (3.2.6) can be written equivalently as

$$0 = \nabla_z L(\bar{z}, \lambda^k)(z^k - \bar{z}) + \sum_{i \in J} (\lambda_i^k - \bar{\lambda}_i) \nabla_y g_i(\bar{z}) + o\left(\|z^k - \bar{z}\|\right).$$

The boundedness of the sequence $\{\tau_k^{-1}(\lambda^k - \bar{\lambda})\}$ is evident. Hence the proof of part (a) is complete. Part (b) follows from (a) and the observation made preceding the proposition.   **Q.E.D.**

The following corollary is immediate from part (a) of the above result.

3.4.4 Corollary. Assume the hypotheses of Proposition **3.4.3**. Then $\bar{z}$ is stationary for the MPEC (3.1.1) if and only if for each $\bar{\lambda} \in M'$, $(\bar{z}, \bar{\lambda})$ is stationary for the KKT constrained MP (1).

3.4.5 Remark. It can be shown that without the SMFCQ or the CRCQ, if $\bar{z}$ is stationary for the MPEC (3.1.1), then for any $\lambda \in M(\bar{z})$, the pair $(\bar{z}, \lambda)$ is stationary for (1). Indeed by the first part of the proof of Proposition **3.4.3** which does not require any constraint qualification, the cone $\mathcal{T}(\bar{z}; \mathcal{F})$ always contains the projection of $\mathcal{T}((\bar{z}, \lambda); \mathcal{F}^{\mathrm{KKT}})$ onto $\Re^{m+n}$ for any $\lambda \in M(\bar{z})$. Nevertheless the reverse implication is in jeopardy because the second part of the proof of this proposition requires the CQs.

For a given $\bar{\lambda} \in M(\bar{z})$ and $dx \in \Re^n$, we establish a basic relation between the set $\mathcal{LS}^{\mathrm{KKT}}_{\bar{w}}(dx)$, where $\bar{w} \equiv (\bar{z}, \bar{\lambda})$, and the linearized cone of the set $\mathrm{Gr}(M)$ at $\bar{w}$, with $\mathrm{Gr}(M)$ considered as defined by a system of nonlinear equations and inequalities and its linearized cone obtained by a standard NLP approach. To begin, note that we can represent vectors in $\mathrm{Gr}(M)$ as solutions to either one of the following two equivalent systems:

$$
\begin{aligned}
L(x, y, \lambda) &= 0, \\
g(x, y) \leq 0, \quad \lambda &\geq 0, \\
\lambda^T g(x, y) &= 0,
\end{aligned}
\tag{5}
$$

or, in the Hadamard form of the complementarity condition $\lambda^T g(x, y) = 0$,

$$
\begin{aligned}
L(x, y, \lambda) &= 0, \\
g(x, y) \leq 0, \quad \lambda &\geq 0, \\
\lambda_i g_i(x, y) &= 0, \quad i = 1, \dots, \ell.
\end{aligned}
$$

Regardless of which system is used and after some simplifications, the linearized cone of $\mathrm{Gr}(M)$ at $\bar{w}$ can be seen to be given by:

$$\mathcal{L}^{\mathrm{NLP}}(\bar{w}; \mathrm{Gr}(M)) = \{(dx, dy, d\lambda) \in \Re^{n+m+\ell} :$$

$$\nabla_x L(\bar{w})dx + \nabla_y L(\bar{w})dy + \sum_{i \in \mathcal{I}(\bar{z})} d\lambda_i \nabla_y g_i(\bar{z}) = 0$$

$$\nabla_x g_i(\bar{z})^T dx + \nabla_y g_i(\bar{z})^T dy \le 0, \ i \in \mathcal{I}_0(\bar{w}),$$

$$\nabla_x g_i(\bar{z})^T dx + \nabla_y g_i(\bar{z})^T dy = 0, \ i \in \mathcal{I}_+(\bar{w}), \tag{6}$$

$$d\lambda_i = 0, \ i \notin \mathcal{I}(\bar{z}),$$

$$d\lambda_i \ge 0, \ i \in \mathcal{I}_0(\bar{w})\}.$$

Thus $(dy, d\lambda) \in \mathcal{LS}_{\bar{w}}^{\mathrm{KKT}}(dx)$ if and only if $(dx, dy, d\lambda) \in \mathcal{L}^{\mathrm{NLP}}(\bar{w}; \mathrm{Gr}(M))$ and $d\lambda_i \nabla g_i(\bar{z})^T dz = 0$ for all $i \in \mathcal{I}_0(\bar{w})$. In other words, for each $dx \in \Re^n$, the section of $\mathcal{L}^{\mathrm{NLP}}(\bar{w}; \mathrm{Gr}(M))$ corresponding to dx is equal to the feasible region of the equivalent mixed complementarity system corresponding to the AVI $(\nabla_x L(\bar{w})dx, \nabla_y L(\bar{w}), \mathcal{K}(\bar{z}, \bar{\lambda}; dx))$. Consequently, we always have

$$\mathrm{Gr}(\mathcal{LS}_{\bar{w}}^{\mathrm{KKT}}) \subseteq \mathcal{L}^{\mathrm{NLP}}(\bar{w}; \mathrm{Gr}(M)),$$

and equality holds if $\bar{w}$ is strictly complementary, i.e., if $\bar{\lambda} - g(\bar{z}) > 0$. Indeed in the latter case, the index set $\mathcal{I}_0(\bar{w})$ is empty and both $\mathrm{Gr}(\mathcal{LS}_{\bar{w}}^{\mathrm{KKT}})$ and $\mathcal{L}^{\mathrm{NLP}}(\bar{w}; \mathrm{Gr}(M))$ become a common linear subspace in $\Re^{n+m+\ell}$. Example **3.1.2** shows that equality could fail to hold between these two cones.

4

Verification of MPEC Hypotheses

In Section **3.3**, we have introduced the full, extreme, and basic CQs for an MPEC. Under any one of these CQs, stationarity of a feasible solution to an MPEC can be characterized; see Corollary **3.3.1**, Theorem **3.3.4**, and Remark **3.3.5**. In this chapter, we consider some important cases of the general MPEC (3.1.1) and introduce certain assumptions in order to verify such CQs, thus completing the task of deriving first-order optimality conditions for the MPEC.

This chapter has four main sections. Section **4.1** considers the AVI constrained mathematical program, i.e., the MPAEC. The main goal of this section is to demonstrate that the extreme CQ must hold without any particular assumptions. This conclusion is consistent with a standard nonlinear program with linear constraints for which the KKT conditions must be necessary for local optimality. The next two sections, **4.2** and **4.3**, discuss two general approaches to deal with the MP with nonlinear VI constraints. The approach in Section **4.2** is of interest because of the bilevel nature of the MPEC and the special significance of the variables x and y, whereas the approach in Section **4.3** is more along the line of traditional nonlinear programming, but with an important modification to deal with the combinatorial feature of the MPEC. The chapter concludes with a discussion of an exact penalization of order 1, using the assumptions set forth in Section **4.2**.

The two approaches discussed in Sections **4.2** and **4.3** together provide a powerful tool for the study of the MPEC. They are based on different assumptions and are therefore applicable to different classes of MPECs. Whereas the implicit programming (IMP) approach in Section **4.2** involves highly technical analysis, the piecewise programming (PCP) approach in Section **4.3** requires only a moderate extension of classical nonlinear programming theory.

4.1 AVI Constrained Mathematical Program

Consider the MPAEC studied in Section **2.4.1**; the constraint function g and the inner VI mapping F are given by (2.4.1) and (2.4.2) respectively, and Z is given by (3.3.3). For ease of reference, we repeat the formulation of this MPEC as follows:

$$
\begin{aligned}
\text{minimize} \quad & f(x, y) \\
\text{subject to} \quad & (x, y) \in Z, \\
& Dx + Ey + b \leq 0, \\
& (y' - y)^T (Px + Qy + q) \geq 0, \\
& \text{for all } y' \text{ such that } Dx + Ey' + b \leq 0.
\end{aligned}
\tag{1}
$$

We also recall the equivalent formulation of this problem where the inner AVI is stated in terms of its KKT conditions:

$$
\begin{aligned}
\text{minimize} \quad & f(x, y) \\
\text{subject to} \quad & (x, y, \lambda) \in Z \times \Re_+^{\ell}, \\
& Dx + Ey + b \leq 0, \\
& Px + Qy + q + E^T \lambda = 0, \\
& \lambda^T (Dx + Ey + b) = 0.
\end{aligned}
\tag{2}
$$

The latter formulation is exactly (2.4.4) without the ball $\mathbb{B}(0, c)$.

Let $\bar{z} \equiv (\bar{x}, \bar{y}) \in \mathcal{F}$ be a given feasible solution to (1). As we have noted before, the SBCQ for the inner AVI holds at $(\bar{x}, \bar{y})$ because g is affine. In fact, the CRCQ holds. Our goal in this subsection is to show that the linearized cone $\mathcal{L}(\bar{z}; \mathcal{F})$ coincides with the tangent cone $\mathcal{T}(\bar{z}; \mathcal{F})$ for the AVI constrained MP. Consequently the full CQ (and indeed the extreme CQ) holds for the MPAEC (1); thus the four statements (a)–(d) in Corollary **3.3.1** must be equivalent; moreover, Theorem **3.3.4** and the remark that follows are valid for (1). Therefore, we have a characterization of a stationary point for this problem, and hence a set of first-order necessary optimality conditions, under no constraint qualification at all. This conclusion is consistent with the optimality conditions for a linearly constrained nonlinear program which are always necessary under no constraint qualifications. Note, however, (1) is not a linearly constrained nonlinear program, due to the complementarity relationship $\lambda^T (Dx + Ey + b) = 0$ which is quadratic. We should stress the two principal reasons why we are able to obtain the first-order necessary conditions for the MPAEC under no constraint qualification: one reason is our special treatment of the tangent cone $\mathcal{T}(\bar{z}; \mathcal{F})$, whereby its piecewise polyhedral nature is preserved; and the other reason is the particular linearization scheme that we have applied to the constraints of the MPAEC.

The following result shows that the converse of Lemma **3.2.1** holds for an AVI constrained MP; specifically the full CQ (3.3.2) and the extreme CQ both hold for the MPAEC (1).

4.1.1 Proposition. For the MPAEC (1) with a polyhedral set Z, assume that $dz \equiv (dx, dy)$ is in $\mathcal{T}(\bar{z}; Z)$ and there exists a $\lambda \in M(\bar{z})$ such that dy solves the AVI $(Pdx, Q, \mathcal{K}(\bar{z}, \lambda; dx))$, then $\bar{z} + \tau dz \in \mathcal{F}$ for all $\tau \geq 0$ sufficiently small; hence $dz \in \mathcal{T}(\bar{z}; \mathcal{F})$. Consequently,

$$\mathcal{T}(\bar{z}; \mathcal{F}) = \mathcal{L}^e(\bar{z}; \mathcal{F}) = \mathcal{L}(\bar{z}; \mathcal{F}).$$

Proof. By the definition (3.2.2) of the directional critical set $\mathcal{K}(\bar{z}, \lambda; dx)$ and by the assumption that dy belongs to $\mathrm{SOL}(Pdx, Q, \mathcal{K}(\bar{z}, \lambda; dx))$, it follows that there exist multipliers $\{d\lambda_i : i \in \mathcal{I}(\bar{z})\}$ with $d\lambda_i \geq 0$ for $i \in \mathcal{I}_0(\bar{z}, \lambda)$, such that

$$Pdx + Qdy + \sum_{i \in \mathcal{I}(\bar{z})} d\lambda_i (E_{i \cdot})^T = 0,$$

$$d\lambda_i (Ddx + Edy)_i = 0, \quad \text{for all } i \in \mathcal{I}_0(\bar{z}, \lambda),$$

where $E_i.$ denotes the i-th row of the matrix E. Moreover, we have

$$(Ddx + Edy)_i \leq 0, \quad \text{for } i \in \mathcal{I}_0(\bar{z}, \lambda),$$

$$(Ddx + Edy)_i = 0, \quad \text{for } i \in \mathcal{I}_+(\bar{z}, \lambda).$$

Since $dz \in \mathcal{T}(\bar{z}; Z)$ and Z is polyhedral, it follows that $z(\tau) \equiv z + \tau dz \in Z$ for all $\tau \geq 0$ sufficiently small. Define $d\lambda_i \equiv 0$ for all $i \notin \mathcal{I}(\bar{z})$ and for $\tau \geq 0$,

$$\lambda(\tau) \equiv \lambda + \tau d\lambda.$$

It is easy to check that $\lambda(\tau) \geq 0$ for all $\tau \geq 0$ sufficiently small. Moreover for all τ, we have

$$Px(\tau) + Qy(\tau) + q + E^T \lambda(\tau) = 0,$$

where $(x(\tau), y(\tau)) = z(\tau)$. We also have

$$(Dx(\tau) + Ey(\tau) + b) \leq 0,$$

for all $\tau \geq 0$ sufficiently small. Finally, we verify that

$$\lambda(\tau)^T (Dx(\tau) + Ey(\tau) + b) = 0$$

for all $\tau > 0$ sufficiently small. Suppose $\lambda(\tau)_i > 0$ for some index i. If $\lambda_i > 0$, then since

$$(Dx + Ey + b)_i = 0 \quad \text{and} \quad (Ddx + Edy)_i = 0,$$

it follows that

$$(Dx(\tau) + Ey(\tau) + b)_i = 0, \tag{3}$$

for all τ. If $\lambda_i = 0$, we must have $d\lambda_i > 0$ and $i \in \mathcal{I}(\bar{z})$. The former condition $d\lambda_i > 0$ implies $(Ddx + Edy)_i = 0$; whereas the latter condition $i \in \mathcal{I}(\bar{z})$ implies $(Dx + Ey + b)_i = 0$. Consequently, (3) holds for this index i. Summarizing the above derivation, we conclude that $z(\tau) \in \mathcal{F}$ for all $\tau \geq 0$ sufficiently small. It follows that $dz \in \mathcal{T}(\bar{z}; \mathcal{F})$ as desired.

To prove the last assertion of the proposition, observe that for g and F given by (2.4.1) and (2.4.2) respectively, the matrices $\nabla_x L(\bar{z}, \lambda)$ and $\nabla_y L(\bar{z}, \lambda)$ are equal to P and Q respectively. Consequently, we have

$$\mathcal{T}(\bar{z}; \mathcal{F}) \subseteq \mathcal{L}^e(\bar{z}; \mathcal{F}) \subseteq \mathcal{L}(\bar{z}; \mathcal{F}) \subseteq \mathcal{T}(\bar{z}; \mathcal{F}),$$

where the first inclusion is just (3.2.12); the second inclusion is always valid by definition of the two cones, and the last inclusion follows from the observation we have just made and what we have proved above.   **Q.E.D.**

Combining Lemma **4.1.1** and Theorem **3.3.4**, we obtain the following stationarity result for an MPAEC.

4.1.2 Theorem. For the problem (1) where Z is given by (3.3.3), a vector $\bar{z} = (\bar{x}, \bar{y}) \in \mathcal{F}$ is a stationary point of (1) if and only if for each vector $\lambda \in M^e(\bar{z})$ and each pair index sets α and $\bar{\alpha}$ that partition

$$\{i : \lambda_i = 0 = (D\bar{x} + E\bar{y} + b)_i\},$$

there exist multipliers $\zeta \in \Re^s$, $\eta \in \Re^\ell$, and $\pi \in \Re^m$ such that

$$\nabla_x f(\bar{z}) + G^T \zeta + D^T \eta = P^T \pi$$

$$\nabla_y f(\bar{z}) + H^T \zeta + E^T \eta = Q^T \pi$$

$$E_{i.}\pi \leq 0, \quad i \in \alpha$$

$$\lambda_i E_{i.}\pi = 0, \quad i \in \mathcal{I}(\bar{z})$$

$$\eta_i \geq 0, \quad i \in \bar{\alpha}$$

$$\eta_i = 0, \quad i \notin \mathcal{I}(\bar{z})$$

$$\zeta \geq 0, \quad \zeta^T(G\bar{x} + H\bar{y} + a) = 0.$$

If Z is a polyhedron, then the feasible region $\mathcal{F}$ of the MPAEC (1) is the union of a finite number of convex polyhedra in $\Re^{n+m}$; i.e., $\mathcal{F}$ is *piecewise polyhedral*. This can easily be seen from the equivalent formulation (2) in terms of complementarity constraints. Based on this observation, we obtain the following property of $\mathcal{T}(\bar{z}; \mathcal{F})$.

4.1.3 Proposition. For the problem (1) where Z is given by (3.3.3), let $\bar{z} \in \mathcal{F}$. There exists an open neighborhood W of $\bar{z}$ such that $dz \in \mathcal{T}(\bar{z}; \mathcal{F})$ if and only if $dz = \tau (z - \bar{z})$ for some $z \in \mathcal{F} \cap W$ and some $\tau > 0$.

Proof. Let $\{\mathcal{P}_1, \ldots, \mathcal{P}_{r+s}\}$ be a family of polyhedra in $\Re^{n+m}$ whose union is equal to $\mathcal{F}$. Let $\{\mathcal{P}_1, \ldots, \mathcal{P}_r\}$ be the subfamily such that

$$\bar{z} \in \bigcap_{i=1}^{r} \mathcal{P}_i$$

and $\bar{z} \notin \mathcal{P}_{r+j}$ for each $j = 1, \ldots, s$. It suffices to choose W to be an open set containing $\bar{z}$ such that W does not intersect any one of these $\mathcal{P}_{r+j}$ for each $j = 1, \ldots, s$. The result now follows from the polyhedrality of each P_i for $i = 1, \ldots, r$.   **Q.E.D.**

The following sufficiency result can be proved easily.

4.1.4 Proposition. Let Z be a polyhedron, and let $f : \Re^{n+m} \to \Re$ be pseudoconvex on $\mathcal{F}$. If $\bar{z} \in \mathcal{F}$ is a stationary point of the MPAEC (1), then $\bar{z}$ is a local minimum of (1).

Proof. By Proposition **4.1.3**, it follows that there exists a neighborhood W of $\bar{z}$ such that for all $z \in W \cap \mathcal{F}$, $z - \bar{z} \in \mathcal{T}(\bar{z}; \mathcal{F})$. Since $\bar{z}$ is stationary, we have

$$\nabla f(\bar{z})^T (z - \bar{z}) \geq 0.$$

By pseudoconvexity of f, it follows easily that $f(z) \geq f(\bar{z})$.   **Q.E.D.**

The validity of the last proposition depends critically on the piecewise polyhedrality of $\mathcal{F}$; in turn, this is a consequence of the linearity of F in (x, y) jointly, among other things. The result does not hold if the upper-level variable x enters nonlinearly in F, even if $F(x, y)$ is linear in y for fixed x and f is convex in (x, y) jointly.

4.1.5 Example. Consider the MPEC in $\Re^2$:

$$
\begin{aligned}
&\text{minimize} \quad f(x, y) \equiv x^2 + x \\
&\text{subject to} \quad -x^2 - x + y \geq 0, \quad y \geq 0, \quad y(-x^2 - x + y) = 0.
\end{aligned}
\tag{4}
$$

The feasible region $\mathcal{F}$ is the following union:

$$\left\{ (x, x + x^2) \in \Re^2 : x \leq -1 \text{ or } x \geq 0 \right\} \cup \left\{ (x, 0) \in \Re^2 : x \in [-1, 0] \right\}.$$

Let $\bar{z} \equiv (0, 0)$. We have

$$\mathcal{T}(\bar{z}; \mathcal{F}) = \{(a, a) : a \geq 0\} \cup \{(-a, 0) : a \geq 0\}.$$

Since $\nabla f(\bar{z}) = 0$, then $\nabla f(\bar{z})^T dz \geq 0$ for all $dz \in \mathcal{T}(\bar{z}, \mathcal{F})$; hence $\bar{z}$ is stationary. Yet $\bar{z}$ is not a local minimum point of (4), because when $x < 0$ is sufficiently near zero, the objective function $f(x, y)$ is negative.

We next give an example to show that in the setting of Proposition **4.1.4** although a stationary point of the MPAEC must be a local minimum, such a point is not necessarily a global minimum.

4.1.6 Example. Consider the MPEC in $\Re^4$:

minimize $\quad f(x_1, x_2, y_1, y_2) \equiv 2y_1 + y_2$

subject to $\quad x \in X \equiv \{(x_1, x_2) : x_1 + x_2 = 0,\ x_1 \in [-1, 1]\}$

and $\qquad (y_1, y_2) \in \operatorname{argmin}\{(y_1 - x_1)^2 + (y_2 - x_2)^2 : (y_1, y_2) \geq 0\}.$

This is an AVI constrained MP with a convex objective function f. The feasible region

$$\mathcal{F} = \{(a, b, a_+, b_+) \in \Re^4 : (a, b) \in X\}.$$

It is easy to verify that $(1, -1, 1, 0)$ is a local minimum point of the given MPAEC; but this point is not equal to the unique global minimum, namely, $(-1, 1, 0, 1)$.

4.2 An Implicit Programming Approach

This section concerns an important special case of the MPEC (3.1.1) in which the upper-level feasible set Z has a simplified structure. Specifically, we consider the following problem:

$$
\begin{aligned}
\text{minimize} \quad & f(x, y) \\
\text{subject to} \quad & (x, y) \in \mathcal{F} \equiv (X \times \Re^m) \cap \operatorname{Gr}(\mathcal{S}),
\end{aligned}
\tag{1}
$$

where $f : \Re^{n+m} \to \Re$ is a continuously differentiable real-valued function, X is a closed convex set in $\Re^n$, and $\mathcal{S} : X \to \Re^m$ is the set-valued solution map of the parametric VI $(F(x, \cdot), C(x))$ for $x \in X$, with $F : \Re^{n+m} \to \Re^m$ being a continuously differentiable vector-valued mapping and $C(x)$ being a closed convex subset of $\Re^m$. The set X is assumed to be defined by constraints that are easy to handle (e.g., X could be a polyhedron, though polyhedrality is not assumed throughout). It is in the setting (1) that we develop sufficient conditions for the assumptions of Corollary **3.3.1** to hold.

The implicit programming or IMP approach to the solution of (1) is motivated by the implicit form (1.3.28) of the MPEC which expresses this problem as an optimization problem in the upper-level variable x alone, by assuming that there is a unique solution $y(x)$ of the VI $(F(x, \cdot), C(x))$. We restate this formulation for ease of presentation:

$$\begin{aligned} \text{minimize} \quad & \tilde{f}(x) \\ \text{subject to} \quad & x \in X, \end{aligned} \tag{2}$$

where $\tilde{f}(x) \equiv f(x, y(x))$. We introduce later an important assumption, much weaker than the strong monotonicity condition discussed in Subsection **1.3.4**, that justifies the problem (2) as being locally (near a given $(\bar{x}, \bar{y}) \in \mathcal{F}$) equivalent to (1).

As shown in Section **4.3**, the IMP approach turns out to be somewhat restrictive, because it requires assumptions that are unnecessarily strong for certain problems. Among its deficiencies, the upper-level feasible region Z is required to be of the form $X \times \Re^m$. Nevertheless, this approach has a natural appeal because of the practical significance of the lower-level variable y as a response (multi)function of the upper-level variable x. In essence, the IMP approach assumes that for x sufficiently close to $\bar{x}$, $y(x)$ is locally single-valued and has a certain first-order directional smoothness property. From an economic point of view, this assumption translates into some well-behaved properties of the follower's rational responses as a function of the leader's strategies. Computationally, the existence of such an implicit function could be put to good use for the design of some iterative descent algorithms for solving the MPEC; see Section **6.3** and the references [69, 138, 215].

This section is exceedingly long and highly technical. To ease the reader, we therefore divide the section into seven subsections. The next subsection is a brief review of various differentiability properties of vector-valued functions; Subsection **4.2.2** states a key conceptual assumption within the IMP approach; Subsection **4.2.3** contains a summary of the main tools (namely, the theory of PC^1 functions and degree theory) for establishing a main result of this section, Theorem **4.2.16**, which is presented in Subsection **4.2.4**. Readers already familiar with the background material in Subsections **4.2.1** and **4.2.3** can safely skip these subsections and proceed directly to the main theorem. The remaining three subsections discuss consequences of the key assumption made in Subsection **4.2.2** and Theorem **4.2.16**. These consequences include the calculation of the directional derivatives of the implicit solution function of the lower-level VIs and the verification of the MPEC CQs required for the validity of the first-order optimality conditions established in Chapter **3**. The seventh and last subsection provides a detailed study of a fundamental assumption in sensitivity and stability analysis of parametric VIs which is the central theme underlying the IMP approach to MPEC. The results presented in this section generalize most known sensitivity results of parametric NLPs and VIs, including Robinson's theory of strongly regular solutions of parametric VIs (specialized from parametric generalized equations) [248], Kojima's theory of strongly stable stationary points of parametric NLPs [140] and the work of many researchers in this area, including [15, 30, 96, 124, 150, 152, 159, 223, 234, 243], to name just a few.

Historical notes on sensitivity analysis

The IMP approach involves a deep understanding of sensitivity and stability analysis of parametric variational inequalities. The latter analysis has its origin in parametric nonlinear programming which was pioneered by Fiacco and McCormick [74]. Under the LICQ, the strict complementarity assumption, and the well-known second-order sufficient condition (SOSC), Fiacco and McCormick considered the KKT system of a parametric NLP and employed the classical implicit function theorem [211] to obtain the existence of a locally unique, F-differentiable solution function. In several seminal papers, Robinson [247, 248, 250] introduced the generalized equations and used them for the study of parametric NLPs and VIs. Among the prominent contributions in these papers was the concept of a strongly regular solution of a generalized equation which has dominated much of the research in this area since its introduction. This theory made no use of the strict complementarity assumption required by Fiacco and McCormick. Thus the implicit function was not F-differentiable but B-differentiable; this important result and indeed the definition of the Bouligand derivative appeared in [251]. Previously Jittorntrum obtained a similar differentiability result for parametric nonlinear programs [124]. At about the same time as Robinson's work [248], Kojima's fundamental paper [140] was published which introduced the strongly stable stationary points of parametric NLPs and provided a complete characterization of such points under the MFCQ and SOSC. Although already sharing a close resemblance in their respective definitions, Robinson's strong regularity concept and Kojima's strong stability concept were not given a precise connection until the work of Jongen, Möbert, Rückmann, and Tammer [126] which showed that for a parametric NLP, these two concepts are equivalent under the LICQ. These two solution concepts are fundamental to some effective computational approaches for solving NLPs [100] and generalized equations [127].

Whereas Robinson's theory of strongly regular generalized equations was built on some fixed-point theorems, Kojima's results were derived using the powerful tool of degree theory in nonlinear analysis. The latter theory is also the main tool in several sensitivity studies of parametric VIs and NCPs; see [96, 103, 141, 161, 227]. Throughout these studies, the MFCQ and SOSC continue to play the central role. It can be said that without them (or their variations), sensitivity analysis would not have been possible.

The development in the following subsections extends the strong regularity theory by relaxing the LICQ to a constant rank condition, thus allowing nonunique multipliers in the lower-level VIs, considered as a family of parametric VIs $(F(x, \cdot), C(x))$ with the upper-level variable x as the parameter. When specialized to parametric NCPs and NLPs, our theory reduces, respectively, to Robinson's theory for parametric generalized equations specialized to NCPs and to the results in [240]. See [31, 32, 60, 150, 261] for results without the CRCQ.

4.2.1 *B-differentiable functions*

The blanket assumption we shall impose on the problem (1) requires the notion of a B-differentiable function. For this reason, we give a brief review of this class of nonsmooth functions and explain the connection of the B-derivative (i.e., the directional derivative) with classical derivatives. We should stress that the differentiability concepts and results discussed below all refer to a function at a given point. Certain properties that hold at this point need not carry over to other points, even those that are near (unless explicitly assumed or stated).

4.2.1 Definition. Let $G : W \subseteq \Re^n \to \Re^m$ be a function defined on the open set W. Let $\bar{x} \in W$ be given. The function G is said to be

(a) *locally Lipschitz* at $\bar{x}$ if G is Lipschitz continuous in a neighborhood of $\bar{x}$;

(b) *directionally differentiable* at $\bar{x}$ if for every vector $d \in \Re^n$ the limit

$$\lim_{\tau \to 0+} \frac{G(\bar{x} + \tau d) - G(\bar{x})}{\tau}$$

exists, which we denote $G'(\bar{x}; d)$ and call the *directional derivative* of G at $\bar{x}$ along the direction d;

(c) *B(ouligand)-differentiable* at $\bar{x}$ if G is both locally Lipschitz and directionally differentiable at $\bar{x}$;

(d) *F(réchet)-differentiable* at $\bar{x}$ if the following limit holds:

$$\lim_{\substack{x \to \bar{x} \\ x \neq \bar{x}}} \frac{G(x) - G(\bar{x}) - \nabla G(\bar{x})(x - \bar{x})}{\|x - \bar{x}\|} = 0,$$

where $\nabla G(\bar{x}) \in \Re^{m \times n}$ is the Jacobian matrix of G at $\bar{x}$;

(e) *strongly F-differentiable* at $\bar{x}$ if

$$\lim_{\substack{(x, x') \to (\bar{x}, \bar{x}) \\ x \neq x'}} \frac{G(x) - G(x') - \nabla G(\bar{x})(x - x')}{\|x - x'\|} = 0.$$

Some authors prefer to define a B-differentiable function at $\bar{x}$ to mean a directionally differentiable function at $\bar{x}$ with the additional requirement that the limit in Definition **4.2.1**(b) be "uniform" in d for all d of bounded magnitude. In place of the latter requirement, we have insisted that the function G be locally Lipschitz continuous near $\bar{x}$. By a result of Shapiro [261] (see part (c) of Proposition **4.2.2**), our definition of B-differentiability implies such a uniform limit condition. We have required the locally Lipschitz condition for B-differentiability because this condition is guaranteed to hold in our context; thus Definition **4.2.1** suffices for our purpose.

We use the 1-dimensional function $G : x \mapsto \min(x, f(x))$, where f is a real-valued function of one variable, to illustrate the various concepts defined above. When f is Lipschitz continuous on $\Re$, then so is G; when f is in addition continuously differentiable near $\bar{x}$, then G is B-differentiable at $\bar{x}$, but not necessarily F-differentiable there unless either $\bar{x} \neq f(\bar{x})$ or $f'(\bar{x}) = 1$; in fact at a point $x = f(x)$, we have $G'(x; d) = \min(d, f'(x)d)$ for $d \in \Re$, by Lemma **2.4.8**.

The following result summarizes some basic properties of the above differentiability concepts. We refer the reader to [219, 252] for its proof.

4.2.2 Proposition. Suppose that the function $G : W \subseteq \Re^n \to \Re^m$ defined on the open set W is B-differentiable at $\bar{x} \in W$. The following statements are valid.

(a) As a function in d, the directional derivative $G'(\bar{x}; d)$ is positively homogeneous; that is, for all $d \in \Re^n$,

$$G'(\bar{x}; \tau\, d) = \tau\, G'(\bar{x}; d), \quad \text{for all } \tau \geq 0.$$

(This property does not require the locally Lipschitz continuity of G.)

(b) As a function in d, the directional derivative $G'(\bar{x}; d)$ is globally Lipschitz with the same modulus as G has in the neighborhood of $\bar{x}$.

(c) The following limit holds:

$$\lim_{\substack{(x,\bar{x})\to(\bar{x},\bar{x})\\ x\neq\bar{x}}} \frac{G(x) - G(\bar{x}) - G'(\bar{x}; x - \bar{x})}{\|x - \bar{x}\|} = 0. \tag{3}$$

(d) The directional derivative $G'(\bar{x}; \cdot)$ is linear in the second argument if and only if the function G is F-differentiable at $\bar{x}$; in this case, we have

$$G'(\bar{x}; d) = \nabla G(\bar{x})d, \quad \text{for all } d \in \Re^n.$$

(e) If G is directionally differentiable at all points in a neighborhood of $\bar{x}$, then G is strongly F-differentiable at $\bar{x}$ if and only if for all $d \in \Re^n$,

$$\lim_{x\to\bar{x}} G'(x; d) = G'(\bar{x}; d),$$

and this limit holds uniformly for all $d \in \Re^n$ on compact sets; or equivalently, if and only if for every $\varepsilon > 0$ there exists a neighborhood V of $\bar{x}$ such that for all $x \in V$ and all $d \in \Re^n$,

$$\| G'(x; d) - G'(\bar{x}; d) \| \leq \varepsilon \|d\|.$$

In particular, if G is strongly F-differentiable at $\bar{x}$, then for every $d \in \Re^n, G'(\cdot; d)$ is a continuous function in its first argument at $\bar{x}$.

We note that the limit condition in part (c) is stronger than the limit condition in the definition of the directional derivative $G'(\bar{x}; d)$. The locally Lipschitz continuity of G at $\bar{x}$ is essential for (3) to hold. In what follows, we give a consequence of Proposition **4.2.2** that will be needed later.

4.2.3 Corollary. Let $G : W \to \Re^m$ be B-differentiable at $\bar{x} \in W \subseteq \Re^n$. If $\{x^k\} \subset W$ and $\{\tau_k\} \subset \Re_{++}$ are sequences of vectors and positive scalars converging respectively to $\bar{x}$ and zero and if

$$\lim_{k \to \infty} \frac{x^k - \bar{x}}{\tau_k} = d,$$

then

$$\lim_{k \to \infty} \frac{G(x^k) - G(\bar{x})}{\tau_k} = G'(\bar{x}; d). \tag{4}$$

Proof. We have

$$\frac{G(x^k) - G(\bar{x}) - \tau_k G'(\bar{x};d)}{\tau_k} = \frac{\|x^k - \bar{x}\|}{\tau_k} \frac{G(x^k) - G(\bar{x}) - \tau_k G'(\bar{x};d)}{\|x^k - \bar{x}\|}$$

$$= \frac{\|x^k - \bar{x}\|}{\tau_k} \frac{G(x^k) - G(\bar{x}) - G'(\bar{x}; x^k - \bar{x})}{\|x^k - \bar{x}\|} + G'(\bar{x}; (x^k - \bar{x})/\tau_k) - G'(\bar{x};d).$$

Taking limits and using Proposition **4.2.2**(b) and (c), we obtain the desired limit (4).   **Q.E.D.**

4.2.2 *Key implicit assumption*

Let $(\bar{x}, \bar{y})$ be a given point in $\mathcal{F}$. The key implicit function assumption for the problem (1) near this point is the following.

(BIF) There exist an open neighborhood $U \times V$ of $(\bar{x}, \bar{y})$ and a Lipschitz continuous function $y : U \cap X \to V$ such that $y(\bar{x}) = \bar{y}$, $y(\cdot)$ is directionally differentiable (thus B-differentiable) at $\bar{x}$, and for each $x \in U \cap X$, $y(x)$ is the unique solution in V of the VI $(F(x, \cdot), C(x))$.

We call the above function $y(x)$ an implicit (solution) function of the parametric VI $(F(x, \cdot), C(x))$ at the pair $(\bar{x}, \bar{y})$. Note that under BIF, the VI $(F(x, \cdot), C(x))$ may still have multiple solutions; BIF only ensures that when x is near $\bar{x}$, this VI has a unique solution that is near $\bar{y}$.

4.2.4 Example. Let $F(x,y) \equiv x + 1 - y$ and $C(x) \equiv \Re_+$ for $(x,y) \in \Re^2$. For each $x \geq -1$, the LCP:

$$y \geq 0, \quad x + 1 - y \geq 0, \quad y(x + 1 - y) = 0,$$

has two solutions: 0 and $x + 1$. At $(\bar{x},\bar{y}) \equiv (0,0)$, $y(x) \equiv 0$ satisfies BIF. At $(x',y') \equiv (-1,0)$, BIF is violated because when x is near x', the above LCP has multiple solutions contained in every neighborhood of (x',y').

We postpone further discussion about BIF, including its validity, until the next section. For the moment, we concentrate on the tangent cone to $\mathcal{F}$ at $(\bar{x},\bar{y})$ and the associated first-order optimality conditions.

Following standard notation, we denote the directional derivative of the implicit function $y(\cdot)$ at $\bar{x}$ along a direction $dx \in \Re^n$ by $y'(\bar{x};dx)$. If the objective function f of (1) is F-differentiable, using Corollary **4.2.3** we can establish the following chain rule of directional differentiation (similar to Corollary **4.2.3**): with $\tilde{f}(x) \equiv f(x,y(x))$,

$$
\tilde{f}'(\bar{x};dx) = \lim_{\tau \to 0+} \frac{f(\bar{x} + \tau\,dx, y(\bar{x} + \tau dx)) - f(\bar{x},\bar{y})}{\tau}
$$

$$
= \nabla_x f(\bar{x},\bar{y})^T dx + \nabla_y f(\bar{x},\bar{y})^T y'(\bar{x};dx). \tag{5}
$$

The following lemma characterizes the tangent cone of a feasible set at the pair $(\bar{x},\bar{y}) \in \mathcal{F}$ in terms of the implicit function $y(\cdot)$ in condition BIF.

4.2.5 Lemma. Let X be convex and $\bar{z} \equiv (\bar{x},\bar{y}) \in \mathcal{F}$ be arbitrary. Under assumption BIF,

$$
\mathcal{T}((\bar{x},\bar{y});\mathcal{F}) = \{(dx,dy) \in \mathcal{T}(\bar{x};X) \times \Re^m : dy = y'(\bar{x};dx)\}. \tag{6}
$$

Consequently, $(\bar{x},\bar{y})$ is a stationary point of (1) if and only if for all vectors $dx \in \mathcal{T}(\bar{x};X)$,

$$
\nabla_x f(\bar{x},\bar{y})^T dx + \nabla_y f(\bar{x},\bar{y})^T y'(\bar{x};dx) \geq 0. \tag{7}
$$

Proof. Write $\mathcal{LS}(\bar{x}) \equiv \{(dx,y'(\bar{x};dx)) : dx \in \mathcal{T}(\bar{x};X)\}$. Let (dx,dy) be in $\mathcal{T}(\bar{z};\mathcal{F})$; so

$$
(dx,dy) = \lim_{k \to \infty} \left(\frac{x^k - \bar{x}}{\tau_k}, \frac{y^k - \bar{y}}{\tau_k} \right)
$$

where $\{(x^k, y^k)\} \subset \mathcal{F}$ converges to $(\bar{x}, \bar{y})$ and $\{\tau_k\} \to 0+$. Then dx must be in $\mathcal{T}(\bar{x}; X)$. Since y^k solves the VI $(F(x^k, \cdot), C(x^k))$ for all k, BIF implies that $y^k = y(x^k)$ for all large k, and $\bar{y} = y(\bar{x})$. So

$$dy = \lim_{k \to \infty} \frac{y(x^k) - y(\bar{x})}{\tau_k} = y'(\bar{x}; dx),$$

by Corollary **4.2.3**. Hence $(dx, dy) \in \mathcal{LS}(\bar{x})$.

Conversely, suppose that $(dx, dy) \in \mathcal{LS}(\bar{x})$. So $dx = \lim_k (x^k - \bar{x})/\tau_k$ for some $\{x^k\} \subset X$ converging to $\bar{x}$ and $\{\tau_k\} \to 0+$; and $dy = y'(\bar{x}; dx)$. By Corollary **4.2.3**, we see that

$$y'(\bar{x}; dx) = \lim_{k \to \infty} \frac{y(x^k) - \bar{y}}{\tau_k};$$

thus $(dx, dy) \in \mathcal{T}(\bar{z}; \mathcal{F})$. The last assertion of the lemma requires no proof. **Q.E.D.**

4.2.6 Remark. If X is polyhedral and the implicit function $y(x)$ is only directionally differentiable at $\bar{x}$ (without $y(x)$ being Lipschitz near $\bar{x}$), then the second part of the proof of the above lemma remains valid (because by the polyhedrality of X, we can replace the sequence $\{x^k\}$ by $\{\bar{x} + \tau_k dx\}$), but the first part is in jeopardy. In other words, with just the directional differentiability of $y(x)$ at $\bar{x}$ (and not the B-differentiability), we still have

$$\mathcal{T}((\bar{x}, \bar{y}); \mathcal{F}) \supseteq \{(dx, dy) \in \mathcal{T}(\bar{x}; X) \times \Re^m : dy = y'(\bar{x}; dx)\},$$

but the equality requires at least some kind of continuity of $y'(\bar{x}; dx)$ in the argument dx (which is guaranteed if $y(x)$ is Lipschitz continuous near $\bar{x}$).

Lemma **4.2.5** has reduced the problem of calculating the tangent cone $\mathcal{T}((\bar{x}, \bar{y}); \mathcal{F})$ to that of calculating the directional derivative function $y'(\bar{x}; \cdot)$ of the implicit function $y(x)$ at $\bar{x}$, if the tangent cone $\mathcal{T}(\bar{x}; X)$ is simple. Under appropriate assumptions, it is not surprising from the results in Section **3.2** that $y'(\bar{x}; dx)$ can be computed by solving certain AVIs; see Subsection **4.2.5** for details.

4.2.3 *Piecewise smooth functions and degree theory*

In the next subsection, we give sufficient conditions for BIF to hold. Actually these conditions will guarantee not only the existence of $y(x)$, but also that $y(x)$ is piecewise smooth, PC^1 to be precise, from which it follows that $y(x)$ is B-differentiable at $\bar{x}$. Moreover, they will yield computational formulas for the directional derivative $y'(\bar{x}; dx)$.

In this subsection, we introduce the tools needed for the derivation of the aforementioned results. Specifically, our ultimate goal is to apply an implicit function theorem to the parametric normal equation that characterizes the VI $(F(x, \cdot), C(x))$ for x near $\bar{x}$; see Subsection **1.3.1** for the definition of the normal map and corresponding equation. As we will show, the normal map is a PC^1 function under an inequality representation of the set $C(x)$ and appropriate assumptions; we find it useful to establish first an implicit function theorem for a general PC^1 parametric equation.

The Habilitation thesis by Scholtes [258] contains an extensive study of the class of piecewise differentiable functions and discusses applications to normal maps. An implicit function theorem for these functions was obtained therein. Using degree theory, we derive a similar theorem but under a different assumption. Further sensitivity analysis results for piecewise smooth functions can be found in [241].

The following definition is the first step to the development of our prerequisite theory.

4.2.7 Definition. A function $G : U \subseteq \Re^n \to \Re^m$ is said to be PC^r for some positive integer r at (or near) a vector $\bar{x} \in U$ if G is continuous at $\bar{x}$ and there exists an open neighborhood $\bar{U} \subseteq U$ of $\bar{x}$ and a finite family of r times continuously differentiable functions (i.e., C^r functions) $G^j : \bar{U} \to \Re^m$, $j = 1, \ldots, p$, such that for each $x \in \bar{U}$, $G(x) = G^j(x)$ for some j. The functions $G^1, \ldots, G^p$ are called the C^r *pieces* of G at $\bar{x}$; we also say that G is a *(continuous) selection* of these pieces near $\bar{x}$. We say that G is PC^r on U if it is PC^r at every vector in U.

Note that the pieces of a PC^r function are not unique, since we can add as many (spurious) C^r functions as we like to an existing family of pieces. In the following discussion, when we say that a PC^r function has a certain property, it is understood that this refers to the existence of a finite family of pieces (the *effective* pieces) for which this property is valid. Note also that a PC^r is completely determined near a point by a finite family of C^r functions whose domains coincide with a single open neighborhood of that point.

The paper by Chaney [45] contains an extensive study of piecewise smooth functions. Every PC^r function is clearly PC^1. Thus PC^1 functions are the most fundamental piecewise smooth functions. Within the class of PC^1 functions are the piecewise affine functions which are formally defined below.

4.2.8 Definition. A PC^1 function G with pieces $\{G^1, \ldots, G^p\}$ is said to be *piecewise linear (affine)* if each G^j is a linear (affine, i.e., linear plus constant) function on its domain U^j; if each G^j is linear with derivative ∇G^j given by the (constant) matrix M^j, we also say that the family of matrices $\{M^1, \ldots, M^p\}$ forms the pieces of G.

PC^1 functions constitute a large subclass of the class of the so-called "semismooth" functions. Inspired by the class of convex functions, Mifflin [205] introduced the class of real-valued semismooth functions. Qi and Sun [233] defined the class of vector-valued semismooth functions; although their definition is in terms of Clarke's generalized Jacobian matrices of a nonsmooth function [51], they also showed that a vector-valued function is semismooth if and only if each of its (real-valued) component functions is a semismooth function as defined by Mifflin. Based on this fact and a result of Chaney [45], it was pointed out in [226] that a PC^1 function must be semismooth.

4.2.9 Example. Let $f_j : U \subseteq \Re^n \to \Re^m$, $j = 1, \ldots, p$, be a finite family of C^r functions defined on the open set U. The function

$$f(x) \equiv \min(f_j(x) : j = 1, \ldots, p), \quad x \in U$$

is PC^r on U. This is easy to see by observing that f is continuous and, for each $x \in U$, $f(x) \in \{f_1(x), \ldots, f_p(x)\}$. If each f^j is affine, then f is piecewise affine.

Another example of a piecewise affine function is the projection map onto a polyhedron; i.e., Π_C where C is polyhedron. We shall discuss projection function further.

If G is a C^1 function near the point $\bar{x} \in \Re^n$, then the directional derivative $G'(\bar{x}; d)$ is a linear function of the direction d. An extension of this fact to a PC^1 function is contained in the following result which gives several important properties of such a function.

4.2.10 Lemma. Let $G : U \subseteq \Re^n \to \Re^m$ be a PC^1 mapping with C^1 pieces $\{G^1, \cdots, G^p\}$ near $\bar{x}$. The following statements hold.

(a) G is B-differentiable at all points near $\bar{x}$.

(b) For every $\varepsilon > 0$, there exists a $\delta > 0$ such that for all $h \in \mathbb{B}(0, \delta)$,

$$\|G'(\bar{x} + h; h) - G'(\bar{x}; h)\| \leq \varepsilon \|h\|. \tag{8}$$

(c) $G'(\bar{x}; \cdot)$ is piecewise linear with pieces $\{\nabla G^1(\bar{x}), \ldots, \nabla G^p(\bar{x})\}$.

Proof. By a result of [104, Theorem 2.3], it is established in [146] that a PC^1 function must be locally Lipschitz continuous. It can also be shown [146] that a PC^1 function is directionally differentiable; a proof can also be found in [258, §4.1]. Thus (a) holds.

Part (b) is a consequence of a result proved in Qi [232]. The third assertion of the lemma is based on the fact that for all vectors $d \in \Re^n$, $G'(\bar{x}; d) \in \{\nabla G^1(\bar{x})d, \ldots, \nabla G^p(\bar{x})d\}$. The proof of the latter fact is elementary, and is contained in the cited references.   **Q.E.D.**

In what follows, we give an example to show that it is possible for a PC^1 function to be F-differentiable at a point but not strongly F-differentiable at this point.

4.2.11 Example. Consider a function $f : \Re^2 \to \Re$ with three pieces, two of which are active in two respective balls which "kiss" at the origin, and the third which is active in the rest of $\Re^2$. Specifically, let B_1 and B_2 denote the closed balls $\mathbb{B}((1, 0), 1)$ and $\mathbb{B}((-1, 0), 1)$ respectively. Define

$$f(x_1, x_2) \equiv \begin{cases} (x_1 - 1)^2 + x_2^2 - 1 & \text{if } (x_1, x_2) \in B_1, \\ -[(x_1 + 1)^2 + x_2^2] + 1 & \text{if } (x_1, x_2) \in B_2, \\ 0 & \text{otherwise.} \end{cases}$$

It is easy to check that f is a PC^1 function. Consider the point $(0, 0)$. The gradients of the two functions on B_1 and B_2 are both equal to $(-2, 0)$ at the origin. Also the tangent cones $\mathcal{T}((0, 0); B_1)$ and $\mathcal{T}((0, 0); B_2)$ cover $\Re^2$. It follows that the function f is F-differentiable at $(0, 0)$, and its gradient there equals $(-2, 0)$.

Now suppose we approach $(0, 0)$ along the line $(0, t)$, where $t \to 0$. Since f is identically zero near $(0, t) \neq (0, 0)$, its gradient there is $(0, 0)$; hence for all $t \neq 0$, $\nabla f(0,t)^T (1,0) = 0$. However, $\nabla f(0,0)^T (1,0) = -2$. So $\nabla f(0,t)^T (1,0)$ does not converge to $\nabla f(0,0)^T (1,0)$ as $t \to 0+$; hence f is not strongly F-differentiable at $(0,0)$ by Proposition **4.2.2**(e).

A very important concept of a PC^1 function mapping $\Re^m$ into itself is defined below.

4.2.12 Definition. If $G : \Re^m \to \Re^m$ is PC^1 near $\bar{x} \in \Re^m$, we say that G is *coherently oriented* at $\bar{x}$ if there exists a family of C^1 pieces $\{G^1, \ldots, G^p\}$ of G near $\bar{x}$ such that sgn $\det \nabla G^j(\bar{x})$ is a nonzero constant independent of j, where "sgn" denotes the sign, $1, -1$, or 0, of a positive, negative, or zero scalar, respectively, and "det" denotes the determinant of a matrix.

The term "coherent orientation" may have been introduced by Kuhn and Löwen [145], for a piecewise affine map, although the property of "constant, nonzero determinantal sign" was recognized much earlier as having an important role to play in the global homeomorphic theory of general piecewise linear maps; see for example [85, 143, 260]. If G is either one of the following two piecewise linear maps

$$z \mapsto Mz^+ - z^- \quad \text{or} \quad x \mapsto \min(x, Mx)$$

for some $m \times m$ matrix M, then G is coherently oriented if and only if M has all principal minors positive, that is, M is a P matrix; see Proposition **1.3.12**.

In several papers [95, 97, 273, 259], Gowda, Sznajder, and Scholtes have studied extensively the piecewise affine mappings and many of its properties, including coherent orientation, zero sets, error bounds, and inverses of these mappings. The results they obtained are by far the most complete for this class of piecewise smooth functions.

Toward the establishment of an implicit function theorem for PC^1 functions, we briefly review the concepts of degree and index of a continuous mapping $G : \Re^m \to \Re^m$; see [162, 211, 119] for a detailed treatment of these concepts and the results given below. Let Ω be an open bounded subset of $\Re^m$ with closure $\text{cl}\,\Omega$ and boundary $\partial\Omega$. Suppose $\bar{a} \in \Re^m$ is such that $G^{-1}(\bar{a}) \cap \partial\Omega = \emptyset$, then the degree of G at $\bar{a}$ with respect to Ω is a well-defined integer which we denote as $\deg(G \,|\, \bar{a}, \Omega)$. Some basic properties of $\deg(G \,|\, \bar{a}, \Omega)$ are as follows.

1. (*Solvability property*) If $\deg(G \,|\, \bar{a}, \Omega) \neq 0$, then there is a solution of the equation $G(v) = \bar{a}$ in Ω.

2. (*Nearness property*) Suppose that $\deg(G \,|\, \bar{a}, \Omega)$ is defined. If $\tilde{G}$ is continuous on $\operatorname{cl} \Omega$ and satisfies

$$\sup_{x \in \Omega} \| G(x) - \tilde{G}(x) \| < \operatorname{dist}(\bar{a}, G(\partial \Omega)),$$

then $\deg(\tilde{G} \,|\, \bar{a}, \Omega)$ is defined and is equal to $\deg(G \,|\, \bar{a}, \Omega)$.

3. (*Homotopy invariance property*) Suppose that $H : [0, 1] \times \operatorname{cl} \Omega \to \Re^m$ is continuous and $\bar{a} \notin H(t, \partial \Omega)$ for all $t \in [0, 1]$. Then $\deg(H(0, \cdot) \,|\, \bar{a}, \Omega)$ and $\deg(H(1, \cdot) \,|\, \bar{a}, \Omega)$ are equal.

4. (*Excision property*) Suppose that $\deg(G \,|\, \bar{a}, \Omega)$ is defined. Let S be a compact subset of Ω such that there are no solutions of the equation $G(v) = \bar{a}$ in S. Then $\deg(G \,|\, \bar{a}, \Omega \backslash S)$ is defined and is equal to $\deg(G \,|\, \bar{a}, \Omega)$.

5. (*Cartesian product property*) Let Ω and Ω' be bounded open subsets of $\Re^m$ and $\Re^{m'}$ respectively. Suppose that $\deg(G \,|\, \bar{a}, \Omega)$ and $\deg(\tilde{G} \,|\, \bar{b}, \Omega')$ are defined. Then the mapping $G \times \tilde{G} : \operatorname{cl} \Omega \times \operatorname{cl} \Omega' \to \Re^m \times \Re^{m'}$ defined by $(G \times \tilde{G})(x, y) \equiv (G(x), \tilde{G}(y))$ has a degree relative to $\Omega \times \Omega'$ given by

$$\deg(G \times \tilde{G} \,|\, (\bar{a}, \bar{b}), \Omega \times \Omega') = \deg(G \,|\, \bar{a}, \Omega) \, \deg(\tilde{G} \,|\, \bar{b}, \Omega').$$

We can also define a local concept of the degree of G near a point $\bar{v}$ in its domain, called the *index* of G at $\bar{v}$. The index of G at $\bar{v}$ can be defined if G has the following property: for some open neighborhood V of $\bar{v}$, $G(x) \neq G(\bar{v})$ for all $x \in V \setminus \{\bar{v}\}$. Indeed, let $\bar{a} = G(\bar{v})$. Then for every open set $\Omega \subseteq V$ with $\bar{v} \in \Omega$, $\bar{a} \notin G(\partial \Omega)$. Thus, $\deg(G \,|\, \bar{a}, \Omega)$ is well defined; moreover, it is independent of Ω as long as $\Omega \subseteq V$ is a neighborhood of $\bar{v}$. This common degree is called the index of G at $\bar{v}$ and is denoted $\operatorname{ind}(G, \bar{v})$. We list two important properties of the index.

1. If G has an F-derivative at $\bar{v}$ and the Jacobian matrix $\nabla G(\bar{v})$ is invertible, then $\operatorname{ind}(G, \bar{v})$ is equal to $\operatorname{sgn} \det \nabla G(\bar{v})$; see [220, Lemma 2]. Thus, $\operatorname{ind}(G, \bar{v}) = \pm 1$.

2. If G (not necessarily differentiable) is injective (i.e., one-to-one) in a neighborhood of $\bar{v}$, then $\operatorname{ind}(G, \bar{v}) = \pm 1$.

The continuous mapping $G : \Omega \subseteq \Re^m \to \Re^m$ is said to be *invertible* on its domain Ω if it is bijective (i.e., injective and onto). By the invariance of domain theorem, it follows that if Ω is open and G is invertible on Ω then $G(\Omega)$ is open. If $\Omega = \Re^m$ and G is invertible on $\Re^m$, then we say that G is *globally invertible*, or just invertible. We say that G is a *(global) Lipschitzian homeomorphism* on Ω if G is invertible and Lipschitz continuous on Ω and G^{-1} is Lipschitz continuous on its domain.

We say that G is *invertible near a point* (or, locally invertible at) $\bar{v} \in \Re^m$ if G is invertible on some open neighborhood V of $\bar{v}$; note that in this case, $G(V)$ is an open neighborhood of $G(\bar{v})$. The concept of "a Lipschitzian homeomorphism near a point" is defined analogously.

The following lemma relates the notions of coherent orientation, index at a point, and invertibility for a PC^1 function and its directional derivative at a point.

4.2.13 Lemma. Let $G : \Re^m \to \Re^m$ be PC^1 near $\bar{v} \in \Re^m$. If G is coherently oriented at $\bar{v}$, then the following three statements are equivalent.

(a) $G'(\bar{v}; \cdot)$ is (globally) invertible;

(b) $\mathrm{ind}(G'(\bar{v}; \cdot), 0)$ is well defined and has value ± 1;

(c) $G'(\bar{v}; \cdot)$ is a global Lipschitzian homeomorphism on $\Re^n$.

Proof. (a) $\Rightarrow$ (b). This has been noted above for any injective function.

(b) $\Rightarrow$ (c). Let $\{G^1, \ldots, G^p\}$ be the C^1 pieces of G near $\bar{v}$ such that $\mathrm{sgn} \det \nabla G^j(\bar{v})$ is a nonzero constant independent of j. By Lemma **4.2.10**, $G'(\bar{v}; \cdot)$ is piecewise linear with pieces $\{\nabla G^1(\bar{v}), \ldots, \nabla G^p(\bar{v})\}$. Hence by [227, Theorem 4], it follows from (b) that $G'(\bar{v}; \cdot)$ is a Lipschitzian homeomorphism near the origin. Since $G'(\bar{v}; \cdot)$ is positively homogeneous, it follows that $G'(\bar{v}; \cdot)$ is a global Lipschitzian homeomorphism on $\Re^n$.

(c) $\Rightarrow$ (a). This is obvious. **Q.E.D.**

An example given in [85] shows that a PC^1 function G can be coherently oriented at a point $\bar{v}$ and yet none of the conditions (a), (b), or (c) in Lemma **4.2.13** holds.

The following is an inverse function theorem for a PC^1 function.

4.2.14 Theorem. Let $G : \Re^m \to \Re^m$ be PC^1 near $\bar{v} \in \Re^m$. If G is coherently oriented at $\bar{v}$ and $G'(\bar{v}; \cdot)$ is (globally) invertible, then the following statements hold with $\bar{a} \equiv G(\bar{v})$.

(a) There is a neighborhood $V \times U$ of $(\bar{v}, \bar{a})$ such that the restricted mapping $\tilde{G} : V \to U$, given by $\tilde{G}(v) \equiv G(v)$ for $v \in V$, is a Lipschitz homeomorphism on its domain, and $\tilde{G}^{-1}$ is PC^1.

(b) G is a Lipschitzian homeomorphism with a PC^1 local inverse, $\tilde{G}^{-1}$, near $\bar{v}$.

(c) For each vector $d \in \Re^m$,

$$(\tilde{G}^{-1})'(\bar{a}; d) = G'(\bar{v}; \cdot)^{-1}(d). \tag{9}$$

Proof. By Lemma **4.2.13**, it follows that $\mathrm{ind}(G'(\bar{v}; \cdot), 0)$ is well defined and has value ± 1. Thus by the nearness property of degree, it follows that $\mathrm{ind}(G, \bar{v})$ exists and has value ± 1; see [220]. The existence of $V \times U$ and the Lipschitzian homeomorphism of $\tilde{G}$ now follow from [227, Theorem 4]. The proof of the quoted result also shows that $V \times U$ can be chosen small enough such that $\tilde{G}^{-1}$ is PC^1. Part (b) follows from (a) in view of the fact that all PC^1 functions are locally Lipschitz.

To show (9), note that $\tilde{G}^{-1}$ is B-differentiable since it is PC^1. Let $d \in \Re^m$, and $v(\tau) \equiv \tilde{G}^{-1}(\bar{a} + \tau d)$ for $\tau > 0$ sufficiently small. Then $v(0) = \bar{v}$ and $v'(0; 1)$ exists and equals $(\tilde{G}^{-1})'(\bar{a}; d)$. Thus

$$d = \lim_{\tau \to 0+} \frac{\tilde{G}(v(\tau)) - \tilde{G}(v(0))}{\tau} = \lim_{\tau \to 0+} \frac{G(v(\tau)) - G(v(0))}{\tau}$$

$$= G'(v(0); v'(0; 1)) = G'(\bar{v}; (\tilde{G}^{-1})'(\bar{a}; d)),$$

where the third equality is a consequence of Corollary **4.2.3**, using the B-differentiability of G and the function $v(\cdot)$. Equation (9) is an immediate consequence of the last equality and invertibility of $G'(\bar{v}; \cdot)$.   **Q.E.D.**

We next apply the above result to a PC^1 function of two arguments, $H : \Re^n \times \Re^m \to \Re^m$, to get an implicit function theorem. If H is PC^1 near $(\bar{x}, \bar{v})$, then it is B-differentiable at $(\bar{x}, \bar{v})$, hence partially B-differentiable with respect to v at $(\bar{x}, \bar{v})$; in particular, the limit

$$H_v'(\bar{x}, \bar{v}; dv) \equiv \lim_{\tau \to 0+} \frac{H(\bar{x}, \bar{v} + \tau dv) - H(\bar{x}, \bar{v})}{\tau}$$

exists and is equal to $H'((\bar{x}, \bar{v}); (0, dv))$. This partial B-derivative plays a central role in the following result.

4.2.15 Theorem. Let $H : \Re^n \times \Re^m \to \Re^m$ be PC^1 near $(\bar{x}, \bar{v}) \in \Re^n \times \Re^m$ with C^1 pieces $H^1, \ldots, H^J$ defined on an open neighborhood of $(\bar{x}, \bar{v})$; let $\bar{a} \equiv H(\bar{x}, \bar{v})$. If sgn $\det \nabla_v H^j(\bar{x}, \bar{v})$ is a nonzero constant independent of $j = 1, \ldots, J$ and $H'_v(\bar{x}, \bar{v}; \cdot)$ is invertible, then:

(a) there exist an open neighborhood $\tilde{U} \times \tilde{V}$ of $(\bar{x}, \bar{v})$ and a PC^1 function $v : \tilde{U} \to \tilde{V}$ such that $v(\bar{x}) = \bar{v}$ and, for each $x \in \tilde{U}$, $v = v(x)$ is the unique solution in $\tilde{V}$ of $H(x, v) = \bar{a}$.

(b) For each $dx \in \Re^n$,

$$v'(\bar{x}; dx) = H'((\bar{x}, \bar{v}); (dx, \cdot))^{-1}(0). \tag{10}$$

Proof. We show that Theorem **4.2.14** can be applied to the mapping $G : (x, v) \mapsto (x, H(x, v))$ at the point $(\bar{x}, \bar{v})$.

First, G is PC^1 near $(\bar{x}, \bar{v})$ with pieces given by

$$G^j(x, v) \equiv (x, H^j(x, v)), \quad \text{for} \quad j = 1, \ldots, J.$$

Second,

$$\nabla G^j(x, v) = \begin{bmatrix} I & 0 \\ \nabla_x H^j(x, v) & \nabla_v H^j(x, v) \end{bmatrix},$$

hence sgn $\det \nabla G^j(\bar{x}, \bar{v}) = $ sgn $\det \nabla_v H^j(\bar{x}, \bar{v})$, which is a nonzero constant independent of j.

The remaining condition to be verified is the invertibility of $G'((\bar{x}, \bar{v}); \cdot)$. Assume this holds. Then according to Theorem **4.2.14**, the restriction $\tilde{G}$ of G to some neighborhood of $(\bar{x}, \bar{v})$ is a Lipschitzian homeomorphism such that $\tilde{G}^{-1}$ is PC^1 and, for each $(dx, da) \in \Re^n \times \Re^m$,

$$(\tilde{G}^{-1})'((\bar{x}, \bar{a}); (dx, da)) = G'((\bar{x}, \bar{v}); \cdot)^{-1}(dx, da). \tag{11}$$

This immediately yields the existence of a neighborhood $\tilde{U} \times \tilde{V}$ of $(\bar{x}, \bar{a})$ such that, for $x \in \tilde{U}$, the restricted mapping $\tilde{H}(x, \cdot) : \tilde{V} \to H(x, \tilde{V})$, defined by $\tilde{H}(x, v) \equiv H(x, v)$, is a Lipschitzian homeomorphism with PC^1 inverse. Indeed given $(x, a) \in \tilde{U} \times \tilde{V}$,

$$\tilde{G}^{-1}(x, a) = (x, \tilde{H}(x, \cdot)^{-1}(a)). \tag{12}$$

Let
$$v(x) \equiv \tilde{H}(x, \cdot)^{-1}(\bar{a}), \quad \text{for} \quad x \in \tilde{U}.$$

Then $v(\cdot)$ satisfies all the requirements except, possibly, equation (10). Using (12) with $(x, a) = (\bar{x}, \bar{a})$, we calculate the B-derivative:

$$(\tilde{G}^{-1})'((\bar{x}, \bar{a}); (dx, 0)) = (dx, v'(\bar{x}; dx)). \tag{13}$$

Also by definition of G,

$$G'((\bar{x}, \bar{v}); (dx, dv)) = (dx, \tilde{H}'((\bar{x}, \bar{v}); (dx, dv))). \tag{14}$$

Thus for any dv,

$$
\begin{aligned}
dv = v'(\bar{x}; dx) \iff & (dx, dv) = (\tilde{G}^{-1})'((\bar{x}, \bar{a}); (dx, 0)) && \text{by (13)} \\
\iff & G'((\bar{x}, \bar{v}); (dx, dv)) = (dx, 0) && \text{by (11)} \\
\iff & \tilde{H}'((\bar{x}, \bar{v}); (dx, dv)) = 0 && \text{by (14)} \\
\iff & dv = \tilde{H}'((\bar{x}, \bar{v}); (dx, \cdot))^{-1}(0),
\end{aligned}
$$

confirming (10).

We now prove that $G'((\bar{x}, \bar{v}); \cdot)$ is invertible. By Lemma **4.2.13**, it suffices to show that $\mathrm{ind}(G'((\bar{x}, \bar{v}); \cdot), 0)$ is well defined with value ± 1. Define two mappings dG and δG for $(dx, dv) \in \Re^n \times \Re^m$:

$$dG(dx, dv) \equiv (dx, H'((\bar{x}, \bar{v}); (dx, dv)))$$

$$\delta G(dx, dv) \equiv (dx, H_v'(\bar{x}, \bar{v}; dv)).$$

So $dG = G'((\bar{x}, \bar{v}); \cdot)$. Also observe that δG is invertible and continuous because $H_v'(\bar{x}, \bar{v}; \cdot)$ is invertible and continuous, hence $\mathrm{ind}(\delta G, 0) = \pm 1$. Therefore from invertibility of δG and the definition of index, it follows that $\deg(\delta G \mid 0, W) = \pm 1$ for any bounded open neighborhood W of $0 \in \Re^{n+m}$. We next use the homotopy invariance property outlined above to show that $\deg(dG \mid 0, W) = \deg(\delta G \mid 0, W)$.

For a point $(dx, dv, t) \in \Re^n \times \Re^m \times [0, 1]$, let

$$
\begin{aligned}
\Gamma(dx, dv, t) \equiv \; & t\, dG(dx, dv) + (1 - t)\, \delta G(dx, dv) \\
= \; & (dx, tH'((\bar{x}, \bar{v}); (dx, dv)) + (1 - t)H_v'(\bar{x}, \bar{v}; dv)).
\end{aligned}
$$

If $\Gamma(dx, dv, t) = 0$ then $dx = 0$, hence $H'((\bar{x}, \bar{v}); (dx, dv)) = H'_v(\bar{x}, \bar{v}; dv)$ and

$$0 = \Gamma(dx, dv, t) = (0, H'_v(\bar{x}, \bar{v}; dv)).$$

Invertibility of $H'_v(\bar{x}, \bar{v}; \cdot)$ implies $dv = 0$. Thus $\Gamma(dx, dv, t)$ is not zero for any (dx, dv) on the boundary of W and $t \in [0, 1]$. It follows from the homotopy invariance property that $\deg(dG \,|\, 0, W)$ is well defined and equal to $\deg(\delta G \,|\, 0, W) = \pm 1$. Furthermore, since W is an arbitrary bounded neighborhood of $0 \in \Re^{n+m}$, taking $t = 1$ shows $(dx, dv) = 0$ is the only solution of $dG(dx, dv) = 0$. Hence the index of dG at 0 must be well defined and equal to ± 1, as desired.   **Q.E.D.**

Piecewise linearity of solutions to parametric QPs

One of the key elements in the next subsection is the demonstration that the projection map $\Pi_{C(x)}(v)$ is a PC^1 function of (x, v) provided that the set $C(x)$ satisfies certain assumptions. Since this result requires nontrivial arguments that might obscure the important ideas, in what follows we consider a simple situation which will highlight the essential steps needed for the subsequent development.

Consider a strictly convex quadratic program:

$$\begin{aligned}
\text{minimize} \quad & c^T x + \tfrac{1}{2} x^T Q x \\
\text{subject to} \quad & Ax \le b,
\end{aligned} \tag{15}$$

where $c \in \Re^n$, $Q \in \Re^{n \times n}$ is symmetric positive definite, $A \in \Re^{m \times n}$, and $b \in \Re^m$. (The notation in this discussion is independent of the notation in the discussion of MPEC.) For simplicity we assume that A has full row rank and that $A\hat{x} < 0$ for some vector $\hat{x} \in \Re^n$. The latter assumption implies that for all vectors $b \in \Re^m$, the above program is feasible. Thus (15) has a unique globally optimal solution and a unique multiplier vector which we denote as $x(\omega)$ and $\lambda(\omega)$ respectively, where $\omega \equiv (b, c)$. When Q is the identity matrix, $x(\omega) = \Pi_C(-c)$ where $C \equiv \{x \in \Re^n : Ax \le b\}$. When Q is not the identity matrix, $x(\omega)$ is the projection of the vector $-Q^{-1}c$ onto the same polyhedron C under the vector norm induced by Q: $\|x\|_Q \equiv (x^T Q x)^{1/2}$; thus in this case the quadratic program (15) can be thought of as a "skewed projection" problem.

We claim that $(x(\omega), \lambda(\omega))$ is a piecewise linear function of ω. For this purpose, we need to demonstrate two things: (i) $(x(\omega), \lambda(\omega))$ is continuous in ω, and (ii) $(x(\omega), \lambda(\omega))$ is composed of finitely many linear pieces for $\omega \in \Re^{n+m}$.

The pair $(x(\omega), \lambda(\omega))$ satisfies the following KKT conditions:

$$0 = c + Qx(\omega) + A^T \lambda(\omega)$$

$$v(\omega) = b - Ax(\omega) \qquad (16)$$

$$(v(\omega), \lambda(\omega)) \geq 0, \quad v(\omega) \circ \lambda(\omega) = 0.$$

Since Q is positive definite and A has full row rank, for each index set $\alpha \subseteq \{1, \ldots, m\}$, the matrix

$$M(\alpha) \equiv \left[\begin{array}{cc} Q & (A_{\alpha \cdot})^T \\ -A_{\alpha \cdot} & 0 \end{array} \right]$$

is nonsingular, where $A_{\alpha \cdot}$ denotes the rows of A indexed by α. For a fixed but arbitrary $\omega = (b, c)$, let

$$\alpha(\omega) \equiv \{i : (Ax(\omega) = b)_i\}.$$

Clearly we have (writing α for $\alpha(\omega)$),

$$\left(\begin{array}{c} x(\omega) \\ \lambda(\omega)_\alpha \end{array} \right) = -M(\alpha)^{-1} \left(\begin{array}{c} c \\ b_\alpha \end{array} \right).$$

Along with the fact that there are only finitely many index subsets α, the above expression shows that $(x(\omega), \lambda(\omega))$ is composed of finitely many linear pieces. It is useful to point out that $x(\omega)$ is actually the unique global solution of the following equality constrained quadratic program whose constraints are obtained by fixing those inequalities in (15) corresponding to the binding constraints at $x(\omega)$ as equations:

$$\text{minimize} \quad c^T x + \tfrac{1}{2} x^T Q x$$

$$\text{subject to} \quad (Ax = b)_{\alpha(\omega)};$$

in addition, $x(\omega)$ remains the unique global solution of the latter program with $\alpha(\omega)$ replaced by any index set α' satisfying

$$\{i : \lambda(\omega)_i > 0\} \subseteq \alpha' \subseteq \alpha(\omega).$$

We now briefly argue that $(x(\omega), \lambda(\omega))$ is a Lipschitz continuous function of ω. Since A has full row rank, we deduce

$$\lambda(\omega) = -(AA^T)^{-1}A(c + Qx(\omega)).$$

By (16), it follows that for $\omega \equiv (b, c)$ and $\omega' \equiv (b', c')$,

$$(x(\omega) - x(\omega'))^T Q(x(\omega) - x(\omega'))$$

$$\leq (x(\omega) - x(\omega'))^T(c' - c) + (\lambda(\omega) - \lambda(\omega'))^T(b' - b)$$

$$= (x(\omega) - x(\omega'))^T(c' - c) + (x(\omega') - x(\omega))^T QA^T(AA^T)^{-1}(b' - b).$$

Thus if $\sigma > 0$ denotes the smallest eigenvalue of Q, then by the Cauchy-Schwartz inequality applied to the right-hand side, we easily deduce

$$\|x(\omega) - x(\omega')\| \leq \sigma^{-1}\sqrt{2}\,\max(1, \|QA^T(AA^T)^{-1}\|)\,\|\omega' - \omega\|,$$

which establishes the Lipschitzian property of the solution function $x(\omega)$; this property in turn implies the same for the multiplier function $\lambda(\omega)$.

Summarizing the above analysis, we conclude that $(x(\omega), \lambda(\omega))$ is a piecewise linear function of ω. The full row rank assumption of A is useful only for the uniqueness, and thus the piecewise linearity property, of the multiplier $\lambda(\omega)$. If A is deficient in row rank, then the optimal solution $x(\omega)$ will remain a piecewise linear function of ω, although the optimal multipliers $\lambda(\omega)$ need no longer be unique. A corollary of this conclusion is that for any matrix $N \in \Re^{n \times n}$ and any polyhedral set $C \subseteq \Re^n$, the normal map $N_C \equiv N \circ \Pi_C + I - N$ is always piecewise affine (piecewise linear if C is cone), regardless of the representation of C.

As far as the KKT system (16) is concerned, the symmetry of Q plays no role; as long as Q is positive definite (not necessarily symmetric) and A has full row rank and the vector $\hat{x}$ exists, the system (16) continues to have a unique solution $(x(\omega), \lambda(\omega))$ which is PC^1 function of ω. (In this case, the scalar σ will be the smallest eigenvalue of the symmetric part of A; the above remark about the full row rank property of A remains relevant.) In the case where A is the negative identity matrix, so that (16) becomes an LCP, the positive definiteness of Q can be weakened to the property of being a P matrix.

The important ideas contained in the above discussion are now summarized. The positive definiteness (or the P-property in the case of the

LCP) of Q and the linear independency of certain rows of A are the two most important factors that contribute to the piecewise linearity of the solution function of the parametric KKT system (16) with (b, c) considered as the parameter. In essence, these two factors remain key to the subsequent analysis that deals with the parametric projection map $\Pi_{C(\cdot)}(\cdot)$. The latter analysis is complicated by the fact that the set-valued map $C(\cdot)$ is defined by nonlinear differentiable (as opposed to simple linear) inequalities that do not necessarily satisfy the LICQ. Due to the generality that our analysis allows, the multipliers in the projection problem are no longer guaranteed unique; this nonuniqueness of the multipliers (which is absent in our above discussion) adds considerable technical difficulty. Finally, since we are no longer dealing with polyhedral sets, piecewise linearity is lost and piecewise smoothness becomes our goal.

4.2.4 *Existence of a piecewise smooth implicit function*

We now return to deal with the assumption BIF. To show the local existence and uniqueness of an implicit PC^1 solution function $y(x)$, we need to introduce some conditions on the lower-level VI $(F(x, \cdot), C(x))$. First assume, for each x in X near $\bar{x} \in X$, that $C(x)$ is given by inequality constraints as in (1.3.3):

$$C(x) \equiv \{y \in \Re^m : g(x, y) \leq 0\},$$

where $g : \Re^{n+m} \to \Re^\ell$ is continuously differentiable, and each component function $g_i(x, \cdot)$ is convex (in y) for each fixed x near $\bar{x}$. Equality affine constraints are omitted only to simplify the exposition. A more general treatment, which includes equality constraints and relaxes the convexity assumption on $C(x)$ by allowing "local" solutions of the lower-level VI, is possible. We also assume that F is a C^1 function, and that $\bar{y}$ solves the VI $(F(\bar{x}, \cdot), C(\bar{x}))$.

In Subsection **1.3.2**, the set of active constraint indices corresponding to (x, y) is denoted $\mathcal{I}(x, y) \equiv \{i : g_i(x, y) = 0\}$. Since we are only interested in the case $y = y(x)$, where existence of the function $y(x)$ has yet to be shown, we simplify the notation here and below by the minor abuse of dropping $y(x)$:

$$\mathcal{I}(x) \equiv \mathcal{I}(x, y(x)).$$

Also recall the definition of the KKT multiplier set $M(x,y)$ from (1.3.4). Consistent with the notation $\mathcal{I}(x)$, let

$$M(x) \equiv M(x, y(x)).$$

Explicitly, $M(x)$ is the set of vectors $\lambda \in \Re^\ell$ such that

$$F(x, y(x)) + \sum_{i=1}^\ell \lambda_i \nabla_y g_i(x, y(x)) = 0,$$
$$\lambda_i \geq 0, \quad g_i(x, y(x)) \leq 0, \quad \lambda_i g_i(x, y(x)) = 0, \quad \text{for } i = 1, \ldots, \ell. \tag{17}$$

Our next two assumptions are the MFCQ and CRCQ at the pair $(\bar{x}, \bar{y})$. In Subsection **1.3.2**, we assume that these CQs hold at all pairs (x, y) in $Z \cap \mathrm{Gr}(C)$; here we only need these two CQs to hold at the given pair $(\bar{x}, \bar{y})$. For ease of reference, we state our assumptions specifically as follows.

(MFCQ) There exists a vector $v \in \Re^n$,

$$\nabla_y g_i(\bar{x}, \bar{y})^T v < 0, \quad \text{for all } i \in \mathcal{I}(\bar{x}).$$

(CRCQ) There is a neighborhood W of $(\bar{x}, \bar{y})$ such that for each $\mathcal{I} \subseteq \mathcal{I}(\bar{x})$, the rank of the family of gradient vectors $\{\nabla_y g_i(x, y) : i \in \mathcal{I}\}$ is a constant independent of (x, y) in W (but dependent on $\mathcal{I}$).

Since $y(x)$, if it exists, solves the VI $(F(x, \cdot), C(x))$, then according to [93], MFCQ holding at $\bar{x}$ implies that the multiplier set $M(x)$ is a nonempty polyhedron that is bounded uniformly for all x near $\bar{x}$; that is, there exist a constant $\gamma > 0$ and a neighborhood V_0 of $\bar{x}$ such that for all $x \in V_0$, $M(x) \neq \emptyset$; moreover, for all $\lambda \in M(x)$, $\|\lambda\| \leq \gamma$.

Our fourth and last assumption, which we call the *strong coherent orientation condition* (SCOC), is the cornerstone of our bid to show the desired properties of $y(x)$. Before stating this assumption, we need to introduce some notation. For each vector a, let $\mathrm{supp}(a) \equiv \{i : a_i \neq 0\}$ be the *support* of a. We now define an important family of index subsets of $\mathcal{I}(\bar{x})$. Specifically, let $\mathcal{B}(\bar{x})$ be the family consisting of index sets $\mathcal{I} \subseteq \mathcal{I}(\bar{x})$ for which there exists a vector $\lambda \in M(\bar{x})$ such that $\mathrm{supp}(\lambda) \subseteq \mathcal{I}$ and the gradient vectors

$$\{\nabla_y g_i(\bar{x}, \bar{y}) : i \in \mathcal{I}\} \tag{18}$$

are linearly independent. Due to its central role in the SCOC, we call $\mathcal{B}(x)$ the *SCOC family of active index sets*. Clearly, $\mathcal{B}(\bar{x})$ is a finite, nonempty family: finite because there are only finitely many constraints; nonempty because any multiplier in the set of extreme points $M^e(\bar{x})$ of $M(\bar{x})$ would easily yield a desired index set $\mathcal{I}$ with the stated properties. Let us label the index sets in $\mathcal{B}(\bar{x})$; specifically, let

$$\mathcal{B}(\bar{x}) \equiv \{\mathcal{I}^1, \ldots, \mathcal{I}^J\}.$$

For each $j = 1, \ldots, J$, the multiplier $\lambda^j \in M(\bar{x})$ satisfying the defining properties of $\mathcal{I}^j$ is unique and must belong to $M^e(\bar{x})$, by the linear independence of the vectors (18). Conversely, given a multiplier $\lambda \in M^e(\bar{x})$, there may be multiple index sets $\mathcal{I} \in \mathcal{B}(\bar{x})$ that contain supp(λ). Hence $J \geq |M^e(\bar{x})|$. In the special case where the pair $(\bar{x}, \bar{y})$ is strongly nondegenerate, then $J = |M(\bar{x})| = 1$. In general, for each $j = 1, \ldots, J$, we define the matrix

$$\Lambda^j \equiv \left[\begin{array}{cc} \nabla_y L(\bar{x}, \bar{y}, \lambda^j) & \nabla_y g_{\mathcal{I}^j}(\bar{x}, \bar{y})^T \\ -\nabla_y g_{\mathcal{I}^j}(\bar{x}, \bar{y}) & 0 \end{array} \right]. \tag{19}$$

The fundamental importance of the matrices Λ^j is well recognized in the literature of parametric nonlinear programming. Kojima, in his seminal paper [140] which considers the strong stability of a KKT triple to a parametric nonlinear program, has made heavy use of these matrices. In particular, by Theorem 3.5 in this reference, it follows that if $\nabla_y F(\bar{x}, \bar{y})$ is symmetric, then for each fixed but arbitrary $j \in \{1, \ldots, J\}$, the sign of det Λ^j is equal to that of det $(B^j)^T \nabla_y L(\bar{x}, \bar{y}, \lambda^j) B^j$, where B^j is a matrix whose columns form a basis of the null space of $\nabla_y g_{\mathcal{I}^j}(\bar{x}, \bar{y})$.

We may now state our last assumption. The requirement of constant determinantal signs in this assumption is reminiscent of the concept of coherent orientation defined in Definition **4.2.12**. We will discuss the connection of this assumption with other conditions in detail in a later subsection, including the reason for calling it SCOC; see Subsection **4.2.7** and [111].

(**SCOC**) The J matrices defined above:

$$\{\Lambda^j : j = 1, \ldots, J\},$$

have the same nonzero determinantal sign.

As noted above, if $(\bar{x}, \bar{y})$ is strongly nondegenerate, then $|J| = 1$; in this case, with $M(\bar{x}) = \{\bar{\lambda}\}$, the SCOC is equivalent to the requirement that the single matrix

$$\begin{bmatrix} \nabla_y L(\bar{x}, \bar{y}, \bar{\lambda}) & \nabla_y g_{\mathcal{I}}(\bar{x}, \bar{y})^T \\ -\nabla_y g_{\mathcal{I}}(\bar{x}, \bar{y}) & 0 \end{bmatrix}, \quad \text{where } \mathcal{I} \equiv \mathcal{I}(\bar{x}), \quad (20)$$

is nonsingular. In turn, the nonsingularity of this matrix is equivalent to two conditions (i) the vectors $\{\nabla_y g_i(\bar{x}, \bar{y}) : i \in \mathcal{I}\}$ are linearly independent (this is the LICQ for the VI $(F(\bar{x}, \cdot), C(\bar{x}))$ at the solution $\bar{y}$), and (ii) the matrix $U^T \nabla_y L(\bar{x}, \bar{y}, \bar{\lambda}) U$ is nonsingular, where the columns of U form an orthonormal basis of the null space of the vectors in (i). Since the strong nondegeneracy of $(\bar{x}, \bar{y})$ already implies (i), it follows that in this special case, the SCOC is equivalent to the nonsingularity of the matrix $U^T \nabla_y L(\bar{x}, \bar{y}, \bar{\lambda}) U$.

As another illustration of the SCOC, we consider the case of the NCP constrained MP which has $g(x, y) \equiv -y$. This case is explored in more detail in the discussion of the "strong regularity condition" in Sections **5.3** and **5.4**. For simplicity, we assume here that $(\bar{x}, \bar{y})$ satisfies

$$F(\bar{x}, \bar{y}) = \bar{y} = 0.$$

In this setting, the SCOC reduces to the condition that for all subsets α of $\{1, \ldots, m\}$, the matrices

$$\begin{bmatrix} \nabla_y F(\bar{x}, \bar{y}) & (I_{\alpha \cdot})^T \\ -I_{\alpha \cdot} & 0 \end{bmatrix}$$

have the same nonzero determinantal sign, where $I_{\alpha \cdot}$ denotes the rows of the identity matrix indexed by α. Since for $\alpha = \{1, \ldots, m\}$, the determinant of the above matrix is equal to 1, and for α equal to a proper subset of $\{1, \ldots, m\}$, the determinant of the above matrix is equal to that of $(\nabla_y F(\bar{x}, \bar{y}))_{\bar{\alpha}\bar{\alpha}}$ where $\bar{\alpha}$ is the complement of α in $\{1, \ldots, m\}$, it follows that SCOC holds in this case if and only if all principal submatrices of $\nabla_y F(\bar{x}, \bar{y})$ have positive determinants; in other words, if and only if $\nabla_y F(\bar{x}, \bar{y})$ is a P-matrix.

The following theorem is the main result of this subsection; it establishes the validity of the key implicit function assumption BIF under the conditions described above. In Subsection **4.2.5** we show how the directional derivative $y'(x; dx)$ can be calculated and discuss the F-differentiability of the implicit function $y(x)$ at $\bar{x}$.

4.2.16 Theorem. Suppose $C(x) \equiv \{y : g(x, y) \leq 0\}$ where g is C^2 and each component function $g_i(x, y)$ is convex in y; the function F is C^1; and $(\bar{x}, \bar{y}) \in \Re^n \times \Re^m$ is such that $\bar{y}$ solves VI $(F(\bar{x}, \cdot), C(\bar{x}))$. If MFCQ, CRCQ and SCOC hold, then there exist a neighborhood $U \times V$ of $(\bar{x}, \bar{y})$ and a PC^1 function $y : U \to V$ such that $y(\bar{x}) = \bar{y}$, and $y(x)$ is the unique solution in V of VI $(F(x, \cdot), C(x))$ for every $x \in U$.

The general idea of the proof of Theorem **4.2.16** is to apply the implicit function result, Theorem **4.2.15**, to the parametric normal equation

$$H(x, v) \equiv F(x, \Pi_{C(x)}(v)) + v - \Pi_{C(x)}(v) = 0 \qquad (21)$$

at $(\bar{x}, \bar{v})$, where $\bar{v} \equiv \bar{y} - F(\bar{x}, \bar{y})$ and $\Pi_{C(x)}$ is the parametric Euclidean projection onto the set $C(x)$. To this end, it would be useful to recall the following equivalence from Proposition **1.3.3**, where it was shown that for any $x \in \Re^n$ and $y, v \in \Re^m$,

$$[v \text{ satisfies } H(x, v) = 0 \text{ and } y = \Pi_{C(x)}(v)]$$

$$\Updownarrow \qquad (22)$$

$$[y \text{ solves VI } (F(x, \cdot), C(x)) \text{ and } v = y - F(x, y)].$$

Below, we use the equivalence (22) rather freely.

Once we have obtained from Theorem **4.2.15** a locally unique, PC^1 solution function $v = v(x)$ of $H(x, v) = 0$ near $(\bar{x}, \bar{v})$, then as shown by Lemma **4.2.22**, this yields a locally unique, PC^1 solution function $y(x) \equiv \Pi_{C(x)}(v(x))$ of VI $(F(x, \cdot), C(x))$ near $(\bar{x}, \bar{y})$, as desired. In turn, to establish the applicability of Theorem **4.2.15** to (21), we need to rely on Lemma **4.2.17** to verify that the Euclidean projector $\Pi_{C(x)}(v)$ is PC^1. Finally, we need to verify the two assumptions in Theorem **4.2.15** for the parametric normal map H; for this verification, we need to identify the C^1 pieces of this map H (see Lemma **4.2.17**) and relate the partial Jacobian matrices of the pieces to the matrices Λ^j in assumption SCOC (see Lemma **4.2.20**).

We start by showing that the projection map $\Pi_{C(x)}(v)$ is PC^1 near $(\bar{x}, \bar{v})$. The study of differentiability properties of the projection map has a rather long history; we refer to [227] for a brief account. Early work has focused on the case where the set $C(x)$ is a constant independent of x. For $C(x)$ given by (3.1.2), results from parametric nonlinear programming theory become applicable [124, 261]. Indeed, for a given pair $(x, v), \Pi_{C(x)}(v)$ is the unique global minimizer of the nonlinear program:

$$\begin{aligned} \text{minimize} \quad & \tfrac{1}{2}\|y - v\|^2 \\ \text{subject to} \quad & g(x, y) \leq 0. \end{aligned} \tag{23}$$

Since the objective function of this program is strictly convex in y for fixed v and each constraint function $g_i(x, \cdot)$ is convex for fixed x, it follows from a fundamental result of Kojima [140, Theorem 7.2] that under the MFCQ at $(\bar{x}, \bar{v})$, the projection map $\Pi_{C(x)}(v)$ is a continuous function of (x, v) near $(\bar{x}, \bar{v})$; moreover, by results of Shapiro [261] and later Dempe [59], $\Pi_{C(x)}(v)$ is directionally differentiable at $(\bar{x}, \bar{v})$. However, the directional derivative at $(\bar{x}, \bar{v})$ is not necessarily continuous let alone Lipschitz in its argument (dx, dv) (see the example on page 642 in [261]), thus making it difficult to use in stationarity analysis of MPEC. These results do not require the CRCQ. In [240], Ralph and Dempe, assuming in addition the latter CQ and extending the earlier work of Pang and Ralph [227], establish the PC^1 property of an optimal solution to a parametric nonlinear program and show how the directional derivative of the solution function can be computed by solving a quadratic program. Instead of just quoting these results of Ralph and Dempe, we give a detailed proof of the lemma below which concerns the special parametric program (23).

4.2.17 Lemma. In the setting of Theorem **4.2.16**, let $\bar{v} \equiv \bar{y} - F(\bar{x}, \bar{y})$. For each $j = 1, \ldots, J$, let $C^j(x)$ be the equality constrained set

$$C^j(x) \equiv \{y \in \Re^m : g_i(x, y) = 0 \text{ for } i \in \mathcal{I}^j\}.$$

The following two statements are valid.

(a) There exist a neighborhood $U \times W$ of $(\bar{x}, \bar{v})$ and a neighborhood V of $\bar{y}$ such that for each $(x, v) \in U \times W$ there is a unique nearest point y in $C^j(x) \cap V$ to v; denote this point y by $\Pi^j(x, v)$. Moreover, Π^j is C^1 near $\bar{x}$ and $\Pi^j(\bar{x}, \bar{v}) = \bar{y}$.

(b) The mapping $(x, v) \mapsto \Pi_{C(x)}(v)$ is PC^1 near $(\bar{x}, \bar{v})$, with C^1 pieces given by Π^j for $j = 1, \ldots, J$.

Proof. Fix an index $j = 1, \ldots, J$. Consider the following equality constrained parametric nonlinear program with y as the primary variable and (x, v) as the parameter:

$$
\begin{aligned}
&\text{minimize} \quad \tfrac{1}{2}\|y - v\|^2 \\
&\text{subject to} \quad g_i(x, y) = 0, \quad \text{for } i \in \mathcal{I}^j.
\end{aligned}
\tag{24}
$$

The Lagrangean system of this program is

$$
\begin{aligned}
&y - v + \sum_{i \in \mathcal{I}^j} \lambda_i \nabla_y g_i(x, y) = 0 \\
&g_i(x, y) = 0, \quad \lambda_i \text{ free }, \quad \text{for } i \in \mathcal{I}^j.
\end{aligned}
\tag{25}
$$

At $(\bar{x}, \bar{v})$, this system is satisfied by $\bar{y}$ and the vector $\lambda^j \in M(\bar{x})$ associated with $\mathcal{I}^j$. Since the gradients $\{\nabla_y g_i(\bar{x}, \bar{y}), i \in \mathcal{I}^j\}$ are linearly independent, it follows that the matrix

$$
\begin{bmatrix}
I + \sum_{i \in \mathcal{I}^j} \lambda_i^j \nabla_{yy}^2 g_i(\bar{x}, \bar{y}) & \nabla_y g_{\mathcal{I}^j}(\bar{x}, \bar{y})^T \\
-\nabla_y g_{\mathcal{I}^j}(\bar{x}, \bar{y}) & 0
\end{bmatrix}
$$

is nonsingular, by the nonnegativity of λ_i^j and the convexity of $g_i(\bar{x}, \cdot)$. Hence, by the classical implicit function theorem for smooth functions [211] applied to the system (25) with y and $\{\lambda_i : i \in \mathcal{I}^j\}$ as the primary variables and (x, v) as the parameter, it follows that there exist a neighborhood $U_0 \times W_0 \times V$ of $(\bar{x}, \bar{v}, \bar{y})$ and a C^1 function $\tilde{y}^j : U_0 \times W_0 \to V$ such that for each $(x, v) \in U_0 \times W_0$, $\tilde{y}^j(x, v)$ is the unique vector in V satisfying (25) for some $\tilde{\lambda}$ close to λ^j. The proof will be complete if we can show that for all (x, v) sufficiently close to $(\bar{x}, \bar{v})$, there exists a j such that $\Pi_{C(x)}(v)$ is equal to $\tilde{y}^j(x, v)$. In fact, it suffices to define for each j, $\Pi^j(x, v) \equiv \tilde{y}^j(x, v)$; then Π^j is C^1 near $(\bar{x}, \bar{v})$ and for all (x, v) sufficiently close to $(\bar{x}, \bar{v})$, $\Pi^j(x, v)$ is a point in $C^j(x) \cap V$ nearest to v.

We consider the projection problem (23) for a fixed pair (x, v) near $(\bar{x}, \bar{v})$. Since MFCQ holds at $(\bar{x}, \bar{y})$, it follows that there is a neighborhood $U_1 \times W_1$ such that for all $(x, v) \in U_1 \times W_1$, the program (23) is equivalent to its KKT system; that is, there exists a multiplier η satisfying

$$
\begin{aligned}
&\Pi_{C(x)}(v) - v + \sum_{i=1}^{\ell} \eta_i \nabla_y g_i(x, \Pi_{C(x)}(v)) = 0 \\
&\left. \begin{array}{c} g_i(x, \Pi_{C(x)}(v)) \le 0, \quad \eta_i \ge 0 \\ \eta_i\, g_i(x, \Pi_{C(x)}(v)) = 0 \end{array} \right\} \ i = 1, \ldots, \ell.
\end{aligned}
\tag{26}
$$

Since $\Pi_{C(x)}(v)$ is continuous in (x, v), it follows that by shrinking $U_1 \times W_1$ if necessary we may assume without loss of generality that $\Pi_{C(x)}(v) \in V$ for all $(x, v) \in U_1 \times W_1$. Similar to the family of index sets $\mathcal{B}(\bar{x})$, we define for each $(x, v) \in U_1 \times W_1$, the finite family of index sets, $\mathcal{B}(x, v)$, which consists of all index sets $\mathcal{I}$ such that

$$\mathcal{I} \subseteq \{i : g_i(x, \Pi_{C(x)}(v)) = 0\}$$

and there exists a vector $\eta \in \Re^\ell$ satisfying (26), $\operatorname{supp}(\eta) \subseteq \mathcal{I}$, and the vectors

$$\{\nabla_y g_i(x, \Pi_{C(x)}(v)) : i \in \mathcal{I}\}$$

are linearly independent. Clearly, if (x, v) is sufficiently close to $(\bar{x}, \bar{v})$, we must have $\mathcal{I} \subseteq \mathcal{I}(\bar{x}, \bar{v})$ for all $\mathcal{I} \in \mathcal{B}(x, v)$.

We now claim that there exists a neighborhood $U_2 \times W_2$ of $(\bar{x}, \bar{v})$ such that for all $(x, v) \in U_2 \times W_2$, $\mathcal{B}(x, v) \subseteq \mathcal{B}(\bar{x})$. Assume, for the sake of contradiction, that there exists a sequence $\{(x^k, v^k)\}$ converging to $(\bar{x}, \bar{v})$ such that for each k, there is an index set $\mathcal{K}_k \in \mathcal{B}(x^k, v^k) \setminus \mathcal{B}(\bar{x})$. Since there are only finitely many such index sets, by working with a subsequence if necessary, we may assume that these index sets $\mathcal{K}_k$ are the same for all k. By letting $\mathcal{K}$ be this common index set, we have $\mathcal{K} \subseteq \mathcal{I}(\bar{x})$; moreover, for each k, the gradients

$$\{\nabla_y g_i(x^k, \Pi_{C(x^k)}(v^k)) : i \in \mathcal{K}\} \tag{27}$$

are linearly independent, and there exists a vector $\eta^k \in \Re^\ell_+$ such that $\operatorname{supp}(\eta^k) \subseteq \mathcal{K}$, and

$$\Pi_{C(x^k)}(v^k) - v^k + \sum_{i \in \mathcal{K}} \eta_i^k \nabla_y g_i(x^k, \Pi_{C(x^k)}(v^k)) = 0$$

$$g_i(x^k, \Pi_{C(x^k)}(v^k)) \le 0, \quad \eta_i^k g_i(x^k, \Pi_{C(x^k)}(v^k)) = 0, \text{ for } i \in \mathcal{K}.$$

Since $\nabla_y g_i(x^k, \Pi_{C(x^k)}(v^k)) \to \nabla_y g_i(\bar{x}, \Pi_{C(\bar{x})}(\bar{v})) = \nabla_y g_i(\bar{x}, \bar{y})$, by (27) and the CRCQ, it follows that the vectors

$$\{\nabla_y g_i(\bar{x}, \Pi_{C(\bar{x})}(\bar{v})) : i \in \mathcal{K}\}$$

are linearly independent; hence, the sequence of vectors $\{\eta_i^k : i \in \mathcal{K}\}$ is bounded. Thus, the full sequence $\{\eta^k\}$ must have an accumulation point, say $\bar{\eta}$, which must necessarily be an element of $M(\bar{x})$. Clearly, $\operatorname{supp}(\bar{\eta}) \subseteq \mathcal{K}$. Consequently, $\mathcal{K} \in \mathcal{B}(\bar{x})$ which is a contradiction.

Now let $U \equiv U_0 \cap U_1 \cap U_2$ and $W \equiv W_0 \cap W_1 \cap W_2$. Take a vector $(x, v) \in U \times W$. Let η be an extreme point of the system (26). Then $\mathrm{supp}(\eta) \in \mathcal{B}(x, v) \subseteq \mathcal{B}(\bar{x})$. So $\mathrm{supp}(\eta) = \mathcal{I}^j$ for some $j = 1, \ldots, J$. It follows that $g_i(x, \Pi_{C(x)}(v)) = 0$ for all $i \in \mathcal{I}^j$, and hence (25) is satisfied with $y \equiv \Pi_{C(x)}(v)$ and $\lambda = \eta$. Since $\Pi_{C(x)}(v) \in V$, by the uniqueness of $y^j(x, v)$ it follows that $\Pi_{C(x)}(v) = y^j(x, v)$ as desired.   **Q.E.D.**

By the definition (21) of the map H, it follows that H is PC^1 if F is C^1. Moreover, by an obvious composition, we obtain the C^1 pieces of H. For each index $j \in \{1, \ldots, J\}$, define

$$G^j \equiv \nabla_y g_{\mathcal{I}^j}(\bar{x}, \bar{y}),$$

$$B^j \equiv \nabla_y L(\bar{x}, \bar{y}, \lambda^j) = \nabla_y F(\bar{x}, \bar{y}) + \sum_{i \in \mathcal{I}^j} \lambda_i^j \nabla_{yy}^2 g_i(\bar{x}, \bar{y}),$$

$$\tilde{B}^j \equiv I + \sum_{i \in \mathcal{I}^j} \lambda_i^j \nabla_{yy}^2 g_i(\bar{x}, \bar{y}),$$

$$\mathcal{N}^j \equiv \{dy \in \Re^m : \nabla_y g_i(\bar{x}, \bar{y})^T dy = 0, \text{ for } i \in \mathcal{I}^j\}.$$

(28)

By properties of $\mathcal{I}^j$, the matrix G^j has full row rank. By convexity of the functions $g_i(x, \cdot)$, it follows that the matrix $\tilde{B}^j$ is symmetric positive definite. Also, $\mathcal{N}^j$ is a linear subspace of $\Re^m$; in fact, $\mathcal{N}^j$ is the null space of the matrix G^j. Let

$$\tilde{G}^j \equiv G^j (\tilde{B}^j)^{-1} (G^j)^T. \tag{29}$$

We have the following easy matrix-theoretic fact. (Note: the notation in the lemma is generic and the matrices should not be confused with the particular matrices defined above.)

4.2.18 Lemma. For any symmetric positive definite matrix $\tilde{B} \in \Re^{m \times m}$ and any matrix $G \in \Re^{p \times m}$ with full row rank, the matrix $\tilde{G} \equiv G \tilde{B}^{-1} G^T$ is positive definite; moreover, with

$$E \equiv \begin{bmatrix} \tilde{B} & G^T \\ -G & 0 \end{bmatrix},$$

we have $\det E > 0$ and

$$E^{-1} = \begin{bmatrix} \tilde{B}^{-1} - \tilde{B}^{-1} G^T \tilde{G}^{-1} G \tilde{B}^{-1} & -\tilde{B}^{-1} G^T \tilde{G}^{-1} \\ \tilde{G}^{-1} G \tilde{B}^{-1} & \tilde{G}^{-1} \end{bmatrix}.$$

Proof. Clearly, $\tilde{G}$ is positive definite. By the the Schur determinantal formula, we have

$$\det E = \det \tilde{B} \det \tilde{G} > 0.$$

The formula for the inverse E^{-1} can be verified by a direct multiplication. **Q.E.D.**

In terms of the matrices defined above, we establish below a result, Lemma **4.2.19**, that is related to [227, Lemma 8]. In particular, part (b) of this result implies that for all vectors $v \in \Re^m$, $\nabla_v \Pi^j(\bar{x}, \bar{v})(v)$ is equal to the unique optimal solution of the following equality constrained, strictly convex quadratic program in the variable u:

$$\text{minimize} \quad \tfrac{1}{2} u^T \tilde{B}^j u - u^T v$$

$$\text{subject to} \quad u \in \mathcal{N}^j.$$

In turn, as described at the end of Subsection **4.2.3**, the latter program can be interpreted as a skewed projection problem of v onto the subspace $\mathcal{N}^j$.

4.2.19 Lemma. In the setting of Theorem **4.2.16**, let $\bar{v} = \bar{y} - F(\bar{x}, \bar{y})$; H be given by (21), and $\Pi^1, \ldots, \Pi^J$ be given by Lemma **4.2.17**(a); let the matrices G^j, B^j, $\tilde{B}^j$, and $\tilde{G}^j$ be given by (28) and (29). The following statements are valid.

(a) H is PC^1 near $(\bar{x}, \bar{v})$ with C^1 pieces given by

$$H^j(x, v) \equiv F(x, \Pi^j(x, v)) + v - \Pi^j(x, v), \quad \text{for } j = 1, \ldots, J.$$

(b) $\nabla_v \Pi^j(\bar{x}, \bar{v}) = (\tilde{B}^j)^{-1} - (\tilde{B}^j)^{-1}(G^j)^T(\tilde{G}^j)^{-1}G^j(\tilde{B}^j)^{-1}.$

(c) $\nabla_v H^j(\bar{x}, \bar{v}) = I + (B^j - \tilde{B}^j) \circ \nabla_v \Pi^j(\bar{x}, \bar{v}).$

Proof. Part (a) is an immediate consequence of Lemma **4.2.17**, because F is C^1. To prove (b), we note that, by the implicit function theorem applied to the system (25),

$$\nabla_v \Pi^j(\bar{x}, \bar{v}) = [I \;\; 0] \begin{bmatrix} \tilde{B}^j & (G^j)^T \\ -G^j & 0 \end{bmatrix}^{-1} \begin{bmatrix} I \\ 0 \end{bmatrix}.$$

By Lemma **4.2.18**, (b) follows.

For part (c), we use part (a) to obtain

$$\nabla_v H^j(\bar{x}, \bar{v}) = \nabla_y F(\bar{x}, \bar{y}) \circ \nabla_v \Pi^j(\bar{x}, \bar{v}) + I - \nabla_v \Pi^j(\bar{x}, \bar{v}),$$

from which the desired formula in part (c) follows easily.   **Q.E.D.**

We can now relate the determinantal sign of $\nabla_v H^j(\bar{x}, \bar{v})$ to that of Λ^j defined in (19).

4.2.20 Lemma. Assume the setting of Lemma **4.2.19**. It holds that for each $j = 1, \ldots, J$,

$$\text{sgn det } \nabla_v H^j(\bar{x}, \bar{v}) = \text{sgn det } \Lambda^j.$$

Proof. For each j, let

$$\tilde{E}^j \equiv \begin{bmatrix} \tilde{B}^j & (G^j)^T \\ -G^j & 0 \end{bmatrix}.$$

By Lemmas **4.2.19** and **4.2.18**, it follows that $\nabla_v \Pi^j(\bar{x}, \bar{v})$ is the upper left-hand block in the inverse of $\tilde{E}^j$. Since

$$\Lambda^j = \tilde{E}^j + \begin{bmatrix} B^j - \tilde{B}^j & 0 \\ 0 & 0 \end{bmatrix},$$

a simple algebraic manipulation yields

$$\Lambda^j (\tilde{E}^j)^{-1} = \begin{bmatrix} \nabla_v H^j(\bar{x}, \bar{v}) & D^j \\ 0 & I \end{bmatrix}$$

for an appropriate matrix D^j, where we have used Lemma **4.2.19**(c). The equality of the signs of the two determinants is now obvious because $\det(\tilde{E}^j)^{-1} > 0$.   **Q.E.D.**

For illustrative purposes, it would be useful for us to state a consequence of Lemmas **4.2.19** and **4.2.20** in a simpler context.

4.2.21 Corollary. Let $A \in \Re^{m \times m}$ be a symmetric positive definite matrix and $G \in \Re^{p \times m}$ with full row rank. Let $\mathcal{N} \subset \Re^m$ denote the null space of G; let $A_{\mathcal{N}}$ be the normal map defined by the pair $(A, \mathcal{N})$; i.e.,

$$A_{\mathcal{N}} \equiv A \circ \Pi_{\mathcal{N}} + I - \Pi_{\mathcal{N}}.$$

(This map is a linear transformation equivalent to a matrix.) Then

$$\text{sgn det } A_{\mathcal{N}} = \text{sgn det } \begin{bmatrix} A & G^T \\ -G & 0 \end{bmatrix}.$$

Proof. Let

$$F(x,y) \equiv Ay, \quad \text{for all } (x,y) \in \Re^{n+m},$$

$$C^j(x) \equiv \mathcal{N}, \quad \text{for all } x \in \Re^m.$$

The desired conclusion is now an easy consequence of the above two lemmas.   **Q.E.D.**

The following result is a refinement of the equivalence (22). Its proof is adopted from [237, Lemma 6] and uses the PC^1 property of the parametric projection map.

4.2.22 Lemma. Assume the setting of Theorem **4.2.16** and let H be given by (21). The following statements are equivalent.

(a) The point $\bar{y}$ equals $\Pi_{C(\bar{x})}(\bar{v})$; and there exist a neighborhood $\tilde{U} \times \tilde{W}$ of $(\bar{x}, \bar{v})$ and a (PC^1) function $v : \tilde{U} \to \tilde{W}$ such that $v(\bar{x}) = \bar{v}$ and, for $x \in \tilde{U}, v = v(x)$ is the unique solution in $\tilde{W}$ of the normal equation $H(x,v) = 0$.

(b) The point $\bar{v}$ equals $\bar{y} - F(\bar{x}, \bar{y})$; and there exist a neighborhood $U \times V$ of $(\bar{x}, \bar{y})$ and a (PC^1) function $y : U \to V$ such that $y(\bar{x}) = \bar{y}$ and, for $x \in U$, $y(x)$ is the unique solution in V of the VI $(F(x,\cdot), C(x))$.

Proof. The main concern of the proof is the uniqueness assertion. We shall prove in detail that statement (a) implies (b) and omit the proof of the converse. We first deal with the two statements without regard to the PC^1 property.

Assume statement (a) holds. Let $\varepsilon > 0$ be such that $\mathbb{B}(\bar{v}, \varepsilon) \subset \tilde{W}$. Then there is $\delta_1 > 0$ such that $\mathbb{B}(\bar{x}, \delta_1) \subset \tilde{U}$ and, for $(x,y) \in \mathbb{B}(\bar{x}, \delta_1) \times \mathbb{B}(\bar{y}, \delta_1)$,

$$\|y - \bar{y}\| + \|F(x,y) - F(\bar{x}, \bar{y})\| < \varepsilon.$$

Let $V \equiv \mathbb{B}(\bar{y}, \delta_1)$. By the continuity of the function $v(x)$ and the continuity of $\Pi_{C(x)}(v)$, there exists $\delta_2 > 0$ such that for all $x \in \mathbb{B}(\bar{x}, \delta_2)$, the vector $y(x) \equiv \Pi_{C(x)}(v(x)) \in V$; moreover, $y(x)$ solves the VI $(F(x,\cdot), C(x))$. Let

$\delta \equiv \min(\delta_1, \delta_2)$ and $U \equiv \mathbb{B}(\bar{x}, \delta)$. Then the function $y : U \to V$ with $y(x)$ as just defined is continuous. It remains to show that for each $x \in U$, $y(x)$ is the only solution of the VI $(F(x, \cdot), C(x))$ that lies in V. For any $(x, y) \in U \times V$, if y solves VI $(F(x, \cdot), C(x))$ then $v \equiv y - F(x, y)$ solves $H(x, v) = 0$, and

$$\|v - \bar{v}\| = \|y - F(x, y) - [\bar{y} - F(\bar{x}, \bar{y})]\|$$

$$\leq \|y - \bar{y}\| + \|F(x, y) - F(\bar{x}, \bar{y})\| < \varepsilon.$$

Hence $v \in \tilde{W}$, giving $v = v(x)$ and $y = \Pi_{C(x)}(v(x)) = y(x)$, i.e., $y(x)$ is the unique solution in V to VI $(F(x, \cdot), C(x))$.

Finally, if $v(x)$ is PC^1, then so is $y(x)$ since the composition of two PC^1 functions is PC^1.   **Q.E.D.**

The last lemma we need is the invertibility of the directional derivative $H_v'(\bar{x}, \bar{v}; \cdot)$.

4.2.23 Lemma. Assume the setting of Lemma **4.2.19**. Then $H_v'(\bar{x}, \bar{v}; \cdot)$ is well defined and invertible on $\Re^m$.

Proof. Since H is PC^1 near $(\bar{x}, \bar{v})$ with pieces $\{H^j\}_1^J$, from Lemma **4.2.10**, it follows that H is B-differentiable at $(\bar{x}, \bar{v})$, so

$$H_v'(\bar{x}, \bar{v}; dv) = H'((\bar{x}, \bar{v}); (0, dv))$$

exists for all $dv \in \Re^m$; and $H_v'(\bar{x}, \bar{v}; \cdot)$ is piecewise linear with pieces $\{\nabla_v H^j(\bar{x}, \bar{v})\}_1^J$. Since Lemma **4.2.20** says that sgn $\det \nabla_v H^j(\bar{x}, \bar{v})$ equals sgn $\det \Lambda^j$, the assumption (SCOC) implies that $H_v'(\bar{x}, \bar{v}; \cdot)$ is coherently oriented. Clearly, $H_v'(\bar{x}, \bar{v}; dv)$ is the directional derivative of the normal map

$$F(\bar{x}, \cdot) \circ \Pi_{C(\bar{x})} + I - \Pi_{C(\bar{x})}$$

at $\bar{v}$ along the direction $dv \in \Re^m$. Now, according to [227, Theorem 9], it follows that $H_v'(\bar{x}, \bar{v}; \cdot)$ is a global Lipschitzian homeomorphism, thus invertible on $\Re^m$.   **Q.E.D.**

With the above lemmas, we are now ready to give the required proof of Theorem **4.2.16**.

Proof of Theorem 4.2.16. Lemmas **4.2.20** and **4.2.23** yield the hypothesis of Theorem **4.2.15**. Hence there is a PC^1 implicit function $v(\cdot)$, whose value $v = v(x)$ is the locally unique solution of $H(x, v) = 0$ for each x near $\bar{x}$. Given Lemmas **4.2.17** and **4.2.22**, the conclusions of Theorem **4.2.16** follow with $y(x) \equiv \Pi_{C(x)}(v(x))$.   **Q.E.D.**

Theorem **4.2.16** does not say anything about the multipliers of the perturbed VI $(F(x, \cdot), C(x))$. By assuming the following

(LICQ) the gradient vectors $\nabla_y g_i(\bar{x}, \bar{y})$, $i \in \mathcal{I}(\bar{x})$, are linearly independent,

which implies both MFCQ and CRCQ, we have the following consequence of Theorem **4.2.16**.

4.2.24 Corollary. Assume the setting of Theorem **4.2.16** with LICQ replacing both MFCQ and CRCQ. Then $M(\bar{x})$ is a singleton, say $\{\bar{\lambda}\}$. Furthermore, there exist a neighborhood $U \times V$ of $(\bar{x}, \bar{y})$ and PC^1 functions $y : U \to V$ and $\lambda : U \to \Re_+^\ell$ such that

$$(y(\bar{x}), \lambda(\bar{x})) = (\bar{y}, \bar{\lambda}),$$

and $(y(x), \lambda(x))$ is the unique solution in $V \times \Re^\ell$ of the KKT conditions (17).

Proof. Theorem **4.2.16** yields the neighborhood $U \times V$ and the PC^1 solution $y : U \to V$ of VI $(F(x, \cdot), C(x))$. The existence and global uniqueness of the KKT multiplier for x near $\bar{x}$ is guaranteed by LICQ. Choose a subset U' of U such that LICQ holds at $(x, y(x))$ for each $x \in U'$; and restrict $y(\cdot)$ to U'. The remaining assertion to be verified is the PC^1 property of the multiplier function $\lambda(x)$. For each x near $\bar{x}$, with $\lambda(x)$ being the unique multiplier corresponding to $(x, y(x))$, we have $\lambda_i(x) = 0$ for all $i \notin \mathcal{I}(\bar{x})$ and

$$F(x, y(x)) + \sum_{i \in \mathcal{I}(\bar{x})} \lambda_i(x) \nabla_y g_i(x, y(x)) = 0.$$

By LICQ, it follows that $\lambda(x)$ is a C^1 function of $(x, y(x))$. Hence, $\lambda(x)$ is a PC^1 function of x.   **Q.E.D.**

The above corollary is a refinement of some frequently cited results in parametric nonlinear programming, such as those of [248].

4.2.5 *Calculation of directional derivatives*

We continue to assume the setting of Theorem **4.2.16**. Our goal in this subsection is to show how to calculate the directional derivative $y'(\bar{x}; dx)$ of the implicit function $y(x)$; in the next subsection, we will establish that the extreme CQ for the MPEC (1) holds under the assumptions of Theorem **4.2.16**. As a consequence, we can give a set of necessary and sufficient conditions for the feasible vector $\bar{z} = (\bar{x}, \bar{y}) \in \mathcal{F}$ to be stationary for the problem (1).

In Section **3.2**, we have defined the directional critical set $\mathcal{K}(\bar{z}, \lambda; dx)$ for each $\lambda \in M(\bar{x}) = M(\bar{z})$ and any given vector $dx \in \Re^n$, and we have established in Proposition **3.2.3** that if $M(\bar{x}) \neq \emptyset$, then $\mathcal{K}(\bar{z}, \lambda; dx)$ is the same nonempty polyhedral set for all critical multipliers λ, i.e., members of $M^c(\bar{z}; dx)$. In particular, under the MFCQ at $\bar{z}$, $M^c(\bar{z}; dx)$ is nonempty and the latter invariance property of $\mathcal{K}(\bar{z}, \lambda; dx)$ holds. For ease of reference, we repeat the definition of $\mathcal{K}(\bar{z}, \lambda; dx)$ and $M^c(\bar{z}; dx)$ as follows:

$$\mathcal{K}(\bar{x}, \lambda; dx) \equiv \{dy \in \Re^m \ :$$

$$dx^T \nabla_x g_i(\bar{x}, \bar{y}) + dy^T \nabla_y g_i(\bar{x}, \bar{y}) \leq 0, \text{ for } i \in \mathcal{I}(\bar{x}) \text{ with } \lambda_i = 0, \quad (30)$$

$$dx^T \nabla_x g_i(\bar{x}, \bar{y}) + dy^T \nabla_y g_i(\bar{x}, \bar{y}) = 0, \text{ for } i \text{ such that } \lambda_i > 0\};$$

$$M^c(\bar{x}; dx) \equiv \operatorname{argmax} \left\{ \sum_{i=1}^{\ell} \lambda_i dx^T \nabla_x g_i(\bar{z}) \ : \ \lambda \in M(\bar{x}) \right\}.$$

By combining Lemmas **3.2.1** and **4.2.5**, it follows that for all vectors $dx \in \mathcal{T}(\bar{x}; X)$ there exists a vector $\lambda \in M(\bar{x})$ such that $y'(\bar{x}; dx)$ solves the AVI $(\nabla_x L(\bar{z}, \lambda) dx, \nabla_y L(\bar{z}, \lambda), \mathcal{K}(\bar{z}, \lambda; dx))$. By Proposition **3.2.3**, it follows that $\lambda \in M^c(\bar{x}; dx)$. The following result shows that any λ in $M^c(\bar{x}; dx)$ can be used to calculate the directional derivative $y'(\bar{x}; dx)$.

4.2.25 Theorem. Assume the setting of Theorem **4.2.16**. For each vector $dx \in \Re^n$ and multiplier $\lambda \in M^c(\bar{x}; dx)$, $y'(\bar{x}; dx)$ is a solution of the

$$\text{AVI } (\nabla_x L(\bar{z}, \lambda) dx, \nabla_y L(\bar{z}, \lambda), \mathcal{K}(\bar{z}, \lambda; dx)); \quad (31)$$

if in addition λ is an extreme multiplier, this AVI has a unique solution.

Proof. Two things need to be proved: one, for each dx in $\Re^n$ and λ in $M^c(\bar{x}; dx)$, $y'(\bar{x}; dx)$ solves the displayed AVI (31); and two, this AVI has a unique solution if in addition $\lambda \in M^e(\bar{x})$. The proof of the first assertion relies on Lemma 3 and the proof of Theorem 3 in [240]. Although the results in the reference refer to the case where $F(x, \cdot)$ is a gradient map, the arguments can nevertheless be modified to suit our present context. For the uniqueness proof, let $\lambda \in M^c(\bar{x}; dx) \cap M^e(\bar{x})$. We need only follow the logic of the proof of Lemma **4.2.23** by considering the normal map associated with the AVI (31):

$$v \mapsto \nabla_x L(\bar{x}, \bar{y}, \lambda) dx + \nabla_y L(\bar{x}, \bar{y}, \lambda)_{\mathcal{K}}(v), \tag{32}$$

where $\mathcal{K} \equiv \mathcal{K}(\bar{x}, \lambda; dx)$. Extending the proof of Lemma **4.2.17** to allow for linear equations in the constraints and using the fact that λ is extreme (which implies that the gradients $\nabla_y g_i(\bar{z})$ for $i \in \text{supp}(\lambda)$ are linearly independent), we can show that the normal map (32) is piecewise affine with affine pieces:

$$v \mapsto \nabla_x L(\bar{x}, \bar{y}, \lambda) dx + \nabla_y L(\bar{x}, \bar{y}, \lambda)_{\mathcal{K}^j}(v), \quad j = 1, \dots, J' \tag{33}$$

for some positive integer J', where

$$\mathcal{K}^j \equiv \{dy \in \Re^m : dx^T \nabla_x g_i(\bar{x}, \bar{y}) + dy^T \nabla_y g_i(\bar{x}, \bar{y}) = 0, \text{ for } i \in K^j\}$$

and K^j is a subset of $\mathcal{I}(\bar{x})$ that contains $\text{supp}(\lambda)$ such that the gradients $\nabla_y g_i(\bar{x}, \bar{y})$, for $i \in K^j$, are linearly independent. (The displayed map (33) is affine because $\mathcal{K}^j$ is an affine subspace.) Each K^j is a SCOC index set; that is, $K^j \in \mathcal{B}(\bar{x})$. By Corollary **4.2.21** and SCOC, it follows that the normal map (32) is coherently oriented. Thus by Theorem 9 in [227] (or [253] since this is actually a normal map induced by a linear transformation and a polyhedral convex set) this map is a global Lipschitzian homeomorphism, thus invertible. Hence by the equivalence of a VI and its normal map (cf. (22)), it follows that the AVI (31) has a unique solution as desired. **Q.E.D.**

There are two important special cases in which we can show, without relying on the result from [240], that $y'(\bar{x}; dx)$ solves (31) for any $\lambda \in M^c(\bar{x}; dx)$. Indeed, as long as the function $\nabla_z L(\bar{z}, \lambda)(dx, \cdot)$ is the same for all $\lambda \in M^c(\bar{x}; dx)$, then the AVI (31) is independent of $\lambda \in M^c(\bar{x}; dx)$, because by Proposition **3.2.3**, the directional critical set $\mathcal{K}(\bar{z}, \lambda; dx)$ is independent of $\lambda \in M^c(\bar{x}; dx)$. Consequently, the first assertion of Theorem **4.2.25** follows without further proof. One special case in which the function $\nabla_z L(\bar{z}, \lambda)(dx, \cdot)$ is independent of $\lambda \in M^c(\bar{x}; dx)$ is when

$$\nabla^2_{yy} g_i(\bar{x}, \bar{y}) = 0 \quad \text{and} \quad \nabla^2_{xy} g_i(\bar{x}, \bar{y}) = 0, \quad \text{for all } i \in \mathcal{I}(\bar{x}).$$

In this case, we have

$$\nabla_z L(\bar{z}, \lambda)(dx, dy) = \nabla_x F(\bar{z}) dx + \nabla_y F(\bar{z}) dy$$

for all $\lambda \in M(\bar{x})$. Another special case is when $M(\bar{x})$ is a singleton, say $M(\bar{x}) = \{\bar{\lambda}\}$, then clearly the same conclusion about $\nabla_z L(\bar{z}, \lambda)$ holds. We recall that $M(\bar{x})$ being a singleton is equivalent to the SMFCQ holding at the pair $(\bar{y}, \bar{\lambda})$ for the VI $(F(\bar{x}, \cdot), C(\bar{x}))$.

If LICQ holds in addition to the hypotheses of Theorem **4.2.16**, then Corollary **4.2.24** says that for all x near $\bar{x}$, $M(x) = \{\lambda(x)\}$ and $\lambda(x)$ is a PC^1 function with the directional derivative calculated as below.

4.2.26 Corollary. Assume the setting of Corollary **4.2.24**. For each vector $dx \in \Re^n$, $(y'(\bar{x}; dx), \lambda'(\bar{x}; dx))$ is the unique pair $(dy, d\lambda)$ that satisfies the KKT system for the AVI (31); that is, $dy \in \mathcal{K}(\bar{x}, \bar{\lambda}; dx)$, $d\lambda_i = 0$ if $i \notin \mathcal{I}(\bar{x})$, and

$$\nabla_z L(\bar{z}, \bar{\lambda})(dx, dy) + \sum_{i \in \mathcal{I}(\bar{x})} d\lambda_i \nabla_y g_i(\bar{x}, \bar{y}) = 0,$$

$$d\lambda_i \geq 0, \quad d\lambda_i \nabla g_i(\bar{x}, \bar{y})^T (dx, dy) = 0, \quad \text{if } i \in \mathcal{I}(\bar{x}) \text{ and } \bar{\lambda}_i = 0,$$

$$d\lambda_i \text{ free}, \quad \text{if } \bar{\lambda}_i > 0.$$

Proof. The uniqueness of the KKT pair $(dy, d\lambda)$ for the AVI (31) follows from Theorem **4.2.25** and the LICQ. It remains to show that the directional derivative

$$\lambda'(\bar{x}; dx) \equiv \lim_{\tau \to 0+} \frac{\lambda(\bar{x} + \tau dx) - \bar{\lambda}}{\tau}$$

exists and satisfies the required properties. For each x sufficiently close to $\bar{x}$, we have

$$F(x, y(x)) + \sum_{i \in \mathcal{I}(\bar{x})} \lambda_i(x) \nabla_y g_i(x, y(x)) = 0.$$

Similar to the proof of Proposition **3.4.3** and using the LICQ, we can show that $\lambda'(\bar{x}; dx)$ exists and satisfies

$$\nabla_z L(\bar{z}, \bar{\lambda})(dx, y'(\bar{x}; dx)) + \sum_{i \in \mathcal{I}(\bar{x})} \lambda_i'(\bar{x}; dx) \nabla_y g_i(\bar{z}) = 0,$$

$$\lambda_i'(\bar{x}; dx) \geq 0, \quad \text{if } i \in \mathcal{I}(\bar{x}) \text{ and } \bar{\lambda}_i = 0,$$

$$\lambda_i'(\bar{x}; dx) = 0, \quad \text{if } i \notin \mathcal{I}(\bar{x}),$$

$$\lambda_i'(\bar{x}; dx) \text{ free}, \quad \text{if } i \in \mathcal{I}(\bar{x}) \text{ and } \bar{\lambda}_i > 0.$$

The only thing left to be shown is the complementarity condition:

$$\lambda_i'(\bar{x}; dx) \nabla g_i(\bar{z})^T (dx, y'(\bar{x}; dx)) = 0$$

for all $i \in \mathcal{I}(\bar{x})$ such that $\bar{\lambda}_i = 0$. Indeed if $\nabla g_i(\bar{z})^T(dx, y'(\bar{x}; dx)) < 0$ for some such i, then we must have $g_i(\bar{x} + \tau dx, y(\bar{x} + \tau dx)) < 0$ for all $\tau > 0$ sufficiently small; by complementarity, this implies $\lambda_i(\bar{x} + \tau dx) = 0$. Hence $\lambda_i'(\bar{x}; dx) = 0$ as desired. **Q.E.D.**

In the rest of this subsection, we discuss the F-differentiability of the implicit solution function $y(\cdot)$ at $\bar{x}$. Specifically, we identify necessary and sufficient conditions for the function $y(x)$ to have this property and give sufficient conditions for $y(x)$ to be strongly F-differentiable at $\bar{x}$. Results of this kind have appeared in [218, 151] under more restrictive assumptions.

Before proceeding, we should mention that $y(x)$, being a PC^1 function and thus Lipschitz continuous in an (open) neighborhood of $\bar{x}$, is F-differentiable at almost all points in this neighborhood (i.e., the set of nondifferentiable points in this neighborhood has zero Lebesque measure), by a renowned result of Rademacher [235]. The result obtained below, Theorem **4.2.28**, can be used to calculate the F-derivative of $y(\cdot)$ at these F-differentiable points. In turn, these F-derivatives of $y(\cdot)$ can be used to compute the Clarke generalized Jacobian $\partial \tilde{f}(\bar{x})$ [51] of the composite objective function $\tilde{f}(x) \equiv f(x, y(x))$ at $\bar{x}$. In essence this is the approach that Outrata [214] has used to obtain the necessary optimality conditions for an MPEC; nevertheless, the underlying framework of Outrata's approach is Robinson's strong regularity theory, which, as shown in Subsection **4.2.7**, is more restrictive than our approach, because the latter is based on the more general SCOC.

For each $dx \in \Re^n$, define

$$\mathcal{E}(\bar{x}, dx) \equiv \{ dy \in \Re^m :$$

$$dx^T \nabla_x g_i(\bar{x}, \bar{y}) + dy^T \nabla_y g_i(\bar{x}, \bar{y}) = 0, \quad \text{for all } i \in \mathcal{I}(\bar{x}) \}.$$

Then $\mathcal{E}(\bar{x}, dx)$ is a (possibly empty) affine subspace. Moreover, for any $\lambda \in M^c(\bar{z}; dx)$, we have

$$\mathcal{E}(\bar{x}, dx) \subseteq \mathcal{K}(\bar{z}, \lambda; dx),$$

and provided that $\mathcal{E}(\bar{x}, dx) \neq \emptyset$, equality holds in the above inclusion if and only if $\mathcal{K}(\bar{z}, \lambda; dx)$ is an affine subspace. By using the maximal elements in the family $\mathcal{B}(\bar{x})$, we may obtain a reduced representation of $\mathcal{E}(\bar{x}, dx)$. Specifically, an index set $\mathcal{I} \in \mathcal{B}(\bar{x})$ is said to be *maximal* if there exists no index set in $\mathcal{B}(\bar{x})$ that properly contains $\mathcal{I}$. Clearly, for each $\lambda \in M^e(\bar{x})$, there must exist at least one maximal element $\mathcal{I} \in \mathcal{B}(\bar{x})$ such that $\text{supp}(\lambda) \subseteq \mathcal{I}$.

4.2.27 Lemma. For each maximal element $\mathcal{I} \in \mathcal{B}(\bar{x})$ and each $dx \in \Re^n$,

$$\mathcal{E}(\bar{x}, dx) \equiv \{ dy \in \Re^m : dx^T \nabla_x g_i(\bar{x}, \bar{y}) + dy^T \nabla_y g_i(\bar{x}, \bar{y}) = 0, \text{ for all } i \in \mathcal{I} \}.$$

Proof. This is obvious.   **Q.E.D.**

Using the above lemma, we obtain several necessary and sufficient conditions for the F-differentiability of $y(x)$ at $\bar{x}$.

4.2.28 Theorem. Assume the setting of Theorem **4.2.16**. The following statements are equivalent.

(a) The implicit solution function $y(x)$ is F-differentiable at $\bar{x}$.

(b) For all $dx \in \Re^n$, $y'(\bar{x}; dx) \in \mathcal{E}(\bar{x}, dx)$.

(c) For all $dx \in \Re^n$ and $\lambda \in M^e(\bar{x})$, $y'(\bar{x}; dx)$ is the unique solution of the

$$\text{AVI } (\nabla_x L(\bar{z}, \lambda) dx, \nabla_y L(\bar{z}, \lambda), \mathcal{E}(\bar{x}, dx)). \tag{34}$$

(d) For all $dx \in \Re^n$, $\lambda \in M^e(\bar{x})$, and maximal elements $\mathcal{I} \in \mathcal{B}(\bar{x})$ with $\text{supp}(\lambda) \subseteq \mathcal{I}$, $y'(\bar{x}; dx)$, along with some $d\lambda_{\mathcal{I}}(\bar{x}; dx)$, is the unique solution $(dy, d\lambda_{\mathcal{I}})$ of the system of linear equations:

$$\nabla_z L(\bar{z}, \lambda)(dx, dy) + \sum_{i \in \mathcal{I}} d\lambda_i \nabla_y g_i(\bar{x}, \bar{y}) = 0,$$

$$\nabla g_i(\bar{x}, \bar{y})^T (dx, dy) = 0, \quad \text{for all } i \in \mathcal{I}. \tag{35}$$

(e) There exist a maximal element $\mathcal{I} \in \mathcal{B}(\bar{x})$ and a multiplier $\lambda \in M^e(\bar{x})$ with supp$(\lambda) \subseteq \mathcal{I}$ such that for all $dx \in \Re^n$, $y'(\bar{x}; dx)$, along with some $d\lambda_{\mathcal{I}}(\bar{x}; dx)$, is the unique solution $(dy, d\lambda_{\mathcal{I}})$ of the system of linear equations (35).

Proof. According to Proposition **4.2.2**(d), the function $y(x)$, being B-differentiable at $\bar{x}$, is F-differentiable at $\bar{x}$ if and only if the directional derivative $y'(\bar{x}; dx)$ is linear in the second argument.

(a) $\Rightarrow$ (b). If (a) holds, then

$$y'(\bar{x}; -dx) = -y'(\bar{x}; dx), \quad \text{for all } dx \in \Re^n. \tag{36}$$

Since $y'(\bar{x}; dx) \in \mathcal{K}(\bar{x}, \lambda; dx)$ for all $\lambda \in M^c(\bar{x}; dx)$, it follows easily from (36) that $y'(\bar{x}; dx) \in \mathcal{E}(\bar{x}, dx)$. Hence (b) holds.

(b) $\Rightarrow$ (c). Since $\mathcal{E}(\bar{x}, dx) \subseteq \mathcal{K}(\bar{x}, \lambda; dx)$ for all $\lambda \in M(\bar{x})$, it follows that any solution of (31) that lies in $\mathcal{E}(\bar{x}, dx)$ must be a solution of (34). Thus (b) implies that $y'(\bar{x}; dx)$ solves (34) for all $dx \in \Re^n$. It remains to show that the latter AVI has a unique solution if $\lambda \in M^e(\bar{x})$. This follows by combining several observations: one, for any $\lambda \in M^e(\bar{x})$, supp(λ) is contained in a maximal index set $\mathcal{I}$ belonging to $\mathcal{B}(\bar{x})$; two, (34) is equivalent to the system of linear equations (35) where $\mathcal{I}$ is any maximal element in $\mathcal{B}(\bar{x})$; and three, for any $\lambda \in M^e(\bar{x})$ and maximal element $\mathcal{I} \in \mathcal{B}(\bar{x})$ that contains supp(λ), the matrix

$$\begin{bmatrix} \nabla_y L(\bar{x}, \bar{y}, \lambda) & \nabla_y g_{\mathcal{I}}(\bar{x}, \bar{y})^T \\ -\nabla_y g_{\mathcal{I}}(\bar{x}, \bar{y}) & 0 \end{bmatrix}$$

is nonsingular, by SCOC. Thus (c) holds.

(c) $\Leftrightarrow$ (d). This has been pointed out in the proof of (b) $\Rightarrow$ (c).

(d) $\Rightarrow$ (e). This is obvious.

(e) $\Rightarrow$ (a). Since (35) is a nonsingular system of linear equations for λ and $\mathcal{I}$ with the properties stated in (e), it follows that $y'(\bar{x}; dx)$ is a linear function of dx. Hence (a) follows. **Q.E.D.**

4.2.29 Remark. It is interesting to note that if $y(x)$ is F-differentiable at $\bar{x}$, then part (b) of the above theorem implies that $\mathcal{K}(\bar{x}, \lambda; dx) \neq \emptyset$ for all $dx \in \Re^n$ and $\lambda \in M(\bar{x})$; in turn, the latter implies that $M^c(\bar{x}; dx) = M(\bar{x})$ for all $dx \in \Re^n$, by Proposition **3.2.3**.

Based on Theorem **4.2.28**, we give two sufficient conditions for the implicit solution function $y(x)$ to be F-differentiable at $\bar{x}$, under the further assumption that LICQ holds.

4.2.30 Proposition. Assume the setting of Corollary **4.2.24**. If either strict complementarity holds at the pair $(\bar{x}, \bar{y})$ (i.e., if $\bar{\lambda} - g(\bar{x}, \bar{y}) > 0$) or the directional critical set $\mathcal{K}(\bar{x}, \bar{\lambda}; dx)$ is an affine subspace of $\Re^m$ for all $dx \in \Re^n$, then the function $y(x)$ is F-differentiable at $\bar{x}$. Moreover, in the former case, $y(x)$ is strongly F-differentiable at $\bar{x}$.

Proof. Under the LICQ, $M(\bar{x})$ is a singleton, say $\{\bar{\lambda}\}$. If either one of the two additional assumptions holds, then $\mathcal{K}(\bar{x}, \bar{\lambda}; dx) = \mathcal{E}(\bar{x}, dx)$. It remains to prove the last assertion. Let $\lambda(x)$ be the multiplier function asserted by Corollary **4.2.24**. By continuity of $(y(x), \lambda(x))$, it follows that the properties below must hold for all x sufficiently close to $\bar{x}$:

(i) strict complementarity of the pair $(y(x), \lambda(x))$; i.e.,

$$\lambda(x) - g(x, y(x)) > 0;$$

(ii) $\mathcal{I}(x) = \mathcal{I}(\bar{x})$;

(iii) the LICQ at $(x, y(x))$, i.e., the gradients

$$\{\nabla_y g_i(x, y(x)) : i \in \mathcal{I}(x)\}$$

remain linearly independent;

(iv) the SCOC at $(x, y(x))$. (Indeed under (i), (ii), and (iii), the SCOC at $(x, y(x))$ is equivalent to the nonsingularity of the matrix

$$\Lambda(x) \equiv \begin{bmatrix} \nabla_y L(x, y(x), \lambda(x)) & \nabla_y g_{\mathcal{I}(\bar{x})}(x, y(x))^T \\ -\nabla_y g_{\mathcal{I}(\bar{x})}(x, y(x)) & 0 \end{bmatrix};$$

the latter matrix is nonsingular because it is a small perturbation of the matrix (20) which is $\Lambda(\bar{x})$; moreover $\lim_{x \to \bar{x}} \Lambda(x)^{-1} = \Lambda(\bar{x})^{-1}$.)

Therefore, the implicit solution function $y(\cdot)$ is F-differentiable at all points x near $\bar{x}$; moreover for all $dx \in \Re^n$, the derivative $y'(x; dx)$, along with an appropriate $d\lambda$, is the unique pair $(dy, d\lambda)$ satisfying the system of linear equations (35) with $(\bar{x}, \bar{y}, \bar{\lambda})$ replaced by $(x, y(x), \lambda(x))$; i.e., with $\mathcal{I} \equiv \mathcal{I}(\bar{x})$, we have

$$
\begin{pmatrix} \nabla_x L(x, y(x), \lambda(x)) \\ \nabla_x g_{\mathcal{I}}(x, y(x)) \end{pmatrix} dx + \Lambda(x) \begin{pmatrix} y'(\bar{x}; dx) \\ d\lambda_{\mathcal{I}}(x) \end{pmatrix} = 0.
$$

Passing to the limit $x \to \bar{x}$ in the above equation and recalling the continuity of $(y(x), \lambda(x))$ at $\bar{x}$, we easily deduce that $\lim_{x \to \bar{x}} y'(x; dx) = y'(\bar{x}; dx)$ and this limit holds uniformly for all $dx \in \Re^n$ on compact sets. Hence by Proposition **4.2.2**, $y(x)$ is strongly F-differentiable at $\bar{x}$.   **Q.E.D.**

Note that it is possible for $\mathcal{K}(\bar{x}, \bar{\lambda}; dx)$ to be an affine subspace without the strict complementarity condition holding. For instance, if $\nabla g_i(\bar{x}, \bar{y})$ is equal to the zero vector for all degenerate indices i, then $\mathcal{K}(\bar{x}, \bar{\lambda}; dx)$ is clearly an affine subspace for all $dx \in \Re^n$.

4.2.6 Verification of CQs for MPEC

In the above subsections, we have provided sufficient conditions for BIF to hold. Unfortunately, as Example **3.4.2** shows (the solution function $y(x)$ therein is C^1 thus B-differentiable near $\bar{x} = (2, 0)$), BIF alone is not sufficient for the extreme or the full CQ (see Section **3.3**) for the MPEC to hold. As it turns out, the assumptions of Theorem **4.2.16** will imply the extreme CQ; in this regard, the CRCQ is particularly relevant.

Adapting the notation from Section **3.2**, we let $\mathrm{Gr}(\mathcal{LS}_{(\bar{x}, \lambda)})$ be the set of pairs (dx, dy) such that dy solves the AVI (31), and

$$
\mathcal{L}^e(\bar{x}; \mathcal{F}) \equiv (\mathcal{T}(\bar{x}; X) \times \Re^m) \bigcap \left(\bigcup_{\lambda \in M^e(\bar{x})} \mathrm{Gr}(\mathcal{LS}_{(\bar{x}, \lambda)}) \right);
$$

for any subset M' of $M(\bar{x})$, let

$$
\mathcal{L}'(\bar{x}; \mathcal{F}) \equiv (\mathcal{T}(\bar{x}; X) \times \Re^m) \bigcap \left(\bigcup_{\lambda \in M'} \mathrm{Gr}(\mathcal{LS}_{(\bar{x}, \lambda)}) \right).
$$

So the extreme CQ holds if $\mathcal{L}^e(\bar{x}; \mathcal{F}) = \mathcal{T}(\bar{x}; \mathcal{F})$, and the basic CQ holds if $\mathcal{L}'(\bar{x}; \mathcal{F}) = \mathcal{T}(\bar{x}; \mathcal{F})$ for some $M' \subseteq M(\bar{x})$.

4.2.31 Theorem. Let $\mathcal{L}^e((\bar{x}, \bar{y}); \mathcal{F})$ and $\mathcal{L}'((\bar{x}, \bar{y}); \mathcal{F})$ be given above. Under the hypotheses of Theorem **4.2.16**, the SBCQ and the extreme CQ for MPEC (1) hold at $(\bar{x}, \bar{y})$; indeed,

$$\mathcal{T}((\bar{x}, \bar{y}); \mathcal{F}) = \mathcal{L}^e(\bar{x}; \mathcal{F}) = \mathcal{L}'(\bar{x}; \mathcal{F}) \tag{37}$$

for any set $M' \subseteq M^e(\bar{x})$ such that

$$M' \cap M^c(\bar{x}; dx) \neq \emptyset, \quad \text{for all } dx \in \Re^n. \tag{38}$$

Proof. Suppose $M^1 \subseteq M^e(\bar{x})$ and $M^2 \subseteq M^e(\bar{x})$ are candidates for M' such that for $i = 1, 2$, and each dx, $M^i \cap M^c(\bar{x}; dx)$ is nonempty. We claim that

$$\bigcup_{\lambda \in M^1} \text{Gr}(\mathcal{LS}_{(\bar{x}, \lambda)}) = \bigcup_{\lambda \in M^2} \text{Gr}(\mathcal{LS}_{(\bar{x}, \lambda)}). \tag{39}$$

To see this, let (dx, dy) be given. If $\lambda^1 \in M^1$ is such that (dx, dy) is in $\text{Gr}(\mathcal{LS}_{(\bar{x}, \lambda^1)})$ then, by Proposition **3.2.3**, $\lambda^1 \in M^c(\bar{x}; dx)$. Hence, by Theorem **4.2.25**, $dy = y'(\bar{x}; dx)$. Since there exists $\lambda^2 \in M^2 \cap M^c(\bar{x}; dx)$, the theorem also says that $y'(\bar{x}; dx)$ is the unique value of dy with (dx, dy) belonging to $\text{Gr}(\mathcal{LS}_{(\bar{x}, \lambda^2)})$. Consequently, it follows by symmetry that there exists $\lambda^1 \in M^1$ such that $(dx, dy) \in \text{Gr}(\mathcal{LS}_{(\bar{x}, \lambda^1)})$ if and only if there exists $\lambda^2 \in M^2$ such that $(dx, dy) \in \text{Gr}(\mathcal{LS}_{(\bar{x}, \lambda^2)})$, and (39) holds.

By Proposition **3.2.2**, $M^e(\bar{x})$ is a candidate for M'. Thus the equality (39) holds with $M^1 \equiv M'$ and $M^2 \equiv M^e(x)$, whence $\mathcal{L}^e(\bar{x}; \mathcal{F}) = \mathcal{L}'(\bar{x}; \mathcal{F})$. Furthermore from equation (3.2.12), $\mathcal{T}((\bar{x}, \bar{y}); \mathcal{F}) \subseteq \mathcal{L}^e(\bar{x}; \mathcal{F})$. Thus it remains only to show that $\mathcal{T}((\bar{x}, \bar{y}); \mathcal{F}) \supseteq \mathcal{L}'(\bar{x}; \mathcal{F})$.

Let $\lambda \in M'$, $dx \in \mathcal{T}(\bar{x}; X)$, and $(dx, dy) \in \text{Gr}(\mathcal{LS}_{(\bar{x}, \lambda)})$. From above, $dy = y'(\bar{x}; dx)$. Thus $(dx, dy) \in \mathcal{T}((\bar{x}, \bar{y}); \mathcal{F})$ from Lemma **4.2.5**. **Q.E.D.**

Primal stationarity conditions, based on Corollary **3.3.1**, can be given using the above decomposition of $\mathcal{T}(\bar{z}; \mathcal{F})$ into finitely many pieces of the form

$$(\mathcal{T}(\bar{x}; X) \times \Re^m) \cap \text{Gr}(\mathcal{LS}_{(\bar{x}, \lambda)})$$

indexed by λ in either $M^e(\bar{x})$ or M'. Next we present primal-dual stationarity conditions which are derived from the specialization of Theorem **3.3.4** to the implicit program.

4.2.32 Corollary. Assume, in addition to the setting of Theorem **4.2.16**, that

$$X = \{x \in \Re^n : Gx + a \leq 0\},$$

where $G \in \Re^{s \times n}$, $a \in \Re^s$. Then the vector $(\bar{x}, \bar{y}) \in \mathcal{F}$ is a stationary point of (4.2.1) if and only if for any $M' \subseteq M^e(\bar{x})$ satisfying (38), for each vector $\lambda \in M'$ and each pair of index sets α and $\bar{\alpha}$ partitioning $\mathcal{I}_0(\bar{x}, \lambda) \equiv \{i \in \mathcal{I}(\bar{x}) : \lambda_i = 0\}$, there exist multipliers $\zeta \in \Re^s$, $\eta \in \Re^\ell$ and $\pi \in \Re^m$ such that

$$\nabla_x f(\bar{z}) + G^T \zeta + \nabla_x g(\bar{z})^T \eta = \nabla_x L(\bar{z}, \lambda)^T \pi$$

$$\nabla_y f(\bar{z}) + \nabla_y g(\bar{z})^T \eta = \nabla_y L(\bar{z}, \lambda)^T \pi$$

$$\pi^T \nabla_y g_i(\bar{z}) \leq 0, \quad i \in \alpha$$

$$\pi^T \nabla_y g_i(\bar{z}) = 0, \quad i \in \mathcal{I}_+(\bar{x}, \lambda)$$

$$\eta_i \geq 0, \quad i \in \bar{\alpha}$$

$$\eta_i = 0, \quad i \notin \mathcal{I}(\bar{x})$$

$$\zeta \geq 0, \quad \zeta^T(G\bar{x} + a) = 0.$$

Proof. The displayed system is essentially (3.3.7) with the simplification that the matrix $H = 0$.   **Q.E.D.**

When $M(\bar{x}) = \{\bar{\lambda}\}$, then $M^c(\bar{x}; dx) = \{\bar{\lambda}\}$ for each dx; hence $y'(\bar{x}; dx)$ is determined by the single AVI $\left(\nabla_x L(\bar{z}, \bar{\lambda})dx, \nabla_y L(\bar{z}, \bar{\lambda}), \mathcal{K}(\bar{z}, \bar{\lambda}; dx)\right)$ for all $dx \in \Re^n$, and (37) holds with $M' = \{\bar{\lambda}\}$. We recall that $M(\bar{x})$ being a singleton is equivalent to the SMFCQ holding at $(\bar{z}, \bar{\lambda})$.

Two other cases in which M' can be chosen to be a singleton $\{\lambda\}$ are given in the following corollary of Lemma **3.2.3**.

4.2.33 Corollary. Assume, in addition to the setting of Theorem **4.2.16**, that either

(a) $g(x, \cdot)$ is independent of x, i.e., $g(x, y) \equiv g(y)$, or

(b) the range of $\nabla_y g_{\mathcal{I}(\bar{x})}(\bar{x}, \bar{y})$ contains the range of $\nabla_x g_{\mathcal{I}(\bar{x})}(\bar{x}, \bar{y})$, i.e.,

$$\{\nabla_y g_{\mathcal{I}(\bar{x})}(\bar{x}, \bar{y})dy : dy \in \Re^m\} \supseteq \{\nabla_x g_{\mathcal{I}(\bar{x})}(\bar{x}, \bar{y})dx : dx \in \Re^n\};$$

then we may take $M' = \{\lambda\}$ for any $\lambda \in M^e(\bar{x})$ in (37).

Proof. Statement (a) implies statement (b). Assume the latter, and let $\lambda \in M^e(x)$. We need to show that $M' \equiv \{\lambda\}$ satisfies the condition (38).

If the range of $\nabla_y g_{\mathcal{I}(\bar{x})}(\bar{x}, \bar{y})$ contains the range of $\nabla_x g_{\mathcal{I}(\bar{x})}(\bar{x}, \bar{y})$, then for any dx there exists dy such that

$$\nabla_x g_{\mathcal{I}(\bar{x})}(\bar{x}, \bar{y})dx + \nabla_y g_{\mathcal{I}(\bar{x})}(\bar{x}, \bar{y})dy = 0.$$

Thus $\mathcal{K}(x, \lambda; dx)$ is nonempty for all dx, and Lemma **3.2.3** says that λ is an element of $M^c(\bar{x}; dx)$.   **Q.E.D.**

Combining Example **3.4.2**, Corollary **4.2.33**, and Theorem **4.2.31**, we can now see the important role of the CRCQ. Under this CQ (and the other hypotheses of Theorem **4.2.16**), the extreme CQ must hold; moreover, under the additional assumptions in Corollary **4.2.33**, we can use any single, extreme multiplier $\lambda \in M^e(\bar{x})$ to completely characterize the tangent cone $\mathcal{T}(\bar{x}; \mathcal{F})$, and hence, the stationarity conditions of the MPEC; see Corollary **4.2.32**. Without the CRCQ, although it might still be possible for the basic CQ to hold for the MPEC, the set of multipliers M' needed for equating the tangent cone $\mathcal{T}(\bar{x}; \mathcal{F})$ with the corresponding linearized cone $\mathcal{L}'(\bar{x}; \mathcal{F})$ becomes difficult to identify. Although we have successfully identified this set M' in Example **3.4.2**, it is not clear to us at the time of this writing what an appropriate set M' should in general be.

4.2.7 *More on strong coherent orientation*

We discuss various conditions that are related to the SCOC; from the results derived herein, we will be able to see why we have called SCOC a strong coherent orientation condition. As a start, we recall that in the simple situation where $(\bar{x}, \bar{y})$ is strongly nondegenerate, and $M(\bar{x}) = \{\bar{\lambda}\}$, the SCOC holds if and only if the single matrix (20) is nonsingular. (This fact has been used in the proof of Proposition **4.2.30**.)

In order to derive necessary and sufficient conditions for SCOC to hold in general, we need to introduce some notation and a matrix-theoretic property associated with pairs of square matrices. To begin, we recall the family $\mathcal{B}(\bar{x})$ introduced in Subsection **4.2.4**, where

$$\mathcal{B}(\bar{x}) = \{\mathcal{I}^1, \ldots, \mathcal{I}^J\}.$$

For each $j = 1, \ldots, J$, there exists a unique multiplier $\lambda^j \in M^e(\bar{x})$ such that $\mathrm{supp}(\lambda^j) \subseteq \mathcal{I}^j$; write $m_j \equiv |\mathcal{I}^j|$. For an arbitrary pair of vectors

$r \equiv (q, p) \in \Re^m \times \Re^{m_j}$ let $\text{AVI}^j(r)$ denote the

$$\text{AVI}\ (q, \nabla_y L(\bar{x}, \bar{y}, \lambda^j), \mathcal{K}^j(p)),$$

where

$$\mathcal{K}^j(p) \equiv \{dy \in \Re^m :\quad dy^T \nabla_y g_i(\bar{x}, \bar{y}) \leq p_i,\ \text{for } i \in \mathcal{I}_0^j,$$

$$dy^T \nabla_y g_i(\bar{x}, \bar{y}) = p_i,\ \text{for } i \in \mathcal{I}_+^j\},$$

with

$$\mathcal{I}_0^j \equiv \{i \in \mathcal{I}^j : \lambda_i^j = 0\}, \quad \text{and} \quad \mathcal{I}_+^j \equiv \{i \in \mathcal{I}^j : \lambda_i^j > 0\};$$

this is the VI defined by the affine map

$$dy \mapsto q + \nabla_y L(\bar{x}, \bar{y}, \lambda^j) dy$$

and the set $\mathcal{K}^j(p)$. With $p \equiv -\nabla_x g(\bar{z}) dx$, $\mathcal{K}^j(p)$ contains the directional critical set $\mathcal{K}(\bar{x}, \lambda^j; dx)$ which involves the constraint $\nabla g_i(\bar{z})^T dz \leq 0$ for all $i \in \mathcal{I}(\bar{x}) \supseteq \mathcal{I}^j$. The KKT system of the $\text{AVI}^j(r)$ is as follows: $dy \in \mathcal{K}^j(p)$ and

$$q + \nabla_y L(\bar{x}, \bar{y}, \lambda^j) dy + \sum_{i \in \mathcal{I}^j} d\lambda_i \nabla_y g_i(\bar{x}, \bar{y}) = 0,$$

$$d\lambda_i \geq 0, \quad d\lambda_i \left(p_i - \nabla_y g_i(\bar{x}, \bar{y})^T dy \right) = 0, \quad \text{if } i \in \mathcal{I}_0^j; \tag{40}$$

we note that this system defines a mixed LCP. We also recall the matrix Λ^j defined in (19) which we partition as

$$\Lambda^j = \begin{bmatrix} \nabla_y L(\bar{x}, \bar{y}, \lambda^j) & \nabla_y g_{\mathcal{I}_+^j}(\bar{x}, \bar{y})^T & \nabla_y g_{\mathcal{I}_0^j}(\bar{x}, \bar{y})^T \\ -\nabla_y g_{\mathcal{I}_+^j}(\bar{x}, \bar{y}) & 0 & 0 \\ -\nabla_y g_{\mathcal{I}_0^j}(\bar{x}, \bar{y}) & 0 & 0 \end{bmatrix}.$$

Also define

$$\tilde{\Lambda}^j \equiv \begin{bmatrix} \nabla_y L(\bar{x}, \bar{y}, \lambda^j) & \nabla_y g_{\mathcal{I}_+^j}(\bar{x}, \bar{y})^T & 0 \\ -\nabla_y g_{\mathcal{I}_+^j}(\bar{x}, \bar{y}) & 0 & 0 \\ -\nabla_y g_{\mathcal{I}_0^j}(\bar{x}, \bar{y}) & 0 & I \end{bmatrix}. \tag{41}$$

Let Γ^j denote the normal map associated with the matrix Λ^j and the cone $\Re^{m+|\mathcal{I}_+^j|} \times \Re_+^{|\mathcal{I}_0^j|}$; thus for $r = (q, p)$, the translated map

$$\begin{pmatrix} dv \\ d\mu_{\mathcal{I}^j} \end{pmatrix} \mapsto r + \Gamma^j(dv, d\mu_{\mathcal{I}^j})$$

is precisely the normal map associated with the KKT system (40). We note that the map Γ^j is piecewise linear.

We introduce some terminology for pairs of square matrices of the same order [286, 272].

4.2.34 Definition. Let C_0 and C_1 be two square matrices of order k. A *column representative* of the pair (C_0, C_1) is a matrix $C \in \Re^{k \times k}$ such that each j-th column of C is equal to either the j-th column of C_0 or C_1. The pair (C_0, C_1) is said to have the $\mathcal{W}$ property if for each column representative C of the pair (C_0, C_1), sgn det C is a nonzero constant.

Among various characterizations of the $\mathcal{W}$ property, the following result was proved in [272].

4.2.35 Lemma. A pair of matrices $C_0, C_1 \in \Re^{k \times k}$ has the $\mathcal{W}$ property if and only if for every vector $r \in \Re^k$, the system

$$C_0 u - C_1 v = r$$

$$(u, v) \geq 0, \quad u^T v = 0,$$

has a unique solution.

The $\mathcal{W}$ property is a generalization of a P matrix; indeed, a matrix $M \in \Re^{k \times k}$ is P if and only if the pair (I, M) has the $\mathcal{W}$ property. In the following lemma, we characterize the $\mathcal{W}$ property for the pair of matrices $(\Lambda^j, \tilde{\Lambda}^j)$ defined above. Subsequently, this lemma will be used to derive necessary and sufficient conditions for the SCOC.

4.2.36 Lemma. Let $j \in \{1, \ldots, J\}$ be fixed but arbitrary. The following statements are equivalent.

(a) There exists an integer $\sigma_j \in \{-1, 1\}$ such that for all index sets $\mathcal{I}$ satisfying $\mathcal{I}_+^j \subseteq \mathcal{I} \subseteq \mathcal{I}^j$,

$$\text{sgn det} \begin{bmatrix} \nabla_y L(\bar{x}, \bar{y}, \lambda^j) & \nabla_y g_{\mathcal{I}}(\bar{x}, \bar{y})^T \\ -\nabla_y g_{\mathcal{I}}(\bar{x}, \bar{y}) & 0 \end{bmatrix} = \sigma_j.$$

(b) The pair of matrices $(\Lambda^j, \tilde{\Lambda}^j)$ has the $\mathcal{W}$ property.

(c) For all vectors $r \equiv (q, p) \in \Re^m \times \Re^{m_j}$, the KKT system (40) has a unique solution $(dy, d\lambda_{\mathcal{I}^j})$.

(d) For all vectors $r \equiv (q, p) \in \Re^m \times \Re^{m_j}$, the $\text{AVI}^j(r)$ has a unique solution.

(e) The normal map Γ^j is a global Lipschitzian homeomorphism.

(f) The normal map Γ^j is coherently oriented.

(g) The matrix

$$\Lambda_+^j \equiv \begin{bmatrix} \nabla_y L(\bar{x}, \bar{y}, \lambda^j) & \nabla_y g_{\mathcal{I}_+^j}(\bar{x}, \bar{y})^T \\ -\nabla_y g_{\mathcal{I}_+^j}(\bar{x}, \bar{y}) & 0 \end{bmatrix}$$

is nonsingular, and the Schur complement of Λ_+^j in Λ^j, i.e., the matrix

$$[\nabla_y g_{\mathcal{I}_0^j}(\bar{x}, \bar{y}) \quad 0](\Lambda_+^j)^{-1} \begin{bmatrix} \nabla_y g_{\mathcal{I}_0^j}(\bar{x}, \bar{y})^T \\ 0 \end{bmatrix}, \tag{42}$$

is a P matrix.

If, in addition, $\nabla_y F(\bar{x}, \bar{y})$ is symmetric, then any one of the conditions (a)–(g) is further equivalent to the condition:

(h) the following implication in the variables $(dy, d\lambda_{\mathcal{I}^j})$ holds:

$$\left. \begin{array}{l} \nabla_y L(\bar{x}, \bar{y}, \lambda^j) dy + \sum_{i \in \mathcal{I}^j} d\lambda_i \nabla_y g_i(\bar{x}, \bar{y}) = 0 \\ \nabla_y g_i(\bar{x}, \bar{y})^T dy = 0, \quad i \in \mathcal{I}_+^j \\ dy \neq 0 \end{array} \right\} \Rightarrow dy^T \nabla_y L(\bar{x}, \bar{y}, \lambda^j) dy > 0.$$

Finally, if $\nabla_y F(\bar{x}, \bar{y})$ is symmetric and $\nabla_y L(\bar{x}, \bar{y}, \lambda^j)$ is positive semidefinite on the null space of $\nabla_y g_{\mathcal{I}^j}(\bar{x}, \bar{y})$, then any one of the conditions (a)–(h) is further equivalent to the condition:

(i) $\nabla_y L(\bar{x}, \bar{y}, \lambda^j)$ is positive definite on the null space of $\nabla_y g_{\mathcal{I}_+^j}(\bar{x}, \bar{y})$.

Proof. (a) $\Leftrightarrow$ (b). For every column representative C of the pair $(\Lambda^j, \tilde{\Lambda}^j)$, let $\mathcal{J}$ be the subset of $\mathcal{I}_0^j$ such that the j-th column of C is equal to the j-th column of Λ^j. Then $\mathcal{I} \equiv \mathcal{J} \cup \mathcal{I}_+^j$ satisfies $\mathcal{I}_+^j \subseteq \mathcal{I} \subseteq \mathcal{I}^j$. Moreover, det C is equal to the determinant of the matrix displayed in (a). Thus the equivalence of (a) and (b) is clear.

Throughout the proof below, the index j will be dropped for simplicity.

(b) $\Leftrightarrow$ (c). By Lemma **4.2.35**, (b) is equivalent to the following system

$$
\Lambda \begin{pmatrix} dy^+ \\ d\lambda^+_{\mathcal{I}_+} \\ d\lambda^+_{\mathcal{I}_0} \end{pmatrix} = r + \tilde{\Lambda} \begin{pmatrix} dy^- \\ d\lambda^-_{\mathcal{I}_+} \\ d\lambda^-_{\mathcal{I}_0} \end{pmatrix}
$$

$$
\begin{pmatrix} dy^+ \\ d\lambda^+_{\mathcal{I}_+} \\ d\lambda^+_{\mathcal{I}_0} \end{pmatrix} \ge 0, \quad \begin{pmatrix} dy^- \\ d\lambda^-_{\mathcal{I}_+} \\ d\lambda^-_{\mathcal{I}_0} \end{pmatrix} \ge 0, \quad \begin{pmatrix} dy^+ \\ d\lambda^+_{\mathcal{I}_+} \\ d\lambda^+_{\mathcal{I}_0} \end{pmatrix}^T \begin{pmatrix} dy^- \\ d\lambda^-_{\mathcal{I}_+} \\ d\lambda^-_{\mathcal{I}_0} \end{pmatrix} = 0,
$$

having a unique solution for all vectors r. In turn it is easy to see that the latter holds if and only if (c) holds.

(c) $\Leftrightarrow$ (d). Since $\mathcal{I} \in \mathcal{B}(\bar{x})$, the gradients $\{\nabla_y g_i(\bar{x}, \bar{y}) : i \in \mathcal{I}\}$ are linearly independent. This implies the uniqueness of the multiplier vector associated with the constraints in the set $\mathcal{K}(p)$ that defines the AVI(r). Thus the equivalence of (c) and (d) is obvious.

(d) $\Leftrightarrow$ (e) $\Leftrightarrow$ (f). The unique solvability of the AVI(r) for all r is equivalent to the bijectivity of the normal map Γ. By a result of Robinson [253] for a normal map induced by a linear transformation, bijectivity is equivalent to global Lipschitzian homeomorphism which in turn is equivalent to coherent orientation. Thus (d), (e), and (f) are all equivalent.

(d) $\Leftrightarrow$ (g). This follows from the theory of mixed LCPs [96, Theorem 4(b)]; see also Robinson [248].

(g) $\Rightarrow$ (h). Suppose $\nabla_y F(\bar{x}, \bar{y})$ is symmetric. Then so is $\nabla_y L(\bar{x}, \bar{y}, \lambda)$ (which is possibly singular). It can be shown, by a continuity argument and the nonsingularity of Λ_+, that the Schur complement (42) is also symmetric. Thus this Schur complement is P if and only if it is positive definite. Let $(dy, d\lambda_{\mathcal{I}})$ satisfy the left-hand conditions in the implication in (g). Since $dy \ne 0$, the nonsingularity of the matrix Λ_+ implies that $d\lambda_{\mathcal{I}_0} \ne 0$. Thus

$$
(d\lambda_{\mathcal{I}_0})^T [\nabla_y g_{\mathcal{I}_0}(\bar{x}, \bar{y}) \ \ 0] \Lambda_+^{-1} \begin{bmatrix} \nabla_y g_{\mathcal{I}_0}(\bar{x}, \bar{y})^T \\ 0 \end{bmatrix} d\lambda_{\mathcal{I}_0} > 0.
$$

By the choice of $(dy, d\lambda_{\mathcal{I}})$, we deduce

$$\Lambda_+^{-1} \left[\begin{array}{c} \nabla_y g_{\mathcal{I}_0}(\bar{x}, \bar{y})^T \\ 0 \end{array} \right] d\lambda_{\mathcal{I}_0(\bar{x})} = - \left(\begin{array}{c} dy \\ d\lambda_{\mathcal{I}_+} \end{array} \right).$$

Consequently, we deduce

$$0 < -(d\lambda_{\mathcal{I}_0})^T \nabla_y g_{\mathcal{I}_0}(\bar{x}, \bar{y}) dy = dy^T \nabla_y L(\bar{x}, \bar{y}, \lambda) dy.$$

Hence (h) holds.

(h) $\Rightarrow$ (g). Suppose (h) holds. Then clearly the matrix Λ_+ must be nonsingular. By reversing the argument given in the last proof, it can be shown that the Schur complement in (g) is positive definite, thus P.

(i) $\Rightarrow$ (h). This is obvious and requires no assumption.

(h) $\Rightarrow$ (i). Assume that $\nabla_y F(\bar{x}, \bar{y})$ is symmetric, $\nabla_y L(\bar{x}, \bar{y}, \lambda)$ is positive semidefinite on the null space of $\nabla_y g_{\mathcal{I}}(\bar{x}, \bar{y})$, and (h) holds. It can easily be shown that $\nabla_y L(\bar{x}, \bar{y}, \lambda)$ must be positive definite on the same subspace. Indeed, if a nonzero dy belongs to the null space of $\nabla_y g_{\mathcal{I}}(\bar{x}, \bar{y})$ and

$$dy^T \nabla_y L(\bar{x}, \bar{y}, \lambda) dy = 0,$$

then $dy \in \operatorname{argmin}\{v^T \nabla_y L(\bar{x}, \bar{y}, \lambda)v : \nabla_y g_i(\bar{x}, \bar{y})^T v = 0, \ \forall i \in \mathcal{I}\}$. Thus there exist $d\lambda_{\mathcal{I}}$ such that the left-hand side of the implication in (h) holds. By (h), we obtain a contradiction.

Let u be a nonzero vector in the null space of $\nabla_y g_{\mathcal{I}_+}(\bar{x}, \bar{y})$. Consider the quadratic program: in the variable $v \in \Re^m$:

$$\text{minimize} \quad \tfrac{1}{2} v^T \nabla_y L(\bar{x}, \bar{y}, \lambda) v$$

$$\text{subject to} \quad \nabla_y g_i(\bar{x}, \bar{y})^T v = 0, \quad i \in \mathcal{I}_+,$$

$$\nabla_y g_i(\bar{x}, \bar{y})^T v = \nabla_y g_i(\bar{x}, \bar{y})^T u, \quad i \in \mathcal{I}_0.$$

The vector u is feasible to this problem; moreover, since $\nabla_y L(\bar{x}, \bar{y}, \lambda)$ is symmetric and positive definite on the null space of $\nabla_y g_{\mathcal{I}}(\bar{x}, \bar{y})$, the objective value of the above quadratic program must be bounded below on its feasible set. Thus this quadratic program has an optimum solution $\bar{v}$ which satisfies

$$u^T \nabla_y L(\bar{x}, \bar{y}, \lambda) u \geq \bar{v}^T \nabla_y L(\bar{x}, \bar{y}, \lambda) \bar{v}.$$

If $\bar{v} = 0$, then u belongs to the null space of $\nabla_y g_{\mathcal{I}}(\bar{x}, \bar{y})$; thus

$$u^T \nabla_y L(\bar{x}, \bar{y}, \lambda) u > 0.$$

If $\bar{v} \neq 0$, then by (h) it follows that $\bar{v}^T \nabla_y L(\bar{x}, \bar{y}, \lambda) \bar{v} > 0$. Consequently, (i) holds.   **Q.E.D.**

The assumption that $\nabla_y L(\bar{x}, \bar{y}, \lambda^j)$ is positive semidefinite on the null space of $\nabla_y g_{\mathcal{I}^j}(\bar{x}, \bar{y})$ is satisfied if $\nabla_y L(\bar{x}, \bar{y}, \lambda^j)$ is *copositive* on the cone $\mathcal{K}^j(0)$; i.e., if

$$v \in \mathcal{K}^j(0) \implies v^T \nabla_y L(\bar{x}, \bar{y}, \lambda^j) v \geq 0.$$

In turn this copositivity condition is derived from the case of a bilevel program where $F(x, y) = \nabla_y \theta(x, y)$ for some real-valued function θ; in this case, the VI $(F(\bar{x}; \cdot), C(\bar{x}))$ is the stationary point problem of the nonlinear program in the variable y:

$$\text{minimize} \quad \theta(\bar{x}, y)$$

$$\text{subject to} \quad y \in C(\bar{x}).$$

It is well known that if $\bar{y}$ is a local minimum of this program where the LICQ holds, then $M(\bar{x})$ is a singleton, say $M(\bar{x}) = \{\lambda^j\}$, and with $\mathcal{I}^j = \mathcal{I}(\bar{x})$, $\mathcal{K}^j(0)$ is equal to the critical cone of the above nonlinear program at $\bar{y}$; thus, the matrix $\nabla_y L(\bar{x}, \bar{y}, \lambda^j)$ must be copositive on $\mathcal{K}^j(0)$; see Remark **5.1.2**.

In what follows, we apply Lemma **4.2.36** to obtain necessary and sufficient conditions for SCOC to hold. Before stating the main result, we recall the fact that although each extreme multipler $\lambda \in M^e(\bar{x})$ may yield multiple elements $\mathcal{I}^j \in \mathcal{B}(\bar{x})$ such that $\text{supp}(\lambda) \subseteq \mathcal{I}^j$, the matrices Λ^j defined by these $\mathcal{I}^j$ all contain Λ_+^j as a principal submatrix. Moreover, Λ_+^j depends only on the given multiplier λ and not on the index set $\mathcal{I}^j$. We call Λ_+^j the *basic matrix* associated with the multiplier $\lambda \in M^e(\bar{x})$. Let $\mathcal{M}(\bar{x})$ denote the family of these basic matrices (not to be confused with the multiplier set $M(\bar{x})$). If the SMFCQ holds at $\bar{x}$, then the family $\mathcal{M}(\bar{x})$ is a singleton, although the family $\mathcal{B}(\bar{x})$ may contain multiple elements.

4.2.37 Theorem. Under the MFCQ at $(\bar{x}, \bar{y})$, the following two statements are equivalent.

(a) The SCOC holds at $(\bar{x}, \bar{y})$.

(b) There is an integer $\sigma \in \{-1, 1\}$ such that for all matrices $\Lambda_+ \in \mathcal{M}(\bar{x})$,

$$\text{sgn det } \Lambda_+ = \sigma;$$

moreover, for each index set $\mathcal{I}^j \in \mathcal{B}(\bar{x})$ with corresponding matrix $\Lambda_+^j \in \mathcal{M}(\bar{x})$, the Schur complement (42) is a P matrix.

Proof. Suppose (a) holds. For each extreme multiplier $\lambda \in M^e(\bar{x})$, $\text{supp}(\lambda) \in \mathcal{B}(\bar{x})$. Hence SCOC implies that sgn det Λ_+ is a nonzero constant for all $\Lambda_+ \in \mathcal{M}(\bar{x})$.

Consider an arbitrary index j and the pair $(\Lambda^j, \tilde{\Lambda}^j)$, where $\tilde{\Lambda}^j$ is defined in (41). For any subset $\mathcal{J} \subseteq \mathcal{I}_0^j$, the SCOC says that the matrix

$$\begin{bmatrix} \nabla_y L(\bar{x}, \bar{y}, \lambda^j) & \nabla_y g_{\mathcal{I}_+^j}(\bar{x}, \bar{y})^T & \nabla_y g_{\mathcal{J}}(\bar{x}, \bar{y})^T \\ -\nabla_y g_{\mathcal{I}_+^j}(\bar{x}, \bar{y}) & 0 & 0 \\ -\nabla_y g_{\mathcal{J}}(\bar{x}, \bar{y}) & 0 & 0 \end{bmatrix} \quad (43)$$

has the same nonzero determinantal sign as the matrix Λ_+^j. If Λ is a column representative of the pair $(\Lambda^j, \tilde{\Lambda}^j)$, let $\mathcal{J}$ denote the indices $j \in \mathcal{I}_0^j$ such that the j-th column of Λ is equal to the j-th column of Λ^j. Then clearly, det Λ is equal to the determinant of the matrix in (43). Thus, the determinantal signs of all column representatives of the pair $(\Lambda^j, \tilde{\Lambda}^j)$ are equal to sgn det Λ_+^j which is a nonzero constant. By Lemma **4.2.36**, (b) follows.

Conversely, suppose (b) holds. By Lemma **4.2.36** again, for each j,

$$\text{sgn det } \Lambda^j = \text{sgn det } \tilde{\Lambda}^j = \text{sgn det } \Lambda_+^j,$$

with the latter being a nonzero constant. Thus the SCOC holds. **Q.E.D.**

Theorem **4.2.37** can be sharpened under the LICQ. This sharpening is possible because of the fact that $\mathcal{I}(\bar{x}) \in \mathcal{B}(\bar{x})$ if LICQ holds at $(\bar{x}, \bar{y})$. The characterization of the SCOC can now be stated in terms of this one index set $\mathcal{I}(\bar{x})$ only.

4.2.38 Corollary. Suppose that LICQ holds at $(\bar{x}, \bar{y})$. Let $M(\bar{x}) \equiv \{\bar{\lambda}\}$ and

$$\mathcal{I}_+ \equiv \{i \in \mathcal{I}(\bar{x}) : \bar{\lambda}_i > 0\}, \quad \mathcal{I}_0 \equiv \{i \in \mathcal{I}(\bar{x}) : \bar{\lambda}_i = 0\}.$$

Let Λ_+ be the basic matrix associated with $\bar{\lambda}$. The following statements are equivalent.

(a) The SCOC holds at $(\bar{x}, \bar{y})$.

(b) The matrix Λ_+ is nonsingular and the Schur complement of Λ_+ in Λ, i.e., the matrix

$$[\nabla_y g_{\mathcal{I}_0}(\bar{x}, \bar{y}) \ \ 0](\Lambda_+)^{-1} \begin{bmatrix} \nabla_y g_{\mathcal{I}_0}(\bar{x}, \bar{y})^T \\ 0 \end{bmatrix}$$

is a P matrix.

(c) For all vectors $r \equiv (q, p) \in \Re^{m+|\mathcal{I}|}$, the KKT system

$$q + \nabla_y L(\bar{x}, \bar{y}, \bar{\lambda}) dy + \sum_{i \in \mathcal{I}(\bar{x})} d\lambda_i \nabla_y g_i(\bar{x}, \bar{y}) = 0,$$

$$\left. \begin{array}{l} d\lambda_i \geq 0, \quad d\lambda_i \left(p_i - \nabla_y g_i(\bar{x}, \bar{y})^T dy \right) = 0 \\ p_i - \nabla_y g_i(\bar{x}, \bar{y})^T dy \geq 0 \end{array} \right\} \text{ if } i \in \mathcal{I}_0,$$

$$d\lambda_i \text{ free}, \quad p_i - \nabla_y g_i(\bar{x}, \bar{y})^T dy = 0, \quad \text{if } i \in \mathcal{I}_+,$$

has a unique solution $(dy, d\lambda_{\mathcal{I}(\bar{x})})$.

(d) For all vectors $r \equiv (q, p) \in \Re^{m+|\mathcal{I}|}$, the

$$\text{AVI } (q, \nabla_y L(\bar{x}, \bar{y}, \bar{\lambda}), \mathcal{K}(p)),$$

where

$$\mathcal{K}(p) \equiv \{ dy \in \Re^m : \ dy^T \nabla_y g_i(\bar{x}, \bar{y}) \leq p_i, \text{ for } i \in \mathcal{I}_0,$$

$$dy^T \nabla_y g_i(\bar{x}, \bar{y}) = p_i, \text{ for } i \in \mathcal{I}_+ \},$$

has a unique solution.

(e) With $\mathcal{C} \equiv \Re^{m+|\mathcal{I}_+|} \times \Re_+^{|\mathcal{I}_0|}$ and

$$\Lambda \equiv \begin{bmatrix} \nabla_y L(\bar{x}, \bar{y}, \bar{\lambda}) & \nabla_y g_{\mathcal{I}_+}(\bar{x}, \bar{y})^T & \nabla_y g_{\mathcal{I}_0}(\bar{x}, \bar{y})^T \\ -\nabla_y g_{\mathcal{I}_+}(\bar{x}, \bar{y}) & 0 & 0 \\ -\nabla_y g_{\mathcal{I}_0}(\bar{x}, \bar{y}) & 0 & 0 \end{bmatrix},$$

the normal map, $\Gamma \equiv \Lambda \circ \Pi_{\mathcal{C}} + I - \Pi_{\mathcal{C}}$, is a global Lipschitzian homeomorphism.

(f) The normal map Γ defined in part (e) is coherently oriented.

If, in addition, $\nabla_y F(\bar{x}, \bar{y})$ is symmetric, then any one of the conditions (a)–(f) is further equivalent to the condition:

(g) The following implication in the variables $(dy, d\lambda_{\mathcal{I}(\bar{x})})$ holds:

$$\left.\begin{array}{l} \nabla_y L(\bar{x}, \bar{y}, \bar{\lambda})dy + \sum_{i \in \mathcal{I}(\bar{x})} d\lambda_i \nabla_y g_i(\bar{x}, \bar{y}) = 0 \\ \nabla_y g_i(\bar{x}, \bar{y})^T dy = 0, \quad i \in \mathcal{I}_+ \\ dy \neq 0 \end{array}\right\} \Rightarrow dy^T \nabla_y L(\bar{x}, \bar{y}, \bar{\lambda})dy > 0.$$

Finally, if $\nabla_y F(\bar{x}, \bar{y})$ is symmetric and $\nabla_y L(\bar{x}, \bar{y}, \bar{\lambda})$ is positive semidefinite on the null space of $\nabla_y g_{\mathcal{I}(\bar{x})}(\bar{x}, \bar{y})$, then any one of the conditions (a)–(g) is further equivalent to the condition:

(h) $\nabla_y L(\bar{x}, \bar{y}, \bar{\lambda})$ is positive definite on the null space of $\nabla_y g_{\mathcal{I}_+}(\bar{x}, \bar{y})$.

Proof. It suffices to note that for every $\mathcal{I}^j \in \mathcal{B}(\bar{x})$, the Schur complement (42) is a principal submatrix of the Schur complement in part (b). Consequently, part (b) of this corollary is equivalent to part (b) of Theorem **4.2.37**. The equivalence of part (b) and of the remaining statements in this corollary follows from Lemma **4.2.36**. **Q.E.D.**

We close this section with some further discussion of the various conditions in the above results. The $\mathcal{W}$ property of pairs of matrices is used again in the next chapter to establish the optimality conditions of an MP with generalized complementarity constraints. Under the LICQ, condition (b) and hence each of the other conditions in Corollary **4.2.38** is a necessary and sufficient condition for Robinson's "strong regularity condition" to hold for the KKT system of the parametric VI $(F(x, \cdot), C(x))$ at the triple $(\bar{x}, \bar{y}, \bar{\lambda})$. The proof of (h) $\Rightarrow$ (i) was borrowed from [223] which in turn was inspired by results in [33].

Lemma **4.2.36** and Corollary **4.2.38** are related to results of many authors, including Kojima [140], Hirabayashi, Jongen, and Shida [111], Jongen, Mobert, Rückmann, and Tammer [126], Jongen, Klatte, and Tammer [125], and Klatte and Tammer [137]. In these references, the authors use conditions similar to SCOC to characterize various stability and regularity properties of KKT triples of parametric nonlinear programs.

Finally, the CRCQ has played an important role in establishing the PC^1 property of the solution function $y(x)$. Some previous results in the area of parametric nonlinear programming [140, 60, 261] have established the existence of a directionally differentiable (but not necessarily Lipschitz continuous) function $y(x)$ under the assumption of MFCQ and a standard strong second-order sufficiency condition, but without CRCQ. The question of whether the basic CQ (and not necessarily the extreme CQ) will hold for MPEC without CRCQ remains to be investigated. We shall provide a partial answer to this question in the next section by using a piecewise approach to deal with the MPEC, which is quite different from the IMP approach that we have used in this section.

4.3　A Piecewise Programming Approach

A major drawback of the class of MPECs studied in the previous section, namely those for which SCOC and other conditions guarantee the existence of a PC^1 implicit solution function $y(x)$ of the inner VI as stated in BIF, and for which $Z = X \times \Re^m$, is that it excludes significant MPECs of interest. To illustrate the restrictiveness of BIF, consider the following very simple case of the MPEC (3.1.1) which has $Z = \Re^{n+m}$ and $C(x) = \Re^m$ for all $x \in \Re^n$:

$$
\begin{aligned}
\text{minimize} \quad & f(x,y) \\
\text{subject to} \quad & F(x,y) = 0.
\end{aligned}
\tag{1}
$$

This is just a standard equality-constrained nonlinear program. Suppose that $\bar{z} \equiv (\bar{x}, \bar{y})$ is a local minimizer of this problem. From classical Lagrangean theory, we know that if the Jacobian matrix $\nabla F(\bar{z})$ has full row rank, then a Lagrange multiplier $\pi \in \Re^m$ exists such that

$$
\nabla f(\bar{z}) + \sum_{i=1}^{m} \pi_i \nabla F_i(\bar{z}) = 0.
$$

The IMP approach would assume nonsingularity of the partial Jacobian matrix $\nabla_y F(\bar{z})$. Needless to say, this assumption is more restrictive than the assumption of full rank of $\nabla F(\bar{z})$, which only implies that near $(\bar{x}, \bar{y})$ there exist some n independent variables among (x, y) (not necessarily x) in terms of which we can solve for the remaining m variables (not necessarily y). In essence, the PCP approach described in this section allows us to directly apply standard nonlinear programming results to various smooth components of the KKT constrained MP; see (3.3.10). We briefly allude to this approach in Subsection **3.3.1**; in this section we explore the PCP approach more fully.

Consider the MPEC formulated via KKT constraints:

$$
\begin{aligned}
\text{minimize} \quad & f(x,y) \\
\text{subject to} \quad & h(x,y) \le 0, \\
& L(x,y,\lambda) = 0, \\
& g(x,y) \le 0, \quad \lambda \ge 0, \quad \lambda^T g(x,y) = 0,
\end{aligned}
\tag{2}
$$

where we assume that $Z \equiv \{(x,y) : h(x,y) \le 0\}$ with $h : \Re^{n+m} \to \Re^q$ being a continuously differentiable function. In what follows, the notation in Sections **3.2** and **3.3** is adhered to. Let $\mathcal{F}^{\mathrm{KKT}} \subseteq \mathcal{F} \times \Re_+^\ell$ denote the feasible region of (2). Recall the "pieces" of this problem, which are defined by (3.3.10) when Z is a polyhedral set. To be specific, for each pair of index sets $\mathcal{J}_1$ and $\mathcal{J}_2$ partitioning $\{1,\dots,\ell\}$, we consider the nonlinear program in the variable (x,y,λ):

$$
\begin{aligned}
\text{minimize} \quad & f(x,y) \\
\text{subject to} \quad & h(x,y) \le 0, \\
& L(x,y,\lambda) = 0, \\
& \left.\begin{array}{l} \lambda_i = 0 \\[4pt] g_i(x,y) \le 0 \end{array}\right\} \forall i \in \mathcal{J}_1, \\
& \left.\begin{array}{l} \lambda_i \ge 0 \\[4pt] g_i(x,y) = 0 \end{array}\right\} \forall i \in \mathcal{J}_2.
\end{aligned}
\tag{3}
$$

Let $\mathcal{F}^{\mathrm{KKT}}_{(\mathcal{J}_1,\mathcal{J}_2)} \subseteq \Re^{n+m} \times \Re_+^\ell$ denote the feasible region of the latter problem. Clearly, we have

$$
\mathcal{F}^{\mathrm{KKT}} = \bigcup_{(\mathcal{J}_1,\mathcal{J}_2)} \mathcal{F}^{\mathrm{KKT}}_{(\mathcal{J}_1,\mathcal{J}_2)},
\tag{4}
$$

where the union ranges over all pairs of index sets $\mathcal{J}_1$ and $\mathcal{J}_2$ partitioning $\{1,\dots,\ell\}$.

Let $\bar{z} \equiv (\bar{x},\bar{y})$ be a given feasible vector of (2) and $\bar{\lambda} \in M(\bar{z})$ be arbitrary. For simplicity, we assume that $h(\bar{z}) = 0$. We recall some notation introduced in Subsection **3.3.1**:

$$\mathcal{I}(\bar{z}) \equiv \{i : g_i(\bar{z}) = 0\},$$

$$\mathcal{I}_0(\bar{z}, \bar{\lambda}) \equiv \{i \in \mathcal{I}(\bar{z}) : \bar{\lambda}_i = 0\}, \quad \text{and} \quad \mathcal{I}_+(\bar{z}, \bar{\lambda}) \equiv \{i \in \mathcal{I}(\bar{z}) : \bar{\lambda}_i > 0\}.$$

Let α and $\bar{\alpha}$ be arbitrary index sets partitioning $\mathcal{I}_0(\bar{z}, \bar{\lambda})$ and define $\mathcal{J}_1$ and $\mathcal{J}_2$ as in (3.3.11); that is,

$$\mathcal{J}_1 \equiv \bar{\alpha} \cup \{i : g_i(\bar{z}) < 0\}, \quad \text{and} \quad \mathcal{J}_2 \equiv \alpha \cup \mathcal{I}_+(\bar{z}, \bar{\lambda}). \tag{5}$$

For such a pair $(\mathcal{J}_1, \mathcal{J}_2)$, we may write down the standard MFCQ specialized to (3) at the triple $(\bar{x}, \bar{y}, \bar{\lambda})$. This says: the matrix

$$\begin{bmatrix} \nabla_z L(\bar{z}, \bar{\lambda}) & \nabla_y g_{\mathcal{J}_2}(\bar{z})^T \\ \nabla g_{\mathcal{J}_2}(\bar{z}) & 0 \end{bmatrix} \tag{6}$$

has full row rank, and there exists a triple $(dx, dy, d\lambda_{\mathcal{J}_2})$ such that

$$
\begin{aligned}
\nabla h_j(\bar{z})^T dz &< 0, \quad j = 1, \ldots, q \\
\nabla_z L(\bar{z}, \bar{\lambda}) dz + \sum_{i \in \mathcal{J}_2} d\lambda_i \nabla_y g_i(\bar{z}) &= 0, \\
\nabla g_i(\bar{z})^T dz &< 0, \quad i \in \bar{\alpha}, \\
d\lambda_i &> 0, \quad i \in \alpha \\
\nabla g_i(\bar{z})^T dz &= 0, \quad i \in \mathcal{J}_2.
\end{aligned}
\tag{7}
$$

Notice that the size of the matrix (6) grows as α becomes larger. In particular, each of these matrices is a submatrix of the following matrix

$$\begin{bmatrix} \nabla_z L(\bar{z}, \bar{\lambda}) & \nabla_y g_{\mathcal{I}(\bar{z})}(\bar{z})^T \\ \nabla g_{\mathcal{I}(\bar{z})}(\bar{z}) & 0 \end{bmatrix}.$$

Under the MFCQ for (3) as described above, we have

$$\mathcal{T}(\bar{z}; Z) = \{dz \in \Re^n : \nabla h_j(\bar{z})^T dz \leq 0, \quad j = 1, \ldots, q\},$$

and the tangent cone of $\mathcal{F}_{(\mathcal{J}_1, \mathcal{J}_2)}^{\mathrm{KKT}}$ at the triple $(\bar{x}, \bar{y}, \bar{\lambda})$, $\mathcal{T}\left((\bar{z}, \bar{\lambda}); \mathcal{F}_{(\mathcal{J}_1, \mathcal{J}_2)}^{\mathrm{KKT}}\right)$, is the set of points $(dz, d\lambda)$ in $\mathcal{T}(\bar{z}; Z) \times \Re^\ell$ such that

$$
\begin{aligned}
\nabla_z L(\bar{z}, \bar{\lambda}) dz + \sum_{i \in \mathcal{J}_2} d\lambda_i \nabla_y g_i(\bar{z}) &= 0, \\
\nabla g_i(\bar{z})^T dz &\leq 0, \quad i \in \bar{\alpha}, \\
\nabla g_i(\bar{z})^T dz &= 0, \quad i \in \mathcal{J}_2 \qquad (8) \\
d\lambda_i &= 0, \quad \text{if } i \in \mathcal{J}_1 \\
d\lambda_i &\geq 0, \quad \text{if } i \in \alpha.
\end{aligned}
$$

Using this representation, we can relate the above tangent cone to the cone $\mathrm{Gr}(\mathcal{LS}_{(\bar{z}, \bar{\lambda})})$ defined in Section **3.2**. Specifically, we have the following result; see also Proposition **3.4.3**.

4.3.1 Lemma. In the above setting, $\mathcal{T}(\bar{z}; Z) \cap \mathrm{Gr}(\mathcal{LS}_{(\bar{z}, \bar{\lambda})})$ is the projection of

$$
\bigcup_{(\mathcal{J}_1, \mathcal{J}_2)} \mathcal{T}\left((\bar{z}, \bar{\lambda}); \mathcal{F}^{\mathrm{KKT}}_{(\mathcal{J}_1, \mathcal{J}_2)} \right)
$$

onto $\Re^{n+m}$, where $(\mathcal{J}_1, \mathcal{J}_2)$ in the above union ranges over all pairs defined by (5) for all $(\alpha, \bar{\alpha})$ partitioning $\mathcal{I}_0(\bar{z}, \bar{\lambda})$.

Proof. Indeed, if $(dz, d\lambda) \in \mathcal{T}\left((\bar{z}, \bar{\lambda}); \mathcal{F}^{\mathrm{KKT}}_{(\mathcal{J}_1, \mathcal{J}_2)} \right)$, then $dz \in \mathcal{T}(\bar{z}; Z)$ and $(dz, d\lambda)$ satisfies (8). Consequently, by recalling the definition of $\mathrm{Gr}(\mathcal{LS}_{(\bar{z}, \bar{\lambda})})$ and by the equivalence of the

$$
\text{AVI } \left(\nabla_x L(\bar{z}, \bar{\lambda}) dx, \nabla_y L(\bar{z}, \bar{\lambda}), \mathcal{K}(\bar{z}, \bar{\lambda}; dx) \right), \qquad (9)
$$

and its formulation using KKT constraints, it follows that dz belongs to $\mathrm{Gr}(\mathcal{LS}_{(\bar{z}, \bar{\lambda})})$.

Conversely, suppose $dz \in \mathcal{T}(\bar{z}; Z) \cap \mathrm{Gr}(\mathcal{LS}_{(\bar{z}, \bar{\lambda})})$. Define

$$
\alpha \equiv \{ i \in \mathcal{I}_0(\bar{z}, \bar{\lambda}) : \nabla g_i(\bar{z})^T dz = 0 \},
$$

and let $\bar{\alpha}$ be the complement of α in $\mathcal{I}_0(\bar{z}, \bar{\lambda})$. Letting $d\lambda_{\mathcal{I}(\bar{z})}$ be the multiplier corresponding to dy as a solution of the AVI (9), and $d\lambda_i \equiv 0$ for $i \notin \mathcal{I}(\bar{z})$, we deduce that $(dz, d\lambda) \in \mathcal{T}\left((\bar{z}, \bar{\lambda}); \mathcal{F}^{\mathrm{KKT}}_{(\mathcal{J}_1, \mathcal{J}_2)} \right)$. **Q.E.D.**

Using the above lemma, we can establish the following main result of this section. See [147, 257] for an interesting, related application of a *piecewise MFCQ* to obtain stationarity conditions in constrained nonsmooth optimization. The diploma thesis [257] also used this CQ within the piecewise programming approach to study exact penalty functions for MPECs. The following result concerns the basic CQ for MPEC and omits the first-order conditions; for the latter, see (3.3.12).

4.3.2 Theorem. Let $Z \equiv \{(x, y) : h(x, y) \le 0\}$. Assume that the SBCQ holds at $\bar{z} \in \mathcal{F}$ and that $\bar{z}$ satisfies $h(\bar{z}) = 0$. Suppose that $g_i(x, \cdot)$ is convex for all x belonging to the projection of Z onto $\Re^n$. If there exists a multiplier set $M' \subseteq M(\bar{z})$ such that $\mathcal{T}(\bar{z}; \mathcal{F}) \subseteq \mathcal{L}'(\bar{z}; \mathcal{F})$ and for every $\bar{\lambda} \in M'$, the following two properties hold for every pair of index sets $(\alpha, \bar{\alpha})$ partitioning $\mathcal{I}_0(\bar{z}, \bar{\lambda})$ and for $\mathcal{J}_2$ defined by (5):

(i) the matrix (6) is nonsingular, and

(ii) there exists a triple $(dx, dy, d\lambda_{\mathcal{J}_2})$ satisfying (7),

then

$$\mathcal{T}(\bar{z}; \mathcal{F}) = \mathcal{L}'(\bar{z}; \mathcal{F});$$

that is, the basic CQ holds for the MPEC (3.1.1) at $\bar{z}$ with this M'.

Proof. The hypotheses imply that for every $\bar{\lambda} \in M'$ and every pair $(\alpha, \bar{\alpha})$ partitioning $\mathcal{I}_0(\bar{z}, \bar{\lambda})$, the MFCQ holds for the set $\mathcal{F}_{(\mathcal{J}_1, \mathcal{J}_2)}^{\text{KKT}}$ at $(\bar{z}, \bar{\lambda})$, where $(\mathcal{J}_1, \mathcal{J}_2)$ is defined in (5).

It suffices to establish

$$\mathcal{T}(\bar{z}; \mathcal{F}) \supseteq \mathcal{L}'(\bar{z}; \mathcal{F}).$$

Let dz belong to the right-hand set; then $dz \in \mathcal{T}(\bar{z}; Z) \cap \text{Gr}(\mathcal{LS}_{(\bar{z}, \bar{\lambda})})$ for some $\bar{\lambda} \in M'$. By Lemma **4.3.1**, it follows that for some pair of index sets $(\mathcal{J}_1, \mathcal{J}_2)$ partitioning $\{1, \dots, \ell\}$ and some vector $d\lambda \in \Re_+^\ell$, $(dz, d\lambda)$ belongs to $\mathcal{T}\left((\bar{z}, \bar{\lambda}); \mathcal{F}_{(\mathcal{J}_1, \mathcal{J}_2)}^{\text{KKT}}\right)$. Since $\mathcal{F}_{(\mathcal{J}_1, \mathcal{J}_2)}^{\text{KKT}}$ is a subset of the feasible region of (2) and since the functions $g_i(x, \cdot)$ are convex, it follows easily that $dz \in \mathcal{T}(\bar{z}; \mathcal{F})$, as desired.   **Q.E.D.**

The following corollary is immediate.

4.3.3 Corollary. In the setting of Theorem **4.3.2**,

(a) if conditions (i) and (ii) hold for all $\bar{\lambda} \in M(\bar{z})$, then the full CQ holds for the MPEC (3.1.1) at $\bar{z}$;

(b) if CRCQ holds at $\bar{z}$ and conditions (i) and (ii) hold for all $\bar{\lambda} \in M^e(\bar{z})$, then the extreme CQ holds for the MPEC (3.1.1) at $\bar{z}$.

Proof. Take M' to be $M(\bar{z})$ in (a) and $M^e(\bar{z})$ in (b); apply Lemma **3.2.1** and Theorem **4.3.2**.   **Q.E.D.**

Although there are only finitely many index subsets of $\{1, \ldots, \ell\}$, there are infinitely many matrices of the form (6) and linear systems (7) if $M(\bar{z})$ is not a singleton. Thus the hypotheses in Corollary **4.3.3** are generally much harder to verify than the SCOC. Under the SMFCQ at the triple $(\bar{x}, \bar{y}, \bar{\lambda})$, there are finitely many matrices (6) and linear systems (7); hence the two conditions (i) and (ii) can be checked in finite time. If $M(\bar{z})$ is not a singleton, the same conclusion holds under the CRCQ.

Theorem **4.3.2** and Corollary **4.3.3** are based on the MFCQ on the component problems (3). Alternatively, other CQs could be imposed on these problems and the two results will remain valid. We leave the details for the readers to explore.

We conclude with a brief comparison of the SCOC, sufficient for the implicit function assumption BIF, and the above MFCQ for appropriate index sets α, $\bar{\alpha}$, and $\mathcal{J}_2$. (For the sake of this discussion, we do not compare SCOC with other CQs imposed on the problems (3)). The SCOC makes a nonzero constant sign (hence nonsingularity) restriction on particular submatrices of (6); see (4.2.19). These submatrices involve only rows and columns from partial derivatives with respect to y of L and g. The piecewise MFCQ requires full row rank of matrices of the form (6), which involves full derivatives of L and g, and thereby seems easier to satisfy. Nevertheless, the full row rank requirement implies in particular that the gradients

$$\{\nabla g_i(\bar{z}) : i \in \mathcal{I}(\bar{z})\} \subset \Re^{n+m}$$

must be linearly independent. These gradients pertain to *all* active constraints at $\bar{z}$; on the contrary, SCOC does not necessarily involve all such constraints; cf. the family $\mathcal{B}(\bar{x})$ of index sets. In addition, condition (ii) in Theorem **4.3.2** imposes existence of certain vectors satisfying (7), which is a kind of interiority assumption on the feasible region of (3). The fact that the PCP approach allows joint upper-level constraints $z \in Z$, where Z is not necessarily of the restrictive form $X \times \Re^m$, is an advantage of this

212   Chapter 4.   Verification of MPEC Hypotheses

approach whose total merit we cannot completely gauge at this point. Finally, the tradeoff between relaxing the coherent orientation of the matrices (4.2.19) and adding conditions like existence of solutions of (7) needs to be investigated further.

We give an example to illustrate the above comparison.

4.3.4 Example. Consider the following MPEC in the variable $(x, y) \in \Re^4$:

$$\text{minimize} \quad f(x, y)$$
$$\text{subject to} \quad (x, y) \in \Re^2 \times \Re^2,$$
$$(y_1, y_2) \in \text{argmin}\{(y_1 - 1)^2 + y_2^2 : y_1^2 \leq 1, \ y_1 \leq 1\}.$$

In this example, the inner problem is independent of x and has a strictly convex quadratic objective function; its unique optimal solution is given by $\bar{y} = (1, 0)$. Thus the feasible set of the MPEC is $\mathcal{F} = \Re^2 \times \{(1, 0)\}$. Let $\bar{x} = (0, 0)$. Then both constraints for the inner problem are active at $\bar{z}$ and the multiplier set at this point is a singleton, $\bar{\lambda} = \{(0, 0)\}$. Thus, $\mathcal{I}(\bar{z}) = \mathcal{I}_0(\bar{z}) = \{1, 2\}$. Moreover, the SMFCQ and CRCQ hold for the inner problem at $\bar{z}$. Since the gradient vectors of the two active constraints at the point $\bar{z}$ are linearly dependent, we have $\mathcal{B}(\bar{x}) = \{\{1\}, \{2\}\}$. It can then be verified that the SCOC condition holds for this MPEC at $\bar{z}$. However, if we let $\alpha = \mathcal{I}_0(\bar{z})$ and consider the following component of the KKT constrained MP:

$$\text{minimize} \quad f(x, y)$$
$$\text{subject to} \quad (x, y) \in \Re^2 \times \Re^2,$$
$$2(y_1 - 1) + 2\lambda_1 y_1 + \lambda_2 = 0,$$
$$2y_2 = 0,$$
$$y_1^2 = 1, \quad y_1 = 1,$$
$$\lambda_1 \geq 0, \quad \lambda_2 \geq 0,$$

we see that the gradient vectors of the constraints $y_1^2 = 1$ and $y_1 = 1$ are linearly dependent at $(\bar{z}, \bar{\lambda})$; hence the MFCQ cannot hold for the above problem at this point.

Although we have used the equality-constrained nonlinear program (1) to illustrate the deficiency of the IMP approach, this program is not representative of a typical MPEC in which the inner VI $(F(x, \cdot), C(x))$ is much more complicated than just a system of equations. For such an MPEC, both the IMP and PCP approach will have their own contributions and neither approach appears to dominate the other.

4.3.1 *Uniqueness of MPEC multipliers*

For a given triple $(\bar{x}, \bar{y}, \bar{\lambda}) \in \mathcal{F}^{\mathrm{KKT}}$, there are multiple nonlinear programs (3) corresponding to various pairs $(\mathcal{J}_1, \mathcal{J}_2)$ each of which is defined by a subset α of $\mathcal{I}_0(\bar{z}, \bar{\lambda})$. In turn, the KKT system of each of these programs contributes to the overall conditions for the triple $(\bar{x}, \bar{y}, \bar{\lambda})$ to be a stationary point of the MPEC (2); see (3.3.12). For ease of reference, we restate this KKT system as follows, where $\alpha \cup \bar{\alpha}$ is some partition of $\mathcal{I}_0(\bar{z}, \bar{\lambda})$ and $\mathcal{J}_1$, $\mathcal{J}_2$ are defined by (5): $(\bar{x}, \bar{y}, \bar{\lambda}) \in \mathcal{F}^{\mathrm{KKT}}_{(\mathcal{J}_1, \mathcal{J}_2)}$ and

$$\nabla_x f(\bar{z}) + \nabla_x h(\bar{z})^T \zeta + \nabla_x g(\bar{z})^T \eta = \nabla_x L(\bar{z}, \bar{\lambda})^T \pi$$

$$\nabla_y f(\bar{z}) + \nabla_y h(\bar{z})^T \zeta + \nabla_y g(\bar{z})^T \eta = \nabla_y L(\bar{z}, \bar{\lambda})^T \pi$$

$$\pi^T \nabla_y g_i(\bar{z}) \leq 0, \quad i \in \alpha$$

$$\pi^T \nabla_y g_i(\bar{z}) = 0, \quad i \in \mathcal{I}_+(\bar{z}, \bar{\lambda}) \qquad (10)$$

$$\eta_i \geq 0, \quad i \in \bar{\alpha}$$

$$\eta_i = 0, \quad i \notin \mathcal{I}(\bar{z})$$

$$\zeta \geq 0, \quad \zeta^T h(\bar{z}) = 0.$$

In general, the MPEC multipliers (ζ, π, η) in the latter conditions depend on the index set α. These systems (10) for different index sets α are not completely unrelated. Exploiting their interconnections, it becomes possible to impose suitable conditions on one particular nonlinear program, called the *relaxed* NLP, under which we can show that there exists a unique MPEC multiplier triple which satisfies the conditions (10) or (3.3.12) simultaneously for *all* $\alpha \subseteq \mathcal{I}_0(\bar{z}, \bar{\lambda})$. This uniqueness result has important algorithmic implications which are discussed in Section **6.4**. As illustrated by a subsequent example, this result does not imply, however, that for a given pair $(\alpha, \bar{\alpha})$, (10) will necessarily have a unique multiplier (ζ, π, η).

Specifically, for a given $(\bar{z}, \bar{\lambda}) \in \mathcal{F}^{\text{KKT}}$, the relaxed nonlinear program is the following one:

$$\begin{aligned}
\text{minimize} \quad & f(x, y) \\
\text{subject to} \quad & h(x, y) \leq 0, \\
& L(x, y, \lambda) = 0,
\end{aligned}$$

$$\left. \begin{aligned}
\lambda_i &= 0 \\
g_i(x, y) &\leq 0
\end{aligned} \right\} \; \forall i \notin \mathcal{I}(\bar{z}), \tag{11}$$

$$\left. \begin{aligned}
\lambda_i &\geq 0 \\
g_i(x, y) &\leq 0
\end{aligned} \right\} \; \forall i \in \mathcal{I}_0(\bar{z}, \bar{\lambda}),$$

$$\left. \begin{aligned}
\lambda_i &\geq 0 \\
g_i(x, y) &= 0
\end{aligned} \right\} \; \forall i \in \mathcal{I}_+(\bar{z}, \bar{\lambda}).$$

Notice that a triple (x, y, λ) feasible to the latter program is not necessarily feasible to (2); this is because no complementarity is enforced between λ_i and $g_i(x, y)$ for $i \in \mathcal{I}_0(\bar{z}, \bar{\lambda})$. Nevertheless, every feasible solution (x, y, λ) of (2) that is sufficiently close to $(\bar{x}, \bar{y}, \bar{\lambda})$ must be feasible to (11). Hence (11) is a local relaxation of (2) near $(\bar{x}, \bar{y}, \bar{\lambda})$.

The KKT system of (11) at the triple $(\bar{x}, \bar{y}, \bar{\lambda})$ is: $(\bar{x}, \bar{y}, \bar{\lambda})$ is feasible for (11) and satisfies

$$\nabla_x f(\bar{z}) + \nabla_x h(\bar{z})^T \zeta + \nabla_x g(\bar{z})^T \eta = \nabla_x L(\bar{z}, \bar{\lambda})^T \pi$$

$$\nabla_y f(\bar{z}) + \nabla_y h(\bar{z})^T \zeta + \nabla_y g(\bar{z})^T \eta = \nabla_y L(\bar{z}, \bar{\lambda})^T \pi$$

$$\left. \begin{aligned}
\pi^T \nabla_y g_i(\bar{z}) &\leq 0 \\
\eta_i &\geq 0
\end{aligned} \right\} \quad \text{for } i \in \mathcal{I}_0(\bar{z}, \bar{\lambda}) \tag{12}$$

$$\pi^T \nabla_y g_i(\bar{z}) = 0, \quad \text{for } i \in \mathcal{I}_+(\bar{z}, \bar{\lambda})$$

$$\eta_i = 0, \quad \text{for } i \notin \mathcal{I}(\bar{z})$$

$$\zeta \geq 0, \quad \zeta^T h(\bar{z}) = 0.$$

The following result clarifies the connection between the latter system with (10) for various index sets $\alpha \subseteq \mathcal{I}_0(\bar{z}, \bar{\lambda})$.

4.3.5 Proposition. A triple (ζ, η, π) satisfies (12) if and only if it satisfies (10) for all index sets $\alpha \subseteq \mathcal{I}_0(\bar{z}, \bar{\lambda})$. Consequently, if the SMFCQ holds at the tuple $(\bar{z}, \bar{\lambda}, \bar{\zeta}, \bar{\eta}, \bar{\pi})$ which satisfies (12), then $(\bar{\zeta}, \bar{\eta}, \bar{\pi})$ is the unique triple which satisfies (10) for all index sets $\alpha \subseteq \mathcal{I}_0(\bar{z}, \bar{\lambda})$.

Proof. Suppose (ζ, η, π) satisfies (12). Then (ζ, η, π) must satisfy (10) for any subset α of $\mathcal{I}_0(\bar{z}, \bar{\lambda})$. Conversely, if (ζ, η, π) satisfies (10) for all subsets α of $\mathcal{I}_0(\bar{z}, \bar{\lambda})$, then with two specific choices, $\alpha \equiv \mathcal{I}_0(\bar{z}, \bar{\lambda})$ and $\alpha \equiv \emptyset$, we easily see that (ζ, η, π) satisfies (12). The second assertion follows easily from the first: it suffices to recall that the assumed SMFCQ is equivalent to $(\bar{\zeta}, \bar{\eta}, \bar{\pi})$ being the unique triple satisfying (12).   **Q.E.D.**

We have noted above that feasible solutions of the NLP (11) are not necessarily feasible to the MPEC (2). In what follows we give two examples to demonstrate the difference between the relaxed NLP (11) and the local MPEC pieces (3). The first example (which has appeared previously) illustrates that the SMFCQ holding for (11) does not necessarily imply that the MFCQ will hold for (3). The second example illustrates that the MFCQ holding for all the MPEC pieces (3) does not necessarily imply that the relaxed NLP (11) will have a unique multiplier.

Example 4.3.4 revisited. Let $f(x, y) \equiv y_2$ in this example. Since the variable x plays no role in this problem, we drop it in the discussion. With $\bar{y} \equiv (1, 0)$ and $\bar{\lambda} \equiv (0, 0)$, the relaxed NLP (11) has the following form:

$$\text{minimize} \quad y_2$$

$$\text{subject to} \quad 2(y_1 - 1) + 2\lambda_1 y_1 + \lambda_2 = 0,$$

$$2y_2 = 0,$$

$$y_1^2 \leq 1, \quad y_1 \leq 1,$$

$$\lambda_1 \geq 0, \quad \lambda_2 \geq 0.$$

One can show that at $(\bar{y}, \bar{\lambda})$, $(\pi, \eta) \equiv (0, -1/2, 0, 0)$ is the unique multiplier associated with the four functional constraints (excluding the nonnegativity constraints on λ_1 and λ_2). Hence the SMFCQ holds for this NLP. Nevertheless, as pointed out before, the MFCQ fails for the MPEC piece obtained by setting the two inequalities $y_1^2 \leq 1$ and $y_1 \leq 1$ as equations.

4.3.6 Example. Consider the following NCP constrained MP in the variables $(x, y) \in \Re^3 \times \Re^2$:

$$\text{minimize} \quad x_3 + y_1 + y_2$$

$$\text{subject to} \quad x_3 \geq 0, \quad 2x_3 \geq 0,$$

$$F(x, y) \equiv (y_1 + x_1, \, y_2 + x_2) \geq 0,$$

$$y \geq 0, \quad y^T F(x, y) = 0.$$

Consider the pair $(\bar{x}, \bar{y}) \equiv (0, 0)$. It can be seen that both NCP constraints are degenerate; thus $\mathcal{I}_0(\bar{x}, \bar{y}) = \{1, 2\}$. The relaxed NLP (11) for this MPEC is given by

$$\text{minimize} \quad x_3 + y_1 + y_2$$

$$\text{subject to} \quad x_3 \geq 0, \quad 2x_3 \geq 0,$$

$$F(x, y) \equiv (y_1 + x_1, \, y_2 + x_2) \geq 0,$$

$$y \geq 0.$$

Let $\mu = (\mu_1, \mu_2) \geq 0$ and $\lambda = (\lambda_1, \lambda_2, \lambda_3, \lambda_4) \geq 0$ be the multiplier vector for the constraints $x_3 \geq 0$, $2x_3 \geq 0$ and $F(x, y) \geq 0$, $y \geq 0$ respectively. Then the KKT condition for the above NLP at the feasible point $(\bar{x}, \bar{y})$ becomes

$$\lambda_1 = 0, \quad \lambda_2 = 0,$$

$$\mu_1 + 2\mu_2 = 1,$$

$$\lambda_3 = 1, \quad \lambda_4 = 1,$$

$$(\mu_1, u_2) \geq 0, \quad (\lambda_1, \lambda_2, \lambda_3, \lambda_4) \geq 0.$$

Clearly, this set of multiplier vectors consists of a bounded line segment. Thus, the SMFCQ does not hold for the above relaxed NLP, although the MFCQ holds. We observe that the condition $(\lambda_1, \lambda_2, \lambda_3, \lambda_4) \geq 0$ is redundant in the above description of the multiplier set. This observation can be used to verify that the multiplier set for each MPEC piece (3) is identical to that of the above NLP. This shows that the MFCQ holds for all the MPEC pieces (3).

If the LICQ holds for (11) at the triple $(\bar{x}, \bar{y}, \bar{\lambda}) \in \mathcal{F}^{\mathrm{KKT}}$ then we can draw a strengthened conclusion in addition to the last statement in Proposition **4.3.5**.

4.3.7 Proposition. Suppose that the LICQ holds for (11) at the triple $(\bar{x}, \bar{y}, \bar{\lambda}) \in \mathcal{F}^{\text{KKT}}$ then the same CQ holds for (3) at the same triple for all pairs $(\mathcal{J}_1, \mathcal{J}_2)$ defined by (5) for arbitrary subsets α of $\mathcal{I}_0(\bar{z}, \bar{\lambda})$. If in addition there exists (ζ, η, π) satisfying (12), then $(\bar{x}, \bar{y}, \bar{\lambda})$ is a stationary point for the MPEC (2).

Proof. It is trivial to observe that, for $(\mathcal{J}_1, \mathcal{J}_2)$ as specified, the active constraints in the two problems (3) and (11) are the same at $(\bar{z}, \bar{\lambda})$. Thus the first assertion of the proposition is obvious. If in addition there is (ζ, η, π) satisfying (12), then by Proposition **4.3.5**, (ζ, η, π) also satisfies (10) for all index sets $\alpha \subseteq \mathcal{I}_0(\bar{z}, \bar{\lambda})$. Since LICQ holds for all these systems, $(\bar{z}, \bar{\lambda})$ is a stationary point for the nonlinear programs (2). The second assertion of the proposition now follows because of the representation (4). **Q.E.D.**

We conclude the discussion in this section with two results, Propositions **4.3.8** and **4.3.9**, that identify another important role of the relaxed NLP (11) in the context of the general MPEC (3.1.1) and the special case with NCP constraints. Roughly speaking, the first result states that the local optimality of $\bar{z} = (\bar{x}, \bar{y})$ to (3.1.1) can be inferred from the local optimality of $(\bar{z}, \bar{\lambda})$ to (11) for all $\bar{\lambda} \in M(\bar{z})$. (This result should not be confused with Proposition **3.4.1** because the NLPs (11) are not the same as (3.4.1) in that proposition; this proposition is nevertheless useful for the proof of Proposition **4.3.8**.) The second result, Proposition **4.3.9**, is a sharpening of the first when the constraints in the MPEC are of the NCP type.

To make the promised results precise, we recall that $M(\bar{z})$ is the set of KKT multipliers $\lambda \in \Re^\ell$ such that $(\bar{z}, \lambda) \in \mathcal{F}^{\text{KKT}}$. If $M(\bar{z})$ is not a singleton, then there will in general be multiple relaxed NLPs (11) corresponding to each $\bar{\lambda} \in M(\bar{x})$ and the given $\bar{z} \in \mathcal{F}$ feasible to the MPEC (3.1.1). If the SMFCQ holds for the inner VI $(F(\bar{x}, \cdot), C(\bar{x}))$ at $(\bar{z}, \bar{\lambda})$, then $M(\bar{z}) = \{\bar{\lambda}\}$ and there will be a unique relaxed NLP corresponding to $\bar{z}$.

4.3.8 Proposition. Let $\bar{z} \equiv (\bar{x}, \bar{y}) \in \mathcal{F}$ be a feasible vector to the MPEC (3.1.1). If $(\bar{x}, \bar{y}, \bar{\lambda})$, where $\bar{\lambda} \in M(\bar{x}, \bar{y})$, is a local minimum of the relaxed NLP (11), then $(\bar{x}, \bar{y}, \bar{\lambda})$ is a local minimum of (2). Thus if the SBCQ holds for the VI $(F(\bar{x}, \cdot), C(\bar{x}))$ at the solution $\bar{y}$ and for every $\bar{\lambda} \in M(\bar{x}, \bar{y})$, $(\bar{x}, \bar{y}, \bar{\lambda})$ is a local minimum of the relaxed NLP (11), then $(\bar{x}, \bar{y})$ is a local minimum of (3.1.1).

Proof. As noted above, (11) is a local relaxation of (2) near $(\bar{x}, \bar{y}, \bar{\lambda})$. Thus the first assertion is obvious. The second assertion of this proposition follows from the "if" part in Proposition **3.4.1** which does not require the convexity of the functions $g_i(x, \cdot)$.   **Q.E.D.**

Next, we specialize the above result to the NCP constrained MP:

$$\text{minimize} \quad f(x, y)$$
$$\text{subject to} \quad (x, y) \in Z \subseteq \Re^{n+m}, \tag{13}$$
$$F(x, y) \geq 0, \quad y \geq 0, \quad y^T F(x, y) = 0,$$

which corresponds to the MPEC (3.1.1) with $g(x, y) \equiv -y$. Let $\bar{z} \equiv (\bar{x}, \bar{y})$ be a feasible solution to this problem. We recall three index sets defined in (3.2.7) that have been used several times:

$$\alpha(\bar{z}) = \{i : F_i(\bar{z}) = 0 < \bar{y}_i\},$$

$$\beta(\bar{z}) = \{i : F_i(\bar{z}) = 0 = \bar{y}_i\},$$

$$\gamma(\bar{z}) = \{i : F_i(\bar{z}) > 0 = \bar{y}_i\}.$$

There is a single relaxed NLP corresponding to $\bar{z}$ and this is

$$\text{minimize} \quad f(x, y)$$
$$\text{subject to} \quad (x, y) \in Z \subseteq \Re^{n+m},$$
$$\left.\begin{array}{c} F_i(x, y) = 0 \\ y_i \geq 0 \end{array}\right\} \quad \forall i \in \alpha(\bar{z}),$$
$$\left.\begin{array}{c} F_i(x, y) \geq 0 \\ y_i \geq 0 \end{array}\right\} \quad \forall i \in \beta(\bar{z}), \tag{14}$$
$$\left.\begin{array}{c} F_i(x, y) \geq 0 \\ y_i = 0 \end{array}\right\} \quad \forall i \in \gamma(\bar{z}).$$

The following result, which is a straightforward consequence of Proposition **4.3.8**, does not require a proof, due to the fact that $M(\bar{x}) = \{F(\bar{z})\}$.

4.3.9 Proposition. Let $\bar{z} \equiv (\bar{x}, \bar{y}) \in \mathcal{F}$ be a feasible vector to (13). If $(\bar{x}, \bar{y})$ is a local minimum of (14), then $(\bar{x}, \bar{y})$ is a local minimum of (13).

4.4 An Exact Penalty Equivalent of Order 1

Besides yielding a piecewise smooth response function $y(x)$ at $\bar{x}$, the setting of Theorem **4.2.16** also allows the derivation of an exact penalty equivalent of order 1 for the MPEC (4.2.1). In what follows, we explain this penalty function. Rewrite (4.2.1) using the normal form of the equilibrium constraints:

$$\text{minimize} \quad f(x, \Pi_{C(x)}(v))$$
$$\text{subject to} \quad x \in X, \quad H(x,v) = 0, \tag{1}$$

where H is defined by (4.2.21). From Lemma **4.2.22** we see that $(\bar{x}, \bar{y})$ is a local solution of the MPEC (1) and $\bar{v} = \bar{y} - F(\bar{x}, \bar{y})$ if and only if $(\bar{x}, \bar{v})$ is a local solution of (1) and $\bar{y} = \Pi_{C(\bar{x})}(\bar{v})$. Our goal is to show that these problems are in turn equivalent to the following exact penalty problem for each $\rho \geq \bar{\rho}$, where $\bar{\rho}$ is a positive constant:

$$\text{minimize} \quad f(x, \Pi_{C(x)}(v)) + \rho \, \|H(x,v)\|$$
$$\text{subject to} \quad (x,v) \in X \times \Re^m. \tag{2}$$

Specifically we show that (1) is equivalent to (2) for certain ρ, using a regularity property of H.

We give below a condition under which a general constraint system of the form

$$x \in X, \quad H(x,v) = 0 \tag{3}$$

is *(metrically) regular* at $(\bar{x}, \bar{v}) \in (X \times \Re^m) \cap H^{-1}(0)$, that is, there exists $\bar{\beta} > 0$ such that for each (x,v) near $(\bar{x}, \bar{v})$ with $x \in X$, and each y near $0 \in \Re^m$, the set $(X \times \Re^m) \cap H^{-1}(y)$ is nonempty and

$$\text{dist}((x,v), (X \times \Re^m) \cap H^{-1}(y)) \leq \bar{\beta} \, \|H(x,v) - y\|.$$

This is an error bound of order 1 on the distance to the set of feasible solutions. See [41] for sufficient conditions for regularity of a nonsmooth system and the relationship between regularity and exact penalty formulations of constrained optimization problems. Observe the relationship between regularity and the inequality (2.1.3) in Theorem **2.1.2**: regularity means we can take $N^* = 1$.

4.4.1 Lemma. Let $H : \Re^{n+m} \to \Re^m$ and $(\bar{x}, \bar{v}) \in (X \times \Re^m) \cap H^{-1}(0)$. Let

$$G(x, v) \equiv (x, H(x, v)), \quad \text{for } (x, v) \in \Re^{n+m}.$$

If G is a Lipschitzian homeomorphism near $(\bar{x}, \bar{v})$ then the system (3) is regular at $(\bar{x}, \bar{v})$.

Proof. Let Θ be a neighborhood of $(\bar{x}, \bar{v})$ such that $G(\Theta)$ is a neighborhood of $(\bar{x}, 0)$; the restriction $\tilde{G}$ of G to Θ, as a mapping from Θ to $G(\Theta)$, is invertible; and $\tilde{G}^{-1}$ is Lipschitz with modulus $\bar{\beta} > 0$. Then $H(\Theta)$ is a neighborhood of $0 \in \Re^m$. Moreover, by the homeomorphism of G, the neighborhood Θ can be chosen with the property that for each x with $(x, v) \in \Theta$ for some v, the equation $H(x, \cdot) = y$ has a solution for all $y \in H(\Theta)$.

For $(x, v) \in \Theta$ with $x \in X$, and $y \in H(\Theta)$, we have

$$\text{dist}((x, v), (X \times \Re^m) \cap H^{-1}(y))$$

$$\leq \text{dist}(v, H(x, \cdot)^{-1}(y))$$

$$\leq \|(x, v) - \tilde{G}^{-1}(x, y)\| \quad = \quad \|\tilde{G}^{-1}(G(x, v)) - \tilde{G}^{-1}(x, y)\|$$

$$\leq \bar{\beta} \|G(x, v) - (x, y)\| \quad = \quad \bar{\beta} \|H(x, v) - y\|,$$

which establishes the regularity of H.   **Q.E.D.**

When H is the parametric normal map (4.2.21), the hypotheses of Theorem **4.2.16** are sufficient for the function G above to be a Lipschitzian homeomorphism near $(\bar{x}, \bar{v})$.

4.4.2 Proposition. Under the hypotheses of Theorem **4.2.16**, $(\bar{x}, \bar{v})$ is a local minimum of (1) if and only if there exists an $\bar{\rho} > 0$ such that for each $\rho \geq \bar{\rho}$, $(\bar{x}, \bar{v})$ is a local minimum of the penalty problem (2).

Proof. By hypothesis $(\bar{x}, \bar{v})$ is feasible to (1). The proof of Theorem **4.2.16** shows that $G(x, v) = (x, H(x, v))$ is a Lipschitzian homeomorphism at $(\bar{x}, \bar{v})$. Given the conclusion of Lemma **4.4.1**, $\bar{\rho}$ can be chosen as any constant larger than the product of $\bar{\beta}$ and the Lipschitz modulus of f near $(\bar{x}, \Pi_{C(\bar{x})}(\bar{v}))$. The proof is by now standard.   **Q.E.D.**

Proposition **4.4.2** does not require the convexity of X. This result can be extended to (global) minimization under compactness of X and boundedness of $C(x)$ for each $x \in X$.

4.4.3 Theorem. Assume the hypotheses of Theorem **4.2.16** hold for each $(\bar{x}, \bar{y}) \in \mathcal{F}$; X is compact; and $C(x)$ is bounded for each $x \in X$. Then there exists a $\bar{\rho} > 0$ such that for each $\rho \geq \bar{\rho}$, a point $(\bar{x}, \bar{v}) \in \Re^n \times \Re^m$ is a minimum of (1) if and only if it is a minimum of (2).

Proof. Let $\tilde{\mathcal{F}}$ be the feasible region of (1). We claim $\tilde{\mathcal{F}}$ is compact. This set is closed, just as $\mathcal{F}$ is closed via SBCQ. From compactness of X, continuity of g, and boundedness of $C(x) \equiv \{y : g(x,y) \leq 0\}$ for each $x \in X$, we deduce that $\mathcal{F} \equiv \{(x,y) \in X \times \Re^m : y \in C(x)\}$ is compact. Similarly, by compactness of $\mathcal{F}$ and continuity of F, we get compactness of $F(\mathcal{F}) \equiv \{F(x,y) : (x,y) \in \mathcal{F}\}$. Now for any $(x,v) \in \tilde{\mathcal{F}}$ and $y = \Pi_{C(x)}(v)$, Proposition **1.3.3** says $(x,y) \in \mathcal{F}$, thus

$$(x,v) = (x,y) - (0, F(x,y)),$$

where the vector on the right is bounded by compactness of $\mathcal{F}$ and $F(\mathcal{F})$. Thus $\tilde{\mathcal{F}}$ is bounded as needed.

    Using compactness of $\tilde{\mathcal{F}}$ and Lemma **4.4.1**, we find a finite family of vectors $\{(x^i, v^i)\} \subset \tilde{\mathcal{F}}$, open neighborhoods $\{\Theta^i\}$ with $\Theta^i \subset \Re^n \times \Re^m$ for each i, and constants $\{\beta_i\} \subset (0, \infty)$ such that first, $\{\Theta^i\}$ covers $\tilde{\mathcal{F}}$, and second, for each i,

$$\text{dist}((x,v), H^{-1}(0)) \leq \beta_i \, \| H(x,v) \| \quad \text{for } (x,v) \in \Theta^i.$$

Thus

$$\text{dist}((x,v), H^{-1}(0)) \leq \beta \, \| H(x,v) \| \quad \text{for } (x,v) \in (X \times \Re^m) \setminus \Theta \quad (4)$$

where $\Theta \equiv \bigcup_i \Theta^i$ and $\beta \equiv \max_i \beta_i$. We also have

$$0 < \delta \equiv \inf\{ \| H(x,v) \| : (x,v) \in (X \times \Re^m) \setminus \Theta \}. \quad (5)$$

For otherwise, there is a sequence $\{(x^k, v^k)\}$ in $(X \times \Re^m) \setminus \Theta$ for which $H(x^k, v^k) \to 0$. By compactness of $\tilde{\mathcal{F}}$, we may assume without loss of generality that $(x^k, v^k) \to (\bar{x}, \bar{v}) \in \tilde{\mathcal{F}} \subset \Theta$. Thus $(\bar{x}, \bar{v})$ is a boundary point of the open set Θ; a contradiction.

    By Lemma **4.2.17**, $\Pi_{C(x)}(v)$ is Lipschitz in (x,v) near each $(\bar{x}, \bar{v}) \in \tilde{\mathcal{F}}$; hence H is continuous on $\tilde{\mathcal{F}}$. Thus, using compactness of $\tilde{\mathcal{F}}$, there exists a solution $(\bar{x}, \bar{v})$ of (1). Let $\bar{y} \equiv \Pi_{C(\bar{x})}(\bar{v})$.

Since $\tilde{\mathcal{F}}$ is bounded, we may assume without loss of generality that Θ is also bounded. Hence $\Pi_{C(x)}(v)$, being locally Lipschitz, is Lipschitz on Θ; let $\gamma_1 > 0$ be the corresponding Lipschitz modulus. Similarly f is Lipschitz on Θ with Lipschitz modulus $\gamma_2 > 0$. Let $C(X) \equiv \bigcup_{x \in X} C(x)$; $C(X)$ is a compact set since $\mathcal{F}$ is compact. Define

$$\phi \equiv \min\{f(x,y) : x \in X, \, y \in C(X)\};$$

ϕ exists by continuity of f and compactness of X and $C(X)$. Choose any scalar $\bar{\rho}$ such that

$$\bar{\rho} > \max\left(\gamma_1\gamma_2\beta, \, \frac{f(\bar{x},\bar{y}) - \phi}{\delta}\right).$$

Let $\rho \geq \bar{\rho}$. On one hand suppose $(x,v) \in (X \times \Re^m) \cap \Theta$, and let $(x',v') \in \tilde{\mathcal{F}}$ be such that $\|(x - x', v - v')\|$ equals $\text{dist}((x,v), \tilde{\mathcal{F}})$. Then the Lipschitz properties of $f(x,v)$ and $\Pi_{C(x)}(v)$ yield

$$f(x, \Pi_{C(x)}(v)) \; \geq \; f(x',v') - \gamma_1\gamma_2\,\text{dist}((x,v), \tilde{\mathcal{F}})$$

$$\geq \; f(\bar{x},\bar{y}) - \gamma_1\gamma_2\beta\,\|H(x,v)\|,$$

using the inequality (4). Hence

$$f(x, \Pi_{C(x)}(v)) + \rho\,\|H(x,v)\| \; \geq \; f(\bar{x},\bar{y}) + (\rho - \gamma_1\gamma_2\beta)\,\|H(x,v)\|$$

$$\geq \; f(\bar{x},\bar{y}),$$

where equality holds throughout only if $\|H(x,v)\| = 0$, i.e., $(x,v) \in \tilde{\mathcal{F}}$.

On the the other hand, if (x,v) in $(X \times \Re^m) \setminus \Theta$, the inequality (5) yields

$$f(x, \Pi_{C(x)}(v)) + \rho\|H(x,v)\| \geq \phi + \rho\delta > f(\bar{x},\bar{y});$$

hence (x,v) cannot be a minimizer of the penalty problem. So the set of minimizers of (1) coincides with the set of minimizers of (2).   **Q.E.D.**

5

Second-Order Optimality Conditions

As for a standard NLP, we can derive some second-order necessary and second-order sufficient conditions for a local minimum of an MPEC. This chapter is a foray into these second-order conditions. For convenience, throughout we require polyhedrality of the upper-level feasible region Z; initially, we also assume the same for the constraint set $C(x)$ in the lower-level VI for all $x \in \text{dom}(C)$. In standard nonlinearly constrained NLPs, second-order conditions at boundary points must generally account for the curvature of the boundary. Such curvature requirement is usually contained in the positive definiteness properties of the partial Hessian matrix of the Lagrangean function of the nonlinear program, which contains not only the Hessian matrix of the objective function, but also the sum of the Hessian matrices of the active constraint functions, using the KKT multipliers as weights; cf. the discussion on SCOC in Subsection **4.2.7**. Initially we confine our interest to polyhedral sets to keep the ideas and analysis relatively simple. Even in this situation, the treatment is complicated by the equilibrium constraint $(x, y) \in \text{Gr}\,\mathcal{S}$ which, as we have seen in the case of the first-order analysis, leads to some combinatorial considerations that are not present in standard nonlinear programming. Indeed, such complications become more pronounced as we work with the second-order conditions in this chapter.

This chapter discusses a multiplier-based approach, an implicit programming approach, and a piecewise programming approach to the derivation of second-order optimality conditions of MPECs; the treatment here extends ideas of the previous two chapters. Overall, the development here is parallel to that of Chapter **4**. Specifically, besides the next section which reviews known NLP theory, there are five sections in the chapter. Section **5.2** deals with the MPAEC; the results therein, Theorems **5.2.1** and **5.2.3**, are quite satisfactory, as they are analogous to the results for a standard, linearly constrained nonlinear program. Proposition **4.1.1** plays an important role in the treatment of the MPAEC. Nevertheless, the results for the NCP constrained MP in Section **5.3**, Theorems **5.3.1** and **5.3.3**, contain some restrictions that ideally should not be present. The fourth section presents a set of second-order conditions for the NCP constrained MP in terms of some "second-order directional derivatives" of the implicit solution function $y(x)$ of the parametric NCP constraints. From the latter conditions, we are able to obtain refinements of the two main results in Section **5.3**; see Theorem **5.4.3**. In the fifth section, we apply the results of Section **5.4** to deal with the MPEC where the inner VIs are formulated in terms of their KKT conditions. Finally, the sixth and last section discusses how the piecewise programming approach can be combined with classical second-order theory for smooth optimization problems to yield some second-order optimality conditions for MPECs.

5.1 Review of Second-Order NLP Optimality Theory

Before describing the results for MPEC, we find it useful to review the standard second-order optimality conditions from nonlinear programming. Consider the following minimization problem:

$$
\begin{aligned}
\text{minimize} \quad & f(x) \\
\text{subject to} \quad & g_i(x) = 0, \ i = 1, ..., \ell, \\
& g_i(x) \leq 0, \ i = \ell + 1, ..., \ell + m,
\end{aligned}
\tag{1}
$$

where f and $g = (g_1, ..., g_{\ell+m})$ are C^2 mappings defined on an open set in $\Re^n$ that contains the feasible region of (1), which we denote $\mathcal{F}^{\text{NLP}}$. Suppose $\bar{x} \in \mathcal{F}^{\text{NLP}}$ is a stationary point of (1) satisfying the KKT conditions with multipliers $\pi \in \Re^\ell \times \Re^m_+$. Let $M(\bar{x})$ be the set of these multipliers. Let $\mathcal{A}(\bar{x})$ denote the set of equality and active inequality constraints at $\bar{x}$; i.e.,

$$\mathcal{A}(\bar{x}) \equiv \{i : g_i(\bar{x}) = 0\} \supseteq \{1, \ldots, \ell\}.$$

The critical cone of (1) at $\bar{x}$, denoted $\mathcal{C}(\bar{x}; \mathcal{F}^{\text{NLP}})$, is defined by

$$\mathcal{C}(\bar{x}; \mathcal{F}^{\text{NLP}}) \equiv \mathcal{T}(\bar{x}; \mathcal{F}^{\text{NLP}}) \cap \nabla f(\bar{x})^\perp,$$

where $\nabla f(\bar{x})^\perp$ denotes the orthogonal complement of the vector $\nabla f(\bar{x})$; that is $\nabla f(\bar{x})^\perp$ consists of vectors in $\Re^n$ that are perpendicular to $\nabla f(\bar{x})$. In general, the critical cone is dependent on the feasible vector $\bar{x}$, the feasible set $\mathcal{F}^{\text{NLP}}$, and the objective function f. However, we have omitted the function f in the notation $\mathcal{C}(\bar{x}; \mathcal{F}^{\text{NLP}})$ for simplicity; with this abuse of notation, the two cones, $\mathcal{C}(\bar{x}; \mathcal{F}^{\text{NLP}})$ and $\mathcal{T}(\bar{x}; \mathcal{F}^{\text{NLP}})$, are therefore consistent in notation.

The paper [251] appears to be the source for the term "critical cone", although this cone was certainly used much earlier; see, e.g., [52] where the cone was used to characterize the local optimality conditions of quadratic programs. Besides these two references, other papers that discuss the fundamental role of the critical cone in nonlinear programming include [250, 41]. As reviewed below, well-known second-order optimality conditions of constrained optimization problems are stated using the critical cone.

Under an appropriate constraint qualification such as the MFCQ, the tangent cone $\mathcal{T}(\bar{x}; \mathcal{F}^{\text{NLP}})$ can be represented explicitly, yielding

$$\mathcal{C}(\bar{x}; \mathcal{F}^{\text{NLP}}) = \{dx \in \Re^n : \nabla f(\bar{x})^T dx = 0,$$

$$\nabla g_i(\bar{x})^T dx = 0, \text{ for } i = 1, \ldots, \ell,$$

$$\nabla g_i(\bar{x})^T dx \leq 0, \text{ for } i \in \mathcal{A}(\bar{x}) \text{ with } i > \ell\}.$$

The critical cone can also be written using any KKT multiplier $\pi \in M(\bar{x})$; specifically,

$$\mathcal{C}(\bar{x}; \mathcal{F}^{\text{NLP}}) = \{dx \in \Re^n : \nabla g_i(\bar{x})^T dx = 0, \text{ for } i \in \mathcal{A}_+(\bar{x}, \pi),$$

$$\nabla g_i(\bar{x})^T dx \leq 0, \text{ for } i \in \mathcal{A}(\bar{x}) \setminus \mathcal{A}_+(\bar{x}, \pi)\},$$

where

$$\mathcal{A}_+(\bar{x}, \pi) \equiv \{1, \dots, \ell\} \cup \{i : i > \ell \text{ with } \pi_i > 0\}.$$

Define the (real-valued) Lagrangean function for the NLP (1):

$$\mathcal{L}^{\mathrm{NLP}}(x, \pi) \equiv f(x) + \sum_{i=1}^{m+\ell} \pi_i g_i(x), \qquad (x, \pi) \in \Re^{n+m+\ell}.$$

The following theorem summarizes the main second-order theory for the standard nonlinear program (1). The proof can be found in [41] which also contains a detailed account of this theory.

5.1.1 Theorem. Assume that the MFCQ holds at the stationary point $\bar{x} \in \mathcal{F}^{\mathrm{NLP}}$. The following two statements are valid.

(a) If $\bar{x}$ is a local minimizer of (1), then for each $dx \in \mathcal{C}(\bar{x}; \mathcal{F}^{\mathrm{NLP}})$,

$$\max \left\{ dx^T \nabla^2_{xx} \mathcal{L}^{\mathrm{NLP}}(\bar{x}, \pi) dx \; : \; \pi \in M(\bar{x}) \right\} \geq 0.$$

(b) If for each $dx \in \mathcal{C}(\bar{x}; \mathcal{F}^{\mathrm{NLP}}) \setminus \{0\}$,

$$\max \left\{ dx^T \nabla^2_{xx} \mathcal{L}^{\mathrm{NLP}}(\bar{x}, \pi) dx \; : \; \pi \in M(\bar{x}) \right\} > 0,$$

then $\bar{x}$ is a strict local minimizer of (1); i.e., there exist a neighborhood $V \subseteq \Re^n$ of $\bar{x}$ and a constant $\gamma > 0$ such that

$$f(x) \geq f(\bar{x}) + \gamma \|x - \bar{x}\|^2$$

for each $x \in \mathcal{F}^{\mathrm{NLP}} \cap V$.

5.1.2 Remark. A sufficient condition for the assumption in (b) to hold is that there exists a multiplier vector $\pi \in M(\bar{x})$ such that the matrix $\nabla^2_{xx} \mathcal{L}^{\mathrm{NLP}}(\bar{x}, \pi)$ is *strictly copositive* on the critical cone $\mathcal{C}(\bar{x}; \mathcal{F}^{\mathrm{NLP}})$; i.e.,

$$dx^T \nabla^2_{xx} \mathcal{L}^{\mathrm{NLP}}(\bar{x}, \pi) dx > 0, \qquad \forall dx \in \mathcal{C}(\bar{x}; \mathcal{F}^{\mathrm{NLP}}) \setminus \{0\}.$$

In this form, part (b) of Theorem **5.1.1** becomes a well-known result in nonlinear programming; see [204, Chapter 10]. In the case where $M(\bar{x}) = \{\bar{\pi}\}$, part (a) of this theorem says that the matrix $\nabla^2_{xx} \mathcal{L}^{\mathrm{NLP}}(\bar{x}, \bar{\pi})$ is *copositive* on the cone $\mathcal{C}(\bar{x}; \mathcal{F}^{\mathrm{NLP}})$. In the general case, the second-order necessary condition in part (a) of the above theorem does not necessarily imply the copositivity of the matrix $\nabla^2_{xx}\mathcal{L}^{\mathrm{NLP}}(\bar{x}, \pi)$ on the cone $\mathcal{C}(\bar{x}; \mathcal{F}^{\mathrm{NLP}})$ for any multiplier $\pi \in M(\bar{x})$; similarly, the requirement in part (b) of the same theorem is in principle less restrictive than the strict copositivity of the matrix $\nabla^2_{xx}\mathcal{L}^{\mathrm{NLP}}(\bar{x}, \pi)$ on the same cone, for any $\pi \in M(\bar{x})$.

5.2 AVI Constrained Mathematical Program

Consider the AVI constrained MP (4.1.1), which for ease of reference, is restated below:

$$\text{minimize} \quad f(x, y)$$

$$\text{subject to} \quad (x, y) \in Z,$$

$$Dx + Ey + b \leq 0, \tag{1}$$

$$(y' - y)^T (Px + Qy + q) \geq 0,$$

$$\text{for all } y' \text{ such that } Dx + Ey' + b \leq 0,$$

where Z is assumed polyhedral. We also recall the equivalent formulation of (1) using the KKT or complementarity formulation of the AVI constraints; see (4.1.2):

$$\text{minimize} \quad f(x, y)$$

$$\text{subject to} \quad (x, y, \lambda) \in Z \times \Re^\ell_+,$$

$$Dx + Ey + b \leq 0, \tag{2}$$

$$Px + Qy + q + E^T \lambda = 0,$$

$$\lambda^T (Dx + Ey + b) = 0.$$

We continue to use the notation introduced in the previous chapters; in particular, we draw attention to the notation defined in Section **3.2**. Let $\mathcal{F}$ denote the feasible region of (1). Let $\bar{z} \equiv (\bar{x}, \bar{y}) \in \mathcal{F}$ be given.

Following the general theory of constrained optimization problems, we define the critical cone of (1) at the feasible solution $\bar{z}$:

$$\mathcal{C}(\bar{z}; \mathcal{F}) \equiv \mathcal{T}(\bar{z}; \mathcal{F}) \cap \nabla f(\bar{z})^\perp.$$

Having defined this cone, we should immediately point out that Theorem **5.1.1** is not applicable to the MPAEC (1) when it is considered as a nonlinear program. Indeed this theorem is crucially dependent on the MFCQ (or an appropriate CQ) holding for the NLP (5.1.1) at a feasible solution. As we recall from Chapter **3**, this CQ fails to hold for the constraints in MPEC because of the special feature of the tangent cone $\mathcal{T}(\bar{z}; \mathcal{F})$. Consequently, although we have used Theorem **5.1.1** to motivate the introduction of the MPEC critical cone $\mathcal{C}(\bar{z}; \mathcal{F})$, we cannot apply this theorem to the treatment of the MPEC (1). Instead we resort to the results in the previous chapters.

From Proposition **4.1.1**, we see that

$$\mathcal{T}(\bar{z}; \mathcal{F}) = \mathcal{L}^e(\bar{z}; \mathcal{F}) = \mathcal{L}(\bar{z}; \mathcal{F}) \equiv \mathcal{T}(\bar{z}; Z) \bigcap \left(\bigcup_{\lambda \in M(\bar{z})} \mathrm{Gr}(\mathcal{LS}_{(\bar{z}, \lambda)}) \right).$$

For each $\lambda \in M(\bar{z})$, define the cone

$$\mathcal{C}(\bar{z}, \lambda) \equiv \mathcal{T}(\bar{z}; Z) \cap \mathrm{Gr}(\mathcal{LS}_{(\bar{z}, \lambda)}) \cap \nabla f(\bar{z})^{\perp} \subseteq \Re^{n+m}.$$

Thus

$$\bigcup_{\lambda \in M(\bar{z})} \mathcal{C}(\bar{z}, \lambda) = \mathcal{C}(\bar{z}; \mathcal{F}), \tag{3}$$

which shows that for each $\lambda \in M(\bar{z})$, $\mathcal{C}(\bar{z}, \lambda)$ is a "component" of the critical cone of the MPAEC (1) at the feasible solution $\bar{z}$.

We are now ready to establish a second-order necessary optimality condition for (1).

5.2.1 Theorem. Let $\bar{z} \equiv (\bar{x}, \bar{y})$ be a local minimum of (1) and let f be twice continuously differentiable in a neighborhood of $\bar{z}$. Suppose that Z is polyhedral. Then in addition to satisfying the first-order stationarity conditions as described in Corollary **3.3.1**, $\bar{z}$ satisfies the following implication for each $\lambda \in M(\bar{z})$:

$$dz \in \mathcal{C}(\bar{z}, \lambda) \implies dz^T \nabla^2 f(\bar{z}) dz \geq 0;$$

thus the Hessian $\nabla^2 f(\bar{z})$ is copositive on the critical cone $\mathcal{C}(\bar{z}; \mathcal{F})$.

Proof. Let $\lambda \in M(x)$ and $dz \equiv (dx, dy) \in \mathcal{C}(\bar{z}, \lambda)$. By Proposition **4.1.1** and polyhedrality of Z, it follows that $z(\tau) \equiv \bar{z} + \tau \, dz \in \mathcal{F}$ for all $\tau > 0$ sufficiently small. Thus we have

$$0 \leq f(z(\tau)) - f(\bar{z}) = \tau \nabla f(\bar{z})^T dz + \frac{\tau^2}{2} dz^T \nabla^2 f(\bar{z}) dz + o(\tau^2),$$

where $o(t)$ denotes some suitable quantity satisfying

$$\lim_{t \to 0} \frac{o(t)}{t} = 0.$$

Since $\nabla f(\bar{z})^T dz = 0$, we obtain, after dividing by τ^2 and passing to the limit $\tau \downarrow 0$,

$$dz^T \nabla^2 f(\bar{z}) dz \geq 0$$

as desired.   **Q.E.D.**

As noted in the preliminary discussion that precedes Theorem **5.2.1**, this theorem is not a consequence of part (a) of Theorem **5.1.1**. An important feature of the MPAEC (2) is the presence of the complementarity constraint:

$$\lambda^T (Dx + Ey + b) = 0, \tag{4}$$

which is a nonlinear condition in (x, y, λ); nevertheless, the second-order derivatives of this constraint are not included in the Hessian matrix $\nabla^2 f(\bar{z})$ when the second-order optimality conditions for (1) are formed. At first thought, this appears to be inconsistent with the second-order theory of a standard nonlinear program. There are two explanations for this phenomenon. One is that as a nonlinear constraint, the condition (4) is not guaranteed to possess a Lagrangean multiplier; see the example to follow. Hence, the well-known optimality theory in traditional nonlinear programming cannot be applied. The other reason is that the constraints of (2), which define a (parametric) mixed linear complementarity problem, are in some sense linear (piecewise linear, to be precise), especially if one replaces (4) by the disjunctions: for $i = 1, \ldots, m$,

$$\lambda_i = 0 \quad \text{or} \quad (Dx + Ey + b)_i = 0.$$

This point of view also offers an explanation for the piecewise polyhedrality structure of the graph of the "tangent" map $\mathcal{LS}_{(\bar{z}, \lambda)}$. Consequently, if one accepts the notion that (1) is a nonlinear program with piecewise linear constraints (assuming that Z is polyhedral also), then it is not surprising that the second-order optimality conditions for this problem would not involve the second derivatives of the constraint (4). This viewpoint is exploited in deriving conditions for NCP constrained MP, below.

In what follows, we give a simple example to illustrate that the constraint (4) does not always possess a Lagrange multiplier in the sense of traditional nonlinear programming. This example reinforces the discussion at the end of Section **3.1** about the distinguished feature of the MPEC.

5.2.2 Example. Consider the KKT constrained MP (3.1.6) appeared in Example **3.1.1**:

$$\text{minimize} \quad f(x,y)$$
$$\text{subject to} \quad y - \lambda = 0, \quad y\lambda = 0, \quad y, \lambda \geq 0.$$

It has been shown that the basic CQ for this problem considered as a standard NLP is violated at $(x, y, \lambda) = (x, 0, 0)$. As a result, no KKT multipliers (in the sense of NLP) exist. However, if we reformulate this example as an optimization problem with disjunctive constraints,

$$\text{minimize} \quad f(x,y)$$
$$\text{subject to} \quad \text{either} \quad y - \lambda = 0, \quad y = 0, \quad y, \lambda \geq 0$$
$$\text{or} \quad y - \lambda = 0, \quad \lambda = 0, \quad y, \lambda \geq 0,$$

since the constraints are now linear, the fundamental CQ for nonlinear programming holds with regard to each of the two feasible subregions. The nonlinear constraint $\lambda y = 0$ has been reduced to the disjunctive linear constraints, $y = 0$ or $\lambda = 0$, which therefore cannot contribute to the Hessian of the nonlinear programming Lagrangean.

By imposing a strict copositivity assumption on the Hessian $\nabla^2 f(\bar{z})$, we can establish a second-order sufficiency result for the MPAEC (1). Specifically, this result states that if $\nabla^2 f(\bar{z})$ is strictly copositive on the critical cone $\mathcal{C}(\bar{z}; \mathcal{F})$, then $\bar{z}$ is a strict local minimizer of (1).

5.2.3 Theorem. Let $\bar{z} \equiv (\bar{x}, \bar{y})$ be a feasible vector of (1) and let f be twice continuously differentiable in a neighborhood of $\bar{z}$. Suppose that Z is polyhedral. If $\bar{z}$ satisfies the first-order stationarity conditions as described in Corollary **3.3.1** and also the following implication for each $\lambda \in M(\bar{z})$:

$$[0 \neq dz \equiv (dx, dy) \in \mathcal{C}(\bar{z}, \lambda)] \implies dz^T \nabla^2 f(\bar{z}) dz > 0,$$

then there exist a neighborhood $V \subseteq \Re^{n+m}$ of $\bar{z}$ and a scalar $\gamma > 0$ such that for all vectors $z' \in \mathcal{F} \cap V$,

$$f(z') - f(\bar{z}) \geq \gamma \|\bar{z} - z'\|^2.$$

Proof. Suppose the claim is false. There exists a sequence $\{z^k\} \subset \mathcal{F}$ converging to $\bar{z}$ such that for each k, $z^k \neq \bar{z}$ and

$$f(z^k) - f(\bar{z}) < \frac{1}{k} \|\bar{z} - z^k\|^2. \tag{5}$$

Write $z^k \equiv (x^k, y^k)$ for each k and consider the normalized sequence

$$\left\{ \frac{z^k - \bar{z}}{\|z^k - \bar{z}\|} \right\}$$

which must have at least one nonzero accumulation point $dz \equiv (dx, dy)$ that belongs to $\mathcal{T}(\bar{z}; \mathcal{F})$. By stationarity of $\bar{z}$, it follows that $\nabla f(\bar{z})^T dz \geq 0$. Also, from Proposition **4.1.1**, $dz \in \mathrm{Gr}(\mathcal{LS}_{(\bar{z}, \lambda)})$ for some $\lambda \in M(\bar{x})$. A Taylor expansion of f in (5) gives

$$\nabla f(\bar{z})^T (z^k - \bar{z}) + o(\|z^k - \bar{z}\|) < \frac{1}{k}\|\bar{z} - z^k\|^2.$$

Dividing by $\|z^k - \bar{z}\|$ and letting $k \to \infty$, we deduce $\nabla f(\bar{z})^T dz = 0$. Hence $dz \in \mathcal{C}(\bar{z}, \lambda)$; by the strict copositivity assumption, it follows that $dz^T \nabla^2 f(\bar{z}) dz > 0$.

Similarly we get from (5), that

$$\nabla f(\bar{z})^T (z^k - \bar{z}) + \tfrac{1}{2}(z^k - \bar{z})^T \nabla^2 f(\bar{z})(z^k - \bar{z}) + o(\|z^k - \bar{z}\|^2)$$
$$< \tfrac{1}{k}\|\bar{z} - z^k\|^2.$$

Now Proposition **4.1.3** tells us that $z^k - \bar{z} \in T(\bar{z}; \mathcal{F})$ for sufficiently large k; so for such k, $\nabla f(\bar{z})^T(z^k - \bar{z}) \geq 0$ and we have

$$\tfrac{1}{2}(z^k - \bar{z})^T \nabla^2 f(\bar{z})(z^k - \bar{z}) + o(\|z^k - \bar{z}\|^2) < \tfrac{1}{k}\|\bar{z} - z^k\|^2.$$

Dividing by $\|z^k - \bar{z}\|^2$ and passing to the limit $k \to \infty$, we conclude that $dz^T \nabla^2 f(\bar{z}) dz \leq 0$, which is a contradiction.   **Q.E.D.**

In classical quadratic programming theory, it is known [52] that for a stationary point of a (general) quadratic program, the copositivity of the

Hessian matrix of the objective function on the critical cone at this point is a necessary and sufficient condition for the point to be a local minimum of the program; similarly, the strict copositivity of the same Hessian matrix on the critical cone at this point is a necessary and sufficient condition for the point to be a strict local minimum of the program. The following result extends these classical facts to the MPAEC with a quadratic upper-level objective function.

5.2.4 Theorem. Let $\bar{z} \equiv (\bar{x}, \bar{y})$ be a feasible point of the MPAEC (1) with a polyhedral Z and a quadratic objective function:

$$f(x,y) \equiv c^T z + \tfrac{1}{2} z^T R z, \quad z \equiv (x,y) \in \Re^{n+m},$$

where $R \in \Re^{(n+m) \times (n+m)}$ and $c \in \Re^{n+m}$. The following two statements are valid:

(a) $\bar{z}$ is a local minimum of (1) if and only if $\bar{z}$ is stationary for (1) and R is copositive on the critical cone $\mathcal{C}(\bar{z}; \mathcal{F})$;

(b) $\bar{z}$ is a strict local minimum of (1) if and only if $\bar{z}$ is stationary for (1) and R is strictly copositive on the critical cone $\mathcal{C}(\bar{z}; \mathcal{F})$.

Proof. We only prove (a) since the proof of (b) is similar. For (a), it suffices to prove the "if" part. We use the notation in Proposition **4.1.3**. Specifically, let $\{\mathcal{P}_1, \ldots, \mathcal{P}_{r+s}\}$ be a family of polyhedra in $\Re^{n+m}$ whose union is equal to $\mathcal{F}$. Let $\{\mathcal{P}_1, \ldots, \mathcal{P}_r\}$ be the subfamily such that

$$\bar{z} \in \bigcap_{i=1}^{r} \mathcal{P}_i$$

and $\bar{z} \notin \mathcal{P}_{r+j}$ for each $j = 1, \ldots, s$. We have

$$\mathcal{T}(\bar{z}; \mathcal{F}) = \bigcup_{j=1}^{r} \mathcal{T}(\bar{z}; \mathcal{P}_j).$$

For each $j = 1, \ldots, r$, consider the quadratic program:

$$\begin{aligned} &\text{minimize} \quad f(x,y) \\ &\text{subject to} \quad (x,y) \in \mathcal{P}_j. \end{aligned} \tag{6}$$

If $\bar{z}$ is stationary for the MPAEC (1), then $\bar{z}$ is stationary for (6). Moreover if the matrix R is copositive on $\mathcal{C}(\bar{z}; \mathcal{F})$, then R remains copositive on the critical cone of (6) at $\bar{z}$. Consequently, by the aforementioned known fact for a quadratic program, it follows that if $\bar{z}$ is stationary for the MPAEC (1) and R is copositive on $\mathcal{C}(\bar{z}; \mathcal{F})$, then $\bar{z}$ is a local minimum of the problem (6). Consequently, for each $j = 1, \ldots, r$, there exists an open neighborhood W_j of $\bar{z}$ such that

$$z \in \mathcal{P}_j \cap W_j \implies f(z) \geq f(\bar{z}).$$

By shrinking each W_j if necessary, we may assume that each W_j does not intersect any one of the P_{r+j} for each $j = 1, \ldots, s$. It then follows that $\bar{z}$ is a minimum of f on $\mathcal{F} \cap W$, where $W \equiv \bigcap_{j=1}^{r} W_j$. This establishes that $\bar{z}$ is a local minimum of the MPAEC (1). **Q.E.D.**

Theorem **5.2.4**(a) complements Proposition **4.1.4**. In the above theorem, the upper-level objective function is a quadratic function and the result characterizes local optimality in terms of stationarity plus a copositivity condition on the objective function. In the previous proposition, the upper-level objective function is assumed to be a pseudoconvex function; the result shows that local optimality is equivalent to stationarity.

5.3 NCP Constrained Mathematical Program

We next extend Theorems **5.2.1** and **5.2.3** to the MPEC with the inner problems defined by a parametric NCP. Further extensions of the results to the general MP with VI constraints can be accomplished by using the KKT formulation of these constraints; see Section **5.5**. The derivation in the NCP case will illustrate the general approach to deal with the KKT case of an inner VI.

Consider the NCP constrained MP:

minimize $f(x, y)$

subject to $x \in X$ (1)

$$F(x, y) \geq 0, \quad y \geq 0, \quad y^T F(x, y) = 0,$$

where $f : \Re^{n+m} \to \Re$ and $F : \Re^{n+m} \to \Re^m$ are assumed twice continuously differentiable in a neighborhood of a given feasible vector (to be specified) and X is a polyhedral subset of $\Re^n$. Our first order of business is to write down the stationarity conditions for this problem in a form that will be useful for the derivation of the second-order conditions. We continue to use $\mathcal{F}$ for the feasible region of (1). In the notation of Subsection **1.3.2**, the problem (1) corresponds to the case where $C(x) = \Re_+^m$, for all x; or equivalently, the function $g(x, y) = -y$.

Fix a vector $\bar{z} \equiv (\bar{x}, \bar{y}) \in \mathcal{F}$. We recall three index sets defined in (3.2.7) that have been used many times:

$$\alpha(\bar{z}) = \{i : F_i(\bar{z}) = 0 < \bar{y}_i\},$$

$$\beta(\bar{z}) = \{i : F_i(\bar{z}) = 0 = \bar{y}_i\},$$

$$\gamma(\bar{z}) = \{i : F_i(\bar{z}) > 0 = \bar{y}_i\}.$$

Throughout the subsequent discussion, we let ∇_β, for any subset β of $\{1, \ldots, m\}$, denote the partial differentiation with respect to the y_β variables. Also, we assume that the SCOC holds at $\bar{z}$. Specializing Corollary **4.2.38** to the case where $g(x, y) \equiv -y$, we see that this assumption is equivalent to the following condition:

(SRC) the partial Jacobian matrix $\nabla_{\alpha(\bar{z})} F_{\alpha(\bar{z})}(\bar{z})$ is nonsingular, and the Schur complement below is a P matrix:

$$\nabla_{\beta(\bar{z})} F_{\beta(\bar{z})}(\bar{z}) - \nabla_{\alpha(\bar{z})} F_{\beta(\bar{z})}(\bar{z}) \left(\nabla_{\alpha(\bar{z})} F_{\alpha(\bar{z})}(\bar{z}) \right)^{-1} \nabla_{\beta(\bar{z})} F_{\alpha(\bar{z})}(\bar{z}). \qquad (2)$$

The abbreviation SRC stands for Strong Regularity Condition. Indeed, SRC as stated above is equivalent to $\bar{y}$ being a strongly regular solution (in the terminology of Robinson [248]) of the following NCP in the variable y (with x fixed at $\bar{x}$):

$$F(\bar{x}, y) \geq 0, \quad y \geq 0, \quad y^T F(\bar{x}, y) = 0.$$

With an abuse of language, we shall say that $\bar{z}$ is a strongly regular feasible solution of (1). In Subsection **4.2.7**, we show that strong regularity implies the implicit function assumption BIF made in Subsection **4.2.2**, so Theorem **4.2.31** says that the full CQ for (1) holds: $\mathcal{T}(\bar{z}; \mathcal{F}) = \mathcal{L}(\bar{z}; \mathcal{F})$.

Hence the stationarity of $\bar{z}$ for (1) is equivalent to:

$$
\left.
\begin{array}{c}
\nabla_x F_i(\bar{z})dx + \nabla_y F_i(\bar{z})dy = 0, \quad i \in \alpha(\bar{z}) \\[2mm]
\left.
\begin{array}{c}
\nabla_x F_i(\bar{z})dx + \nabla_y F_i(\bar{z})dy \geq 0 \\[1mm]
dy_i \geq 0 \\[1mm]
dy_i\, (\nabla_x F_i(\bar{z})dx + \nabla_y F_i(\bar{z})dy) = 0
\end{array}
\right\} \; i \in \beta(\bar{z}) \\[4mm]
dx \in \mathcal{T}(x;X), \quad dy_i = 0, \quad i \in \gamma(\bar{z})
\end{array}
\right\}
\tag{3}
$$

$$
\implies \nabla f(\bar{z})^T dz \geq 0,
$$

where $dz \equiv (dx, dy)$. Indeed, the set of vectors (dx, dy) satisfying the left-hand conditions in the above implication constitutes the tangent cone of $\mathcal{F}$ at $\bar{z}$. In what follows, we assume that the set X is polyhedral:

$$
X \equiv \{ x \in \Re^n : Gx + a \leq 0 \}
$$

where $G \in \Re^{s \times n}$ and $a \in \Re^s$; we shall further assume, for simplicity, that the given vector $\bar{x}$ satisfies all constraints in X as equations. Thus,

$$
\mathcal{T}(\bar{x}; X) = \{ dx \in \Re^n : G\, dx \leq 0 \}.
$$

Under these assumptions on the set X, we may apply Theorem **3.3.6**. Specifically, for each subset $\beta \subseteq \beta(\bar{z})$ (take $\beta \equiv \beta(\bar{z}) \setminus \beta_1$ where β_1 appears in the theorem), there exist MPEC multipliers $\zeta \in \Re^s$, π_i, $i \in \alpha(\bar{z}) \cup \beta(\bar{z})$, and η_i, $i \in \beta$, such that

$$
\begin{array}{rcl}
\nabla_x f(\bar{z}) & = & -G^T \zeta + \sum_{i \in \alpha(\bar{z}) \cup \beta(\bar{z})} \pi_i \nabla_x F_i(\bar{z}) \\[3mm]
\nabla_{\alpha(\bar{z})} f(\bar{z}) & = & \sum_{i \in \alpha(\bar{z}) \cup \beta(\bar{z})} \pi_i \nabla_{\alpha(\bar{z})} F_i(\bar{z}) \\[3mm]
\nabla_\beta f(\bar{z}) & = & \eta_\beta + \sum_{i \in \alpha(\bar{z}) \cup \beta(\bar{z})} \pi_i \nabla_\beta F_i(\bar{z})
\end{array}
\tag{4}
$$

$$
\zeta, \; \pi_{\beta(\bar{z}) \setminus \beta}, \; \eta_\beta \geq 0.
$$

Let $P(\bar{z}; \beta)$ denote the set of vectors $\pi \in \Re^{|\beta(\bar{z})| + |\alpha(\bar{z})|}$ satisfying (4), and let

$$
P(\bar{z}) \equiv \bigcup_{\beta \subseteq \beta(\bar{z})} P(\bar{z}; \beta).
$$

As expected, the MPEC multipliers π and η_β play an important role in the second-order conditions. In addition, so does the critical cone

$$
\mathcal{C}(\bar{z}; \mathcal{F}) \equiv \mathcal{T}(\bar{z}; \mathcal{F}) \cap \nabla f(\bar{z})^\perp.
\tag{5}
$$

The following result gives a second-order necessary condition for a local minimum of (1).

5.3.1 Theorem. Let X be a polyhedron in $\Re^n$ and $\bar{z} \equiv (\bar{x}, \bar{y})$ be a local minimum of (1). Suppose that $\bar{z}$ is strongly regular and f and F are twice continuously differentiable in a neighborhood of $\bar{z}$. Then in addition to satisfying the first-order stationarity conditions (3), $\bar{z}$ satisfies the following second-order necessary optimality condition: for all vectors $dz \equiv (dx, dy)$ belonging to $\mathcal{C}(\bar{z}; \mathcal{F})$ and satisfying

$$dy_i + \nabla F_i(\bar{z})^T dz > 0, \quad \text{for all } i \in \beta(\bar{z}), \tag{6}$$

there exists a vector $\pi \in P(\bar{z})$ such that

$$dz^T \left(\nabla^2 f(\bar{z}) - \sum_{i \in \alpha(\bar{z}) \cup \beta(\bar{z})} \pi_i \nabla^2 F_i(\bar{z}) \right) dz \geq 0, \tag{7}$$

where $\alpha(\bar{z})$ and $\beta(\bar{z})$ are defined above.

Proof. Let the vector dz be as given. In particular, dz satisfies the left-hand conditions in (3). By the strong regularity assumption of $\bar{z}$, it follows that BIF holds. That is to say, in a neighborhood of $\bar{x}$, there exists an implicit solution function $y(x)$ of the parametric NCP:

$$y \geq 0, \quad F(x, y) \geq 0, \quad y^T F(x, y) = 0,$$

that is B-differentiable at $\bar{x}$. Moreover $dy = y'(\bar{x}; dx)$ by Lemma **4.2.5**. Consequently there exists a sequence of positive scalars $\{\tau_k\}$ such that with $x^k \equiv \bar{x} + \tau_k dx$, $y^k \equiv y(x^k)$, and $z^k \equiv (x^k, y^k)$ for all k, we have

$$\lim_{k \to \infty} z^k = \bar{z}, \quad \lim_{k \to \infty} \frac{z^k - \bar{z}}{\tau_k} = dz,$$

and for each k, $(Gx^k)_i + a_i = 0$ whenever $(G\, dx)_i = 0$, and

$$F(x^k, y^k) \geq 0, \quad y^k \geq 0, \quad (y^k)^T F(x^k, y^k) = 0.$$

Define the index set

$$\beta \equiv \{i \in \beta(\bar{z}) : dy_i > 0 = \nabla_x F_i(\bar{z})^T dx + \nabla_y F_i(\bar{z})^T dy\}. \tag{8}$$

Since $dy_i + \nabla F_i(\bar{z})dz > 0$ for all $i \in \beta(\bar{z})$, we may write

$$\beta_1 \equiv \beta(\bar{z}) \setminus \beta = \{i \in \beta(\bar{z}) : dy_i = 0 < \nabla_x F_i(\bar{z})^T dx + \nabla_y F_i(\bar{z})^T dy\}.$$

The following two statements must hold for all k sufficiently large:

- for all $i \in \gamma(\bar{z}) \cup \beta_1$, $F_i(z^k) > 0$ which implies, by complementarity, $y_i^k = 0$;

- for all $i \in \alpha(\bar{z}) \cup \beta$, $y_i^k > 0$ which implies, by complementarity, $F_i(z^k) = 0$.

Substituting the expressions (4) for the various components of $\nabla f(\bar{z})$ and noticing that $dy_i = 0$ for all $i \notin (\alpha(\bar{z}) \cup \beta)$, we obtain

$$
\begin{aligned}
0 &= dz^T \nabla f(\bar{z}) \\
&= -\zeta^T G \, dx + \sum_{i \in \alpha(\bar{z}) \cup \beta(\bar{z})} \pi_i dz^T \nabla F_i(\bar{z}) + \eta_\beta^T dy_\beta.
\end{aligned}
$$

Since each summand in the right-hand side of this equation is either zero or nonnegative, we deduce that each summand is equal to zero. Thus, we have

$$\eta_\beta = 0, \quad \text{and} \quad \zeta_i (G \, dx)_i = 0, \quad \text{for all } i,$$

and

$$\pi_i = 0, \quad \text{for all } i \in \beta_1.$$

Since $\bar{z}$ is a local minimum of (1), we have for all k sufficiently large,

$$
\begin{aligned}
0 &\leq f(z^k) - f(\bar{z}) \\
&= \nabla f(\bar{z})^T (z^k - \bar{z}) + \tfrac{1}{2}(z^k - \bar{z})^T \nabla^2 f(\bar{z})(z^k - \bar{z}) + o(\|z^k - \bar{z}\|^2).
\end{aligned}
$$

Substituting (4) for $\nabla f(\bar{z})$, using the fact that $\eta_\beta = 0$ and

$$y_i^k = \bar{y}_i = 0, \quad \text{for all } i \in \gamma(\bar{z}) \cup \beta_1,$$

we obtain

$$
\begin{aligned}
0 \leq \ & -\zeta^T G(x^k - \bar{x}) + \sum_{i \in \alpha(\bar{z}) \cup \beta(\bar{z})} \pi_i \nabla F_i(\bar{z})^T (z^k - \bar{z}) \\
& + \tfrac{1}{2}(z^k - \bar{z})^T \nabla^2 f(\bar{z})(z^k - \bar{z}) + o(\|z^k - \bar{z}\|^2).
\end{aligned}
$$

For each i for which $\zeta_i > 0$, we have $(G\,dx)_i = 0$, which implies $(Gx^k)_i = a_i$. Consequently, by our simplifying assumption that $G\bar{x} = a$, we deduce

$$\zeta^T G(x^k - \bar{x}) = 0.$$

For each $i \in \alpha(\bar{z}) \cup \beta$, we have

$$
\begin{aligned}
0 &= F_i(z^k) \\
&= \nabla F_i(\bar{z})^T(z^k - \bar{z}) + \tfrac{1}{2}(z^k - \bar{z})^T \nabla^2 F_i(\bar{z})(z^k - \bar{z}) + o(\|z^k - \bar{z}\|^2),
\end{aligned}
$$

which implies

$$\nabla F_i(\bar{z})^T(z^k - \bar{z}) = -\tfrac{1}{2}(z^k - \bar{z})^T \nabla^2 F_i(\bar{z})(z^k - \bar{z}) + o(\|z^k - \bar{z}\|^2).$$

Hence, it follows that (since $\pi_i = 0$ for all $i \in \beta_1$)

$$0 \le \tfrac{1}{2}(z^k - \bar{z})^T \left(\nabla^2 f(\bar{z}) - \sum_{i \in \alpha(\bar{z}) \cup \beta(\bar{z})} \pi_i \nabla^2 F_i(\bar{z}) \right)(z^k - \bar{z}) + o(\|z^k - \bar{z}\|^2).$$

Dividing by τ_k^2 and passing to the limit as $k \to \infty$, we obtain the desired inequality (7).   **Q.E.D.**

We illustrate the above theorem with the following example.

5.3.2 Example. Consider the problem in 2 variables:

$$
\begin{aligned}
&\text{minimize}\quad \tfrac{1}{2}(x^2 - y^2) \\
&\text{subject to}\quad (x,y) \ge 0 \\
&\qquad\qquad\quad y^4 + y + x \ge 0, \quad y(y^4 + y + x) = 0.
\end{aligned}
$$

It is easy to verify that $\mathcal{F} = \Re_+ \times \{0\}$ and that $\bar{z} \equiv (0,0)$ is the unique global minimizer of this problem; moreover, this point satisfies the SRC and $\mathcal{T}(\bar{z}; \mathcal{F}) = \mathcal{F}$. We have $\nabla f(\bar{z}) = 0$,

$$\beta(\bar{z}) = \{1\}, \quad \text{and} \quad \alpha(\bar{z}) = \gamma(\bar{z}) = \emptyset.$$

It is easy to see that for all scalars π,

$$dz^T \left(\nabla^2 f(\bar{z}) - \pi \nabla^2 F(\bar{z}) \right) dz \ge 0$$

for all vectors $dz \in \mathcal{T}(\bar{z}; \mathcal{F})$; hence, Theorem **5.3.1** is verified.

The $(n + m) \times (n + m)$ matrix

$$\nabla^2 f(\bar{z}) - \sum_{i \in \alpha(\bar{z}) \cup \beta(\bar{z})} \pi_i \nabla^2 F_i(\bar{z})$$

that appears in the condition (7) and also in several analogous conditions throughout this section can be related to the MPEC Lagrangean function of the problem (1); see Subsection **3.3.1** where this function was introduced for the general MP with VI constraints. Specifically, modifying (3.3.9) to suit the NCP constrained MP (1), we define the MPEC Lagrangean function for this problem: for $(x, y, \zeta, \pi) \in \Re^{n+m+s+m}$,

$$\mathcal{L}^{\mathrm{MPEC}}(x, y, \zeta, \pi) \equiv f(x, y) + \zeta^T (Gx + a) - \pi^T F(x, y),$$

with ζ and π being the MPEC multipliers corresponding to the constraints $Gx + a \le 0$ and $F(x, y) \ge 0$ respectively. Setting $\pi_i = 0$ for all $i \in \gamma(\bar{z})$, (this is consistent with the first-order stationarity conditions for (1); see Theorem **3.3.6**), we see that, with $z \equiv (x, y)$,

$$\nabla^2_{zz} \mathcal{L}^{\mathrm{MPEC}}(\bar{x}, \bar{y}, \zeta, \pi) = \nabla^2 f(\bar{z}) - \sum_{i \in \alpha(\bar{z}) \cup \beta(\bar{z})} \pi_i \nabla^2 F_i(\bar{z}). \tag{9}$$

Consequently, under this identification, the second-order necessary condition in Theorem **5.3.1** states that the partial Hessian matrix of the MPEC Lagrangean function with respect to the variable z satisfies a "restricted copositivity" condition on the critical cone $\mathcal{C}(\bar{z}; \mathcal{F})$. This restriction is the imperfection of this theorem; it has to to with the assertion that the inequality (7) is valid, not for all vectors in the critical cone $\mathcal{C}(\bar{z}; \mathcal{F})$, but only for those critical vectors dz that also satisfy the strict complementarity condition (6). Ideally, we would like (7) to hold for all critical vectors. At this time, we are not sure whether this is right or wrong. In the next section, we shall derive an extension of Theorem **5.3.1** that will provide a significant relief of the restriction (6) on dz.

Another remark about Theorem **5.3.1** is that the polyhedrality of X is really not an essential restriction to the validity of the result. In principle, we could generalize this theorem to the case where X is defined by a finite system of smooth inequalities satisfying some standard constraint qualifications. We shall omit the details of this generalization.

As we did in Theorem **5.2.3**, we next establish a second-order sufficiency result for the problem (1).

5.3.3 Theorem. Let X be a polyhedron in $\Re^n$ and let $\bar{z} \equiv (\bar{x}, \bar{y})$ be a feasible vector of (1). Assume that f and F are twice continuously differentiable in a neighborhood of $\bar{z}$. If $\bar{z}$ satisfies the first-order stationarity implication (3) and also the following second-order sufficiency condition: for every nonzero critical vector $dz \in \mathcal{C}(\bar{z}; \mathcal{F})$ and every $\pi \in P(\bar{z})$,

$$dz^T \left(\nabla^2 f(\bar{z}) - \sum_{i \in \alpha(\bar{z}) \cup \beta(\bar{z})} \pi_i \nabla^2 F_i(\bar{z}) \right) dz > 0, \qquad (10)$$

where $\alpha(\bar{z})$ and $\beta(\bar{z})$ are defined in (7), then there exist a neighborhood $V \subseteq \Re^{n+m}$ of $\bar{z}$ and a scalar $\gamma > 0$ such that for all vectors $z' \in \mathcal{F} \cap V$,

$$f(z') - f(\bar{z}) \geq \gamma \|\bar{z} - z'\|^2.$$

Proof. As in the proof of Theorem **5.2.3**, assume that the claim is false. Then there exists a sequence $\{z^k\} \subseteq \mathcal{F}$ converging to $\bar{z}$ such that for each k, $z^k \neq \bar{z}$ and

$$f(z^k) - f(\bar{z}) < \frac{1}{k} \|\bar{z} - z^k\|^2. \qquad (11)$$

Write $z^k \equiv (x^k, y^k)$ for each k and consider the normalized sequence

$$\left\{ \frac{z^k - \bar{z}}{\|z^k - \bar{z}\|} \right\}$$

which must have at least one accumulation point $dz \equiv (dx, dy) \neq 0$. Clearly, $dz \in \mathcal{T}(\bar{z}; \mathcal{F})$. In particular, dz satisfies the left-hand conditions in (3). Hence, it follows that $\nabla f(\bar{z})^T dz \geq 0$. From (11), we can show that $\nabla f(\bar{z})^T dz \leq 0$. Thus $\nabla f(\bar{z})^T dz = 0$. Consequently, (10) is applicable to this vector dz. By an inductive argument, we can deduce that there exist an infinite index set κ and a subset β of $\beta(\bar{z})$ such that

$$F_i(z^k) = 0, \quad \text{for all } i \in \beta \text{ and all } k \in \kappa;$$

we certainly have

$$F_i(z^k) > 0, \quad \text{for all } i \in \beta(\bar{z}) \setminus \beta \text{ and all } k.$$

Let $\pi \in P(\bar{z}; \beta)$ be arbitrary, and let η_β and ζ be such that (4) holds. We have

$$f(z^k) - f(\bar{z}) =$$

$$\nabla f(\bar{z})^T (z^k - \bar{z}) + \tfrac{1}{2}(z^k - \bar{z})^T \nabla^2 f(\bar{z})(z^k - \bar{z}) + o(\|z^k - \bar{z}\|^2).$$

Assuming for simplicity, as above, that $G\bar{x} = a$, we may use (4) to substitute for $\nabla f(\bar{z})$. By the fact that for all $k \in \kappa$ sufficiently large,

$$y_i^k = \bar{y}_i = 0, \quad \text{for all } i \in \gamma(\bar{z}) \cup (\beta(\bar{z}) \setminus \beta)$$

we obtain

$$f(z^k) - f(\bar{z}) =$$
$$-\zeta^T G(x^k - \bar{x}) + \eta_\beta^T y_\beta^k + \sum_{i \in \alpha(\bar{z}) \cup \beta(\bar{z})} \pi_i \nabla F_i(z^k)^T (z^k - \bar{z})$$
$$+ \tfrac{1}{2}(z^k - \bar{z})^T \nabla^2 f(\bar{z})(z^k - \bar{z}) + o(\|z^k - \bar{z}\|^2).$$

Since $G\bar{x} = a$, we have $\zeta^T G(x^k - \bar{x}) \leq 0$; also $\eta_\beta^T y_\beta^k \geq 0$. Hence,

$$\begin{aligned} f(z^k) - f(\bar{z}) \;\geq\; & \sum_{i \in \alpha(\bar{z}) \cup \beta(\bar{z})} \pi_i \nabla F_i(z^k)^T (z^k - \bar{z}) \\ & + \tfrac{1}{2}(z^k - \bar{z})^T \nabla^2 f(\bar{z})(z^k - \bar{z}) + o(\|z^k - \bar{z}\|^2). \end{aligned} \tag{12}$$

For each $i \in \alpha(\bar{z}) \cup \beta$, we have for all $k \in \kappa$,

$$\begin{aligned} 0 \;=\; & F_i(z^k) \\ =\; & \nabla F_i(\bar{z})^T (z^k - \bar{z}) + \tfrac{1}{2}(z^k - \bar{z})^T \nabla^2 F_i(\bar{z})(z^k - \bar{z}) + o(\|z^k - \bar{z}\|^2), \end{aligned}$$

which implies

$$\nabla F_i(\bar{z})^T (z^k - \bar{z}) = -\tfrac{1}{2}(z^k - \bar{z})^T \nabla^2 F_i(\bar{z})(z^k - \bar{z}) + o(\|z^k - \bar{z}\|^2).$$

For each $i \in \beta(\bar{z}) \setminus \beta$, we have $\pi_i \geq 0$ and for all $k \in \kappa$,

$$\begin{aligned} 0 \;\leq\; & F_i(z^k) \\ =\; & \nabla F_i(\bar{z})^T (z^k - \bar{z}) + \tfrac{1}{2}(z^k - \bar{z})^T \nabla^2 F_i(\bar{z})(z^k - \bar{z}) + o(\|z^k - \bar{z}\|^2), \end{aligned}$$

which implies

$$\nabla F_i(\bar{z})^T (z^k - \bar{z}) \geq -\tfrac{1}{2}(z^k - \bar{z})^T \nabla^2 F_i(\bar{z})(z^k - \bar{z}) + o(\|z^k - \bar{z}\|^2).$$

Substituting these expressions into (12) and using (11), we deduce

$$\tfrac{1}{k}\|\bar{z} - z^k\|^2 > f(z^k) - f(\bar{z}) \geq$$
$$\tfrac{1}{2}(z^k - \bar{z})^T \left(\nabla^2 f(\bar{z}) - \sum_{i \in \alpha(\bar{z}) \cup \beta(\bar{z})} \pi_i \nabla^2 F(\bar{z}) \right) (z^k - \bar{z}) + o(\|z^k - \bar{z}\|^2).$$

Normalizing and taking limits, we obtain

$$dz^T \left(\nabla^2 f(\bar{z}) - \sum_{i \in \alpha(\bar{z}) \cup \beta(\bar{z})} \pi_i \nabla^2 F(\bar{z}) \right) dz \leq 0$$

which contradicts the assumption.   **Q.E.D.**

The requirement that (10) hold for all multipliers $\pi \in P(\bar{z})$ is rather strong. Ideally, we would like to require that (10) hold only for some π in $P(\bar{z}; \beta)$ for an appropriate β. The proof of the above theorem suggests that this might be possible. We treat this issue in more depth in the next section. In particular, we identify the required β.

5.4 Implicit Programming Based Results

The proofs of the two results, Theorems **5.3.1** and **5.3.3**, are very much in line with the derivation of the classical second-order conditions in nonlinear programming; see [204]. In order to sharpen these results, we need to go beyond standard derivations. Specifically, we turn to the IMP formulation of the problem (5.3.1).

Under SRC at $\bar{z}$, it follows that there exists a PC^1 implicit solution function $y(x)$ of the parametric NCP:

$$F(x, y) \geq 0, \quad y \geq 0, \quad y^T F(x, y) = 0$$

for all x near $\bar{x}$. Moreover, for each $dx \in \Re^n$, the directional derivative $y'(\bar{x}; dx)$ is the unique solution of the following mixed LCP in the variable dy:

$$\nabla_x F_i(\bar{z})^T dx + \nabla_y F_i(\bar{z})^T dy = 0, \quad i \in \alpha(\bar{z}),$$

$$\left. \begin{array}{c} \nabla_x F_i(\bar{z})^T dx + \nabla_y F_i(\bar{z})^T dy \geq 0 \\[2mm] dy_i \geq 0 \\[2mm] dy_i \left(\nabla_x F_i(\bar{z})^T dx + \nabla_y F_i(\bar{z})^T dy \right) = 0 \end{array} \right\} \quad i \in \beta(\bar{z}),$$

$$dy_i = 0, \quad i \in \gamma(\bar{z}).$$

Our goal is to gain a deeper understanding of the implicit function $y(x)$ near $\bar{x}$. For this purpose, we introduce, similar to the triple $(\alpha(\bar{z}), \beta(\bar{z}), \gamma(\bar{z}))$, three index sets that depend on the direction dx and the directional derivative $d\bar{y} \equiv y'(\bar{x}; dx)$ as follows:

$$d\beta_a(\bar{x}; dx) \equiv \{i \in \beta(\bar{z}) : \nabla_x F_i(\bar{z})^T dx + \nabla_y F_i(\bar{z})^T d\bar{y} = 0 < d\bar{y}_i\}$$

$$d\beta_b(\bar{x}; dx) \equiv \{i \in \beta(\bar{z}) : \nabla_x F_i(\bar{z})^T dx + \nabla_y F_i(\bar{z})^T d\bar{y} = 0 = d\bar{y}_i\}$$

$$d\beta_g(\bar{x}; dx) \equiv \{i \in \beta(\bar{z}) : \nabla_x F_i(\bar{z})^T dx + \nabla_y F_i(\bar{z})^T d\bar{y} > 0 = d\bar{y}_i\}.$$

The strict complementarity restriction (5.3.6) on $(dx, y'(\bar{x}; dx))$ that is required in Theorem **5.3.1** corresponds to the case where $d\beta_b(\bar{x}; dx) = \emptyset$. The following result establishes a kind of second-order directional derivative of the function $y(x)$ at $\bar{x}$. This result is the key to the generalization of this theorem and also the sufficiency Theorem **5.3.3** without assuming the emptiness of $d\beta_b(\bar{x}; dx)$.

5.4.1 Proposition. Suppose that SRC holds at $\bar{z}$. If F is C^2 in a neighborhood of $\bar{z}$, then for every vector $dx \in \Re^n$,

(a) the limit

$$\lim_{\tau \to 0+} \frac{y(\bar{x} + \tau \, dx) - \bar{y} - \tau \, y'(\bar{x}; dx)}{\tau^2} \tag{1}$$

exists, which we denote $\frac{1}{2} y^{(2)}(\bar{x}; dx)$;

(b) with $dz \equiv (dx, y'(\bar{x}; dx))$, $y^{(2)}(\bar{x}; dx)$ is the unique vector $d^{(2)}y$ that satisfies the mixed LCP system:

$$dz^T \nabla^2 F_i(\bar{z}) dz + \nabla_y F_i(\bar{z})^T d^{(2)}y = 0, \quad i \in \alpha(\bar{z}) \cup d\beta_a(\bar{x}; dx),$$

$$\left. \begin{array}{c} dz^T \nabla^2 F_i(\bar{z}) dz + \nabla_y F_i(\bar{z})^T d^{(2)}y \geq 0 \\ \\ d^{(2)}y_i \geq 0 \\ \\ d^{(2)}y_i \left(dz^T \nabla^2 F_i(\bar{z}) dz + \nabla_y F_i(\bar{z})^T d^{(2)}y \right) = 0 \end{array} \right\} i \in d\beta_b(\bar{x}; dx), \tag{2}$$

$$d^{(2)}y_i = 0, \quad i \in \gamma(\bar{z}) \cup d\beta_g(\bar{x}; dx);$$

(c) $y^{(2)}(\bar{x}; \cdot)$ is positively homogeneous of degree 2; that is

$$y^{(2)}(\bar{x}; \tau \, dx) = \tau^2 y^{(2)}(\bar{x}; dx), \quad \text{for all } \tau \geq 0;$$

(d) for every sequence of vectors $\{x^k\}$ converging to $\bar{x}$ and every sequence of positive scalars $\{\tau_k\}$ converging to zero such that

$$\lim_{k \to \infty} \frac{x^k - \bar{x}}{\tau_k} = dx,$$

it holds that

$$\lim_{k \to \infty} \frac{y(x^k) - \bar{y} - y'(\bar{x}; x^k - \bar{x})}{\tau_k^2} = \tfrac{1}{2} y^{(2)}(\bar{x}; dx).$$

Proof. To simplify the notation, we drop the vectors $\bar{z}$, $\bar{x}$, and dx from the index sets and Hessian matrices. We first show that the mixed LCP system (2) has a unique solution for each fixed $dx \in \Re^n$. By a fundamental result in mixed LCP theory [219, Proposition 2], it suffices to show that (i) the matrix

$$\begin{bmatrix} \nabla_\alpha F_\alpha & \nabla_{d\beta_a} F_\alpha \\ \nabla_\alpha F_{d\beta_a} & \nabla_{d\beta_a} F_{d\beta_a} \end{bmatrix} \tag{3}$$

is nonsingular, and (ii) the Schur complement of this matrix in

$$\begin{bmatrix} \nabla_\alpha F_\alpha & \nabla_{d\beta_a} F_\alpha & \nabla_{d\beta_b} F_\alpha \\ \nabla_\alpha F_{d\beta_a} & \nabla_{d\beta_a} F_{d\beta_a} & \nabla_{d\beta_b} F_{d\beta_a} \\ \nabla_\alpha F_{d\beta_b} & \nabla_{d\beta_a} F_{d\beta_b} & \nabla_{d\beta_b} F_{d\beta_b} \end{bmatrix}$$

is a P matrix. Since $\nabla_\alpha F_\alpha$ is nonsingular and the Schur complement of this matrix in (3) is a principal submatrix of the P matrix (5.3.2), it follows that (3) is nonsingular. Moreover, the determinantal sign of the latter matrix, which we denote σ, is equal to that of $\nabla_\alpha F_\alpha$. To prove (ii), let $\beta' \subseteq d\beta_b$ be arbitrary, we need to show that

$$\nabla_{\beta'} F_{\beta'} - \begin{bmatrix} \nabla_\alpha F_{\beta'} & \nabla_{d\beta_a} F_{\beta'} \end{bmatrix} \begin{bmatrix} \nabla_\alpha F_\alpha & \nabla_{d\beta_a} F_\alpha \\ \nabla_\alpha F_{d\beta_a} & \nabla_{d\beta_a} F_{d\beta_a} \end{bmatrix}^{-1} \begin{bmatrix} \nabla_{\beta'} F_\alpha \\ \nabla_{\beta'} F_{d\beta_a} \end{bmatrix}$$

has positive determinant. By the Schur formula, the determinant of the latter matrix is equal to

$$
\frac{\det \begin{bmatrix} \nabla_\alpha F_\alpha & \nabla_{d\beta_a} F_\alpha & \nabla_{\beta'} F_\alpha \\ \nabla_\alpha F_{d\beta_a} & \nabla_{d\beta_a} F_{d\beta_a} & \nabla_{\beta'} F_{d\beta_a} \\ \nabla_\alpha F_{\beta'} & \nabla_{d\beta_a} F_{\beta'} & \nabla_{\beta'} F_{\beta'} \end{bmatrix}}{\det \begin{bmatrix} \nabla_\alpha F_\alpha & \nabla_{d\beta_a} F_\alpha \\ \nabla_\alpha F_{d\beta_a} & \nabla_{d\beta_a} F_{d\beta_a} \end{bmatrix}}.
$$

As for the matrix in (3), it can be shown that the numerator in the above ratio of determinants has the same sign as σ, which is the sign of the denominator. This establishes our two claims (i) and (ii) as well as the existence and uniqueness of solution to the system (2).

We now consider the limit (1). Write

$$
x(\tau) \equiv \bar{x} + \tau \, dx, \quad y(\tau) \equiv y(x(\tau)),
$$

and

$$
dz(\tau) \equiv (dx, dy(\tau)) \equiv \left(dx, \frac{y(\tau) - \bar{y}}{\tau} \right).
$$

Notice that $dz(\tau) \to dz$ as $\tau \to 0+$. Clearly, for all $i \in \gamma$, we have

$$
y_i(\tau) = \bar{y}_i = y_i'(\bar{x}; dx) = 0.
$$

Moreover, by continuity of $y(\cdot)$ and F, we must have for each $i \in d\beta_g$,

$$
F_i(x(\tau), y(\tau)) > 0,
$$

for all $\tau > 0$ sufficiently small; hence for these i and τ, it follows that $y_i(\tau) = 0$. Consequently, we have for all $i \in \gamma \cup d\beta_g$,

$$
\lim_{\tau \to 0+} \frac{y_i(\tau) - \bar{y}_i - \tau \, y_i'(\bar{x}; dx)}{\tau^2} = 0 = \tfrac{1}{2} y_i^{(2)}(\bar{x}; dx).
$$

By a similar argument, we can deduce that $F_i(x(\tau), y(\tau)) = 0$ for all i in $d\beta_a$. Consider an index $i \in \alpha \cup d\beta_a$. We have, for a small enough $\tau > 0$,

$$
0 = F_i(x(\tau), y(\tau))
$$
$$
= F_i(\bar{z}) + \tau \, \nabla F_i(\bar{z})^T dz(\tau) + \tfrac{\tau^2}{2} \, dz(\tau)^T \nabla^2 F_i(\bar{z}) dz(\tau) + o(\tau^2),
$$

which implies

$$0 = \nabla F_i(\bar{z})^T dz(\tau) + \frac{\tau}{2} dz(\tau)^T \nabla^2 F_i(\bar{z}) dz(\tau) + o(\tau).$$

Since

$$0 = \nabla_x F_i(\bar{z})^T dx + \nabla_y F_i(\bar{z}) y'(\bar{x}; dx),$$

subtracting, we obtain

$$0 = \nabla_y F_i(\bar{z})^T (dy(\tau) - y'(\bar{x}; dx)) + \frac{\tau}{2} dz(\tau)^T \nabla^2 F_i(\bar{z}) dz(\tau) + o(\tau),$$

for all $i \in \alpha \cup d\beta_a$. Similarly, we can show that for all $i \in d\beta_b$,

$$0 \le \nabla_y F_i(\bar{z})^T (dy(\tau) - y'(\bar{x}; dx)) + \frac{\tau}{2} dz(\tau)^T \nabla^2 F_i(\bar{z}) dz(\tau) + o(\tau),$$

$$0 \le (dy(\tau) - y'(\bar{x}; dx))_i,$$

and at least one of these two inequalities must hold as an equation. By the claims (i) and (ii) proved above it can be deduced easily that for each $i \in \alpha \cup d\beta_a \cup d\beta_b$, and each sequence of positive scalars $\{\tau_k\}$ converging to zero,

$$\sup_k \frac{|dy_i(\tau_k) - y'_i(\bar{x}; dx)|}{\tau_k} < \infty.$$

(This follows from the fact that for any subset $\beta' \subseteq d\beta_b$, the matrix

$$\begin{bmatrix} \nabla_\alpha F_\alpha & \nabla_{d\beta_a} F_\alpha & \nabla_{\beta'} F_\alpha \\ \nabla_\alpha F_{d\beta_a} & \nabla_{d\beta_a} F_{d\beta_a} & \nabla_{\beta'} F_{d\beta_a} \\ \nabla_\alpha F_{\beta'} & \nabla_{d\beta_a} F_{\beta'} & \nabla_{\beta'} F_{\beta'} \end{bmatrix}$$

is nonsingular.) Moreover, any accumulation point of the sequence

$$\left\{ \frac{dy(\tau_k) - y'(\bar{x}; dx)}{\tau_k} \right\}$$

must satisfy the mixed LCP system (2). By the uniqueness of the solution to the latter system, it follows that the limit (1) must exist and is equal to $\frac{1}{2}$ times this solution. Thus (b) holds. The proof of (d) is analogous to that of part (b); the details are omitted. Part (c) follows from the following observations: for $s = a, b, g$,

$$d\beta_s(\bar{x}; \tau\, dx) = d\beta_s(\bar{x}; dx), \quad \text{for all } \tau > 0;$$

$dz^T \nabla^2 F_i(\bar{z}) dz$ is positively homogeneous of degree 2 in dx; and $y^{(2)}(\bar{x}; 0)$ is clearly equal to zero.   **Q.E.D.**

The vector $y^{(2)}(\bar{x}; dx)$ is a kind of second-order directional derivative of the nonsmooth function $y(x)$ at $\bar{x}$. We explain this observation by considering the strictly complementary case where $\beta(\bar{z}) = \emptyset$. In this case, for all x sufficiently near $\bar{x}$, $y(x)$ is the implicit solution function of the parametrized system of nonlinear equations with y as the primary variable and x the parameter:

$$F_{\alpha(\bar{z})}(x, y) = 0, \quad y_{\gamma(\bar{z})} = 0.$$

The SRC at $\bar{z}$ reduces to the single condition that the matrix $\nabla_{\alpha(\bar{z})} F_{\alpha(\bar{z})}(\bar{z})$ is nonsingular. By the classical implicit function theorem [3, Theorem 2.3], it follows that $y(x)$ is C^2 near $\bar{x}$; hence parts (a), (c), and (d) of Proposition **5.4.1** are clearly valid. Moreover we have

$$\begin{pmatrix} y^{(2)}_{\alpha(\bar{z})}(\bar{x}; dx) \\ y^{(2)}_{\gamma(\bar{z})}(\bar{x}; dx) \end{pmatrix} = \begin{pmatrix} -(\nabla_{\alpha(\bar{z})} F_{\alpha(\bar{z})}(\bar{z}))^{-1} (dz^T \nabla^2 F_i(\bar{z}) dz)_{i \in \alpha(\bar{z})} \\ 0 \end{pmatrix}$$

which is the assertion of part (b), taking into account the assumption that all the beta index sets are empty. In regard to the general case when y is not C^2, part (d) is the second-order analog of Corollary **4.2.3** for the function $y(x)$. Indeed with some additional work, a rigorous connection could be established between $y^{(2)}(\bar{x}; dx)$ and such a second derivative. In spite of the independent interest and importance of this issue, we will not explore it further herein; see Liu [159] who has studied the high-order directional differentiability of the solution function of parametric variational inequalities under the MFCQ, CRCQ, and a strong second-order sufficiency condition. Instead, we state below the second-order optimality conditions for (5.3.1) in terms of $y^{(2)}(\bar{x}; dx)$.

5.4.2 Theorem. Let X be a polyhedron in $\Re^n$ and $\bar{z} \equiv (\bar{x}, \bar{y})$ be a given stationary point of (5.3.1). Suppose that $\bar{z}$ is strongly regular and f and F are twice continuously differentiable in a neighborhood of $\bar{z}$.

(a) If $\bar{z}$ is a local minimum of (5.3.1), then for all critical vectors dz in $\mathcal{C}(\bar{z}; \mathcal{F})$, where $\mathcal{C}(\bar{z}; \mathcal{F})$ is given by (5.3.5),

$$dz^T \nabla^2 f(\bar{z}) dz + \nabla_y f(\bar{z})^T y^{(2)}(\bar{x}; dx) \geq 0. \tag{4}$$

(b) Conversely, if

$$dz^T \nabla^2 f(\bar{z}) dz + \nabla_y f(\bar{z})^T y^{(2)}(\bar{x}; dx) > 0,$$

for all $0 \neq dz \equiv (dx, dy) \in C(\bar{z}; \mathcal{F})$, then the conclusion of Theorem **5.3.3** holds.

Proof. To prove (a), let $dz = (dx, dy)$ be an arbitrary critical vector. Let $x(\tau)$, $y(\tau)$, and $dz(\tau)$ be as defined in the proof of Proposition **5.4.1**. Also let

$$z(\tau) \equiv (x(\tau), y(\tau)) = \bar{z} + \tau \, dz(\tau).$$

By Lemma **4.2.5**, we have $dy = y'(\bar{x}; dx)$. For all $\tau > 0$, we have

$$\begin{aligned}
0 &\leq f(z(\tau)) - f(\bar{z}) \\
&= \nabla f(\bar{z})^T (z(\tau) - \bar{z}) + \tfrac{1}{2}(z(\tau) - \bar{z})^T \nabla^2 f(\bar{z})(z(\tau) - \bar{z}) + o(\tau^2),
\end{aligned}$$

where we have used the fact that $y(\tau)$ is Lipschitz near $\tau = 0$ to obtain $o(\|z(\tau) - \bar{z}\|^2) = o(\tau^2)$. Since

$$\nabla_x f(\bar{z})^T dx + \nabla_y f(\bar{z})^T y'(\bar{x}; dx) = 0,$$

we deduce

$$0 \leq \nabla_y f(\bar{z})^T (dy(\tau) - y'(\bar{x}; dx)) + \frac{\tau}{2} dz(\tau)^T \nabla^2 f(\bar{z}) dz(\tau) + o(\tau).$$

Dividing by $\tau > 0$ and taking the limit $\tau \to 0$, we deduce the desired inequality (4), using Proposition **5.4.1**(a).

To prove (b), we proceed as in the proof of Theorem **5.3.3**. Assume for contradiction that there exists a sequence $\{z^k\} \subset \mathcal{F}$ converging to $\bar{z}$ such that the normalized sequence

$$\left\{ \frac{z^k - \bar{z}}{\|z^k - \bar{z}\|} \right\}$$

converges to a nonzero vector $dz \in C(\bar{z}; \mathcal{F})$ and for each k, $z^k = (x^k, y(x^k))$, and

$$f(z^k) - f(\bar{z}) < \frac{1}{k} \|z^k - \bar{z}\|^2.$$

We have

$$f(z^k) - f(\bar{z}) = \\
\nabla f(\bar{z})^T (z^k - \bar{z}) + \tfrac{1}{2}(z^k - \bar{z})^T \nabla^2 f(\bar{z})(z^k - \bar{z}) + o(\|z^k - \bar{z}\|^2).$$

Since X is polyhedral, it follows that for all k sufficiently large, the pair $(x^k - \bar{x}; y'(\bar{x}; x^k - \bar{x})) \in \mathcal{T}(\bar{z}; \mathcal{F})$. Since $\bar{z}$ is stationary, it follows that

$$0 \leq \nabla_x f(\bar{z})^T (x^k - \bar{x}) + \nabla_y f(\bar{z})^T y'(\bar{x}; x^k - \bar{x}).$$

Consequently,

$$\tfrac{1}{k} \|z^k - \bar{z}\|^2 > f(z^k) - f(\bar{z}) \geq$$

$$\nabla_y f(\bar{z})^T (y^k - \bar{y} - y'(\bar{x}; x^k - \bar{x})) + \tfrac{1}{2} (z^k - \bar{z})^T \nabla^2 f(\bar{z})(z^k - \bar{z})$$

$$+ o(\|z^k - \bar{z}\|^2).$$

Dividing by $\|z^k - \bar{z}\|^2$, passing to the limit $k \to \infty$, and using Proposition **5.4.1**(d), we obtain

$$0 \geq \nabla_y f(\bar{z})^T y^{(2)}(\bar{x}; dx) + dz^T \nabla^2 f(\bar{z}) dz$$

which contradicts the assumption in (b).   **Q.E.D.**

Compared to Theorems **5.3.1** and **5.3.3**, Theorem **5.4.2** does not involve the multipliers π in the set $P(\bar{z})$. By applying the first-order optimality conditions (5.3.4) with an appropriate choice of the index set β, we may derive a refinement of Theorems **5.3.1** and **5.3.3**. To state this refinement, we introduce for each vector $dx \in \Re^n$ the following index set associated with the LCP conditions in the system (2):

$$d^{(2)}\alpha(\bar{x}; dx) \equiv \{i \in d\beta_b(\bar{x}; dx) : y_i^{(2)}(\bar{x}; dx) > 0\}.$$

5.4.3 Theorem. Let X be a polyhedron in $\Re^n$ and $\bar{z} \equiv (\bar{x}, \bar{y})$. Assume the setting of Theorem **5.4.2**.

(a) If $\bar{z}$ is a local minimum of (5.3.1), then for each vector dz belonging to $\mathcal{C}(\bar{z}; \mathcal{F})$, each index set $\beta_0 \subseteq d\beta_b(\bar{x}; dx)$ with complement

$$\bar{\beta}_0 \equiv d\beta_b(\bar{x}; dx) \setminus \beta_0,$$

and each pair (π, η_β) satisfying (5.3.4), where

$$\beta \equiv d\beta_a(\bar{x}; dx) \cup \beta_0, \tag{5}$$

it holds that

$$dz^T \left(\nabla^2 f(\bar{z}) - \sum_{i \in \alpha(\bar{z}) \cup \beta(\bar{z})} \pi_i \nabla^2 F_i(\bar{z}) \right) dz + (\eta_{\beta_0})^T y_{\beta_0}^{(2)}(\bar{x}; dx)$$

$$+ \left(\nabla_{\bar{\beta}_0} f(\bar{z}) - \sum_{i \in \alpha(\bar{z}) \cup \beta(\bar{z})} \pi_i \nabla_{\bar{\beta}_0} F_i(\bar{z}) \right)^T y_{\bar{\beta}_0}^{(2)}(\bar{x}; dx) \qquad (6)$$

$$+ \sum_{i \in d\beta_b(\bar{x}; dx)} \pi_i \left(\nabla_y F_i(\bar{z})^T y^{(2)}(\bar{x}; dx) + dz^T \nabla^2 F_i(\bar{z}) dz \right) \geq 0.$$

(b) Conversely, if for each nonzero critical vector dz there exist an index set $\beta_0 \subseteq d\beta_b(\bar{x}; dx)$ with complement $\bar{\beta}_0$ as given above and a multiplier pair (π, η_β) satisfying (5.3.4), where β is defined in (5), such that the inequality in (6) holds strictly, then the conclusion of Theorem **5.3.3** holds. In particular, the same conclusion holds if for each nonzero critical vector dz, there exists a multiplier pair (π, η_β) satisfying (5.3.4), where

$$\beta \equiv d\beta_a(\bar{x}; dx) \cup d^{(2)}\alpha(\bar{x}; dx),$$

such that

$$dz^T \left(\nabla^2 f(\bar{z}) - \sum_{i \in \alpha(\bar{z}) \cup \beta(\bar{z})} \pi_i \nabla^2 F_i(\bar{z}) \right) dz > 0.$$

Proof. Let dz be a fixed but arbitrary critical vector and let (π, η_β) be an arbitrary multiplier pair satisfying (5.3.4), where β is defined by (5). As in the proof of Theorem **5.3.1**, it can be shown that

$$\eta_i = 0, \quad \text{for all } i \in d\beta_a(\bar{x}; dx), \qquad (7)$$

$$\pi_i = 0, \quad \text{for all } i \in d\beta_g(\bar{x}; dx). \qquad (8)$$

Since, from Proposition **5.4.1**(b), $y_i^{(2)}(\bar{x}; dx) = 0$ for all i belonging to $\gamma(\bar{z}) \cup d\beta_g(\bar{z}; dx)$, it follows that

$$\nabla_y f(\bar{z})^T y^{(2)}(\bar{x}; dx) =$$

$$\nabla_{\alpha(\bar{z})} f(\bar{z})^T y_{\alpha(\bar{z})}^{(2)}(\bar{x}; dx) + \nabla_{d\beta_a(\bar{x}; dx)} f(\bar{z})^T y_{d\beta_a(\bar{x}; dx)}^{(2)}(\bar{x}; dx) +$$

$$\nabla_{\beta_0} f(\bar{z})^T y_{\beta_0}^{(2)}(\bar{x}; dx) + \nabla_{\bar{\beta}_0} f(\bar{z})^T y_{\bar{\beta}_0}^{(2)}(\bar{x}; dx).$$

Using the conditions (5.3.4) to substitute for $\nabla_{\alpha(\bar{z})} f(\bar{z})$ and $\nabla_\beta f(\bar{z})$, and noticing (7), we obtain

$$\nabla_y f(\bar{z})^T y^{(2)}(\bar{x}; dx) = (\eta_{\beta_0})^T y_{\beta_0}^{(2)}(\bar{x}; dx)$$

$$+ \sum_{i \in \alpha(\bar{z}) \cup \beta(\bar{z})} \pi_i \nabla_y F_i(\bar{z})^T y^{(2)}(\bar{x}; dx)$$

$$+ \left(\nabla_{\bar{\beta}_0} f(\bar{z}) - \sum_{i \in \alpha(\bar{z}) \cup \beta(\bar{z})} \pi_i \nabla_{\bar{\beta}_0} F_i(\bar{z}) \right)^T y_{\bar{\beta}_0}^{(2)}(\bar{x}; dx).$$

Since for all $i \in \alpha(\bar{z}) \cup d\beta_a(\bar{x}; dx)$,

$$dz^T \nabla^2 F_i(\bar{z}) dz + \nabla_y F_i(\bar{z})^T y^{(2)}(\bar{x}; dx) = 0,$$

we obtain

$$\nabla_y f(\bar{z})^T y^{(2)}(\bar{x}; dx) = (\eta_{\beta_0})^T y_{\beta_0}^{(2)}(\bar{x}; dx)$$

$$- \sum_{i \in \alpha(\bar{z}) \cup \beta(\bar{z})} \pi_i \, dz^T \nabla^2 F_i(\bar{z}) dz$$

$$+ \left(\nabla_{\bar{\beta}_0} f(\bar{z}) - \sum_{i \in \alpha(\bar{z}) \cup \beta(\bar{z})} \pi_i \nabla_{\bar{\beta}_0} F_i(\bar{z}) \right)^T y_{\bar{\beta}_0}^{(2)}(\bar{x}; dx)$$

$$+ \sum_{i \in d\beta_b} \pi_i \left(\nabla_y F_i(\bar{z})^T y^{(2)}(\bar{x}; dx) + dz^T \nabla^2 F_i(\bar{z}) dz \right).$$

Consequently, we deduce

$$dz^T \nabla^2 f(\bar{z}) dz + \nabla_y f(\bar{z})^T y^{(2)}(\bar{x}; dx) =$$

$$dz^T \left(\nabla^2 f(\bar{z}) - \sum_{i \in \alpha(\bar{z}) \cup \beta(\bar{z})} \pi_i \nabla^2 F_i(\bar{z}) \right) dz + (\eta_{\beta_0})^T y_{\beta_0}^{(2)}(\bar{x}; dx)$$

$$+ \left(\nabla_{\bar{\beta}_0} f(\bar{z}) - \sum_{i \in \alpha(\bar{z}) \cup \beta(\bar{z})} \pi_i \nabla_{\bar{\beta}_0} F_i(\bar{z}) \right)^T y_{\bar{\beta}_0}^{(2)}(\bar{x}; dx) \tag{9}$$

$$+ \sum_{i \in d\beta_b(\bar{x}; dx)} \pi_i \left(\nabla_y F_i(\bar{z})^T y^{(2)}(\bar{x}; dx) + dz^T \nabla^2 F_i(\bar{z}) dz \right).$$

If $\bar{z}$ is a local minimum of (5.3.1), then by Theorem **5.4.2**, the left-hand term in (9) is nonnegative; hence the inequality (6) holds.

The first assertion in part (b) follows from Theorem **5.4.2** and the expression (9). To prove the second assertion, let $\beta_0 \equiv d^{(2)} \alpha(\bar{x}; dx)$. Since

$$\eta_i \geq 0, \qquad \text{for all } i \in \beta_0,$$

$$y_i^{(2)}(\bar{x}; dx) = 0, \quad \text{for all } i \in \bar{\beta}_0,$$

and for all $i \in d\beta_b(\bar{x}; dx)$,

$$\pi_i \left(\nabla_y F_i(\bar{z})^T y^{(2)}(\bar{x}; dx) + dz^T \nabla^2 F_i(\bar{z}) dz \right) \geq 0,$$

by (9), we must have

$$dz^T \nabla^2 f(\bar{z}) dz + \nabla_y f(\bar{z})^T y^{(2)}(\bar{x}; dx) \geq$$

$$dz^T \left(\nabla^2 f(\bar{z}) - \sum_{i \in \alpha(\bar{z}) \cup \beta(\bar{z})} \pi_i \nabla^2 F_i(\bar{z}) \right) dz.$$

Consequently, the desired conclusion follows as before.   **Q.E.D.**

In the remainder of this section, we discuss several special cases of part (a) of Theorem **5.4.3**. Throughout this discussion, we assume that $\bar{z}$ is a local minimum of (5.3.1). One special case yields Theorem **5.3.1** in which we restrict the choice of the critical vector dz such that $d\beta_b(\bar{x}; dx)$ is empty. Another special case occurs when we take $\beta_0 \equiv d^{(2)} \alpha(\bar{x}; dx)$, as in the proof of part (b) of Theorem **5.4.3**. With this choice of β_0, we have

$$y^{(2)}_{\beta_0}(\bar{x}; dx) = 0;$$

thus (6) becomes

$$dz^T \left(\nabla^2 f(\bar{z}) - \sum_{i \in \alpha(\bar{z}) \cup \beta(\bar{z})} \pi_i \nabla^2 F_i(\bar{z}) \right) dz + \sum_{i \in d^{(2)} \alpha(\bar{x}; dx)} \eta_i y_i^{(2)}(\bar{x}; dx)$$

$$+ \sum_{i \in d^{(2)} \gamma(\bar{x}; dx)} \pi_i \left(\nabla_y F_i(\bar{z})^T y^{(2)}(\bar{x}; dx) + dz^T \nabla^2 F_i(\bar{z}) dz \right) \geq 0, \qquad (10)$$

where

$$d^{(2)} \gamma(\bar{x}; dx) \equiv \{ \ i \in d\beta_b(\bar{x}; dx) \ :$$

$$0 < dz^T \nabla^2 F_i(\bar{z}) dz + \nabla_y F_i(\bar{z})^T y^{(2)}(\bar{x}; dx) \}.$$

In other words, (10) must necessarily hold for all vectors $dz \in \mathcal{C}(\bar{z}; \mathcal{F})$. Noting (5.3.9), the mixed LCP system (2) that is satisfied by $y^{(2)}(\bar{x}; dx)$, and the two expressions (7) and (8), and by letting

$$\eta_i = 0 \text{ for all } i \in \alpha(\bar{z}) \quad \text{and} \quad \pi_i = 0 \text{ for all } i \in \gamma(\bar{z}),$$

we may rewrite the condition (10) in the more compact form:

$$dz^T \nabla^2_{zz} \mathcal{L}^{\mathrm{MPEC}}(\bar{z}, \zeta, \pi) dz + \eta^T y^{(2)}(\bar{x}; dx)$$
$$+ \pi^T \left(\nabla F(\bar{z}) y^{(2)}(\bar{x}; dx) + dz^T \nabla^2 F(\bar{z}) dz \right) \geq 0, \tag{11}$$

where $dz^T \nabla^2 F(\bar{z}) dz$ denotes the m-vector with components $dz^T \nabla^2 F_i(\bar{z}) dz$, $i = 1, \ldots, m$. We formalize this observation in the following special result.

5.4.4 Corollary. Let X be a polyhedron in $\Re^n$ and $\bar{z} \equiv (\bar{x}, \bar{y})$ Assume the setting of Theorem **5.4.2**. If $\bar{z}$ is a local minimum of (5.3.1), then (11) holds for each $dz \in \mathcal{C}(\bar{z}; \mathcal{F})$ and each pair (π, η_β) satisfying (5.3.4), where

$$\beta \equiv d\beta_a(\bar{x}; dx) \cup d^{(2)}\alpha(\bar{x}; dx).$$

Conversely, if for each nonzero $dz \in \mathcal{C}(\bar{z}; \mathcal{F})$, there exists a pair (π, η_β) satisfying (5.3.4), where β is as defined above, such that (11) holds with strict inequality, then $\bar{z}$ is a strict local minimum of (5.3.1).

Further special cases of Theorem **5.4.3** are stated in the following result.

5.4.5 Corollary. Let X be a polyhedron in $\Re^n$ and $\bar{z} \equiv (\bar{x}, \bar{y})$ Assume the setting of Theorem **5.4.2**. If $\bar{z}$ is a local minimum of (5.3.1), then (5.3.7), with π as specified in Theorem **5.4.3**, holds for all critical vectors $dz \in \mathcal{C}(\bar{z}; \mathcal{F})$ satisfying for all $i \in d\beta_b(\bar{x}; dx)$,

$$y_i^{(2)}(\bar{x}; dx) = dz^T \nabla^2 F_i(\bar{z}) dz + \nabla_y F_i(\bar{z})^T y^{(2)}(\bar{x}; dx) = 0.$$

In particular, if $\bar{z}$ is a local minimum of (5.3.1) and if

$$dz \in \mathcal{C}(\bar{z}; \mathcal{F}) \implies$$
$$dz^T \nabla^2 F_i(\bar{z}) dz = 0, \quad \forall i \in \alpha(\bar{z}) \cup d\beta_a(\bar{x}; dx) \cup d\beta_b(\bar{x}; dx), \tag{12}$$

then (5.3.7), with π as specified in Theorem **5.4.3**, holds for all critical vectors dz.

Proof. The first assertion follows from (10). To prove the second assertion, observe that by the uniqueness of the solution to the mixed LCP system (5.4.2), if (12) holds, then $y^{(2)}(\bar{x}; dx) = 0$ for all $dz \in \mathcal{C}(\bar{z}; \mathcal{F})$. Hence the second assertion follows from the first.   **Q.E.D.**

If each $F_i(x, y)$ is an affine function for all i, then clearly (12) holds. In this case, we recover the results of Section **5.2** specialized to an LCP constrained MP. More generally, if

$$\mathcal{C}(\bar{z}; \mathcal{F}) \subseteq \bigcap_{i \in \alpha(\bar{z}) \cup d\beta_a(\bar{x}; dx) \cup d\beta_b(\bar{x}; dx)} \text{null space of } \nabla^2 F_i(\bar{z}),$$

then (12), and thus the second conclusion of Corollary **5.4.5** also holds.

5.5 KKT Constrained Mathematical Program

In this section we consider the following MPEC in the variables (x, y, λ):

$$
\begin{aligned}
\text{minimize} \quad & f(x, y) \\
\text{subject to} \quad & Gx + a \leq 0, \\
& 0 = L(x, y, \lambda) \equiv F(x, y) + \nabla_y g(x, y)^T \lambda, \\
& g(x, y) \leq 0, \quad \lambda \geq 0, \quad \lambda^T g(x, y) = 0,
\end{aligned}
\tag{1}
$$

which is the problem (3.3.8) with the matrix $H = 0$. We refer to Subsection **3.3.1** for the dimensions of the variables and functions in this problem and to Chapter **3** in general for the notation and first-order results used freely throughout this section. In particular, we let $\mathcal{F}^{\text{KKT}} \subseteq \Re^{n+m+\ell}$ denote the feasible region of (1).

Our goal in this section is to derive, using the implicit programming approach, some second-order conditions for a local minimizer of (1). In light of Proposition **3.4.1** and Corollary **3.4.4** which relate the MPEC (1.1.1) and the KKT constrained MP (1) in terms of their respective local minima and stationary points, the main result of this section, Theorem **5.5.1**, can be used to produce a corresponding set of second-order optimality conditions for a local minimizer of (1.1.1); these will be omitted.

Fix a triple $(\bar{x}, \bar{y}, \bar{\lambda}) \in \mathcal{F}^{\text{KKT}}$ which we assume is a stationary point of (1). We further assume that f and F are C^2 and g is C^3 in a neighborhood of $(\bar{x}, \bar{y})$. Finally, we assume that the LICQ holds at $(\bar{x}, \bar{y})$ and that any one of the conditions (a)–(f) in Corollary **4.2.38** holds. By this corollary and Corollary **4.2.24**, it follows that a PC^1 implicit function $(y(x), \lambda(x))$ exists such that $\bar{y} = y(\bar{x})$, $\bar{\lambda} = \lambda(\bar{x})$, and the KKT constraints in (1) hold for all x near $\bar{x}$. Moreover, Corollary **4.2.26** identifies the directional derivative

$(y'(\bar{x}; dx), \lambda'(\bar{x}; dx))$ as the unique solution $(dy, d\lambda)$ of the following affine KKT system for each fixed but arbitrary $dx \in \Re^n$:

$$\nabla_x L(\bar{z}, \dot{\lambda}) dx + \nabla_y L(\bar{z}, \bar{\lambda}) dy + \nabla_y g_{\mathcal{I}(\bar{x})}(\bar{z})^T d\lambda_{\mathcal{I}(\bar{x})} = 0,$$

$$\left.\begin{array}{l} \nabla_x g_i(\bar{z})^T dx + \nabla_y g_i(\bar{z})^T dy \le 0, \quad d\lambda_i \ge 0 \\ d\lambda_i \left(\nabla_x g_i(\bar{z})^T dx + \nabla_y g_i(\bar{z})^T dy\right) = 0 \end{array}\right\} \quad i \in \mathcal{I}_0,$$

$$\nabla_x g_i(\bar{z})^T dx + \nabla_y g_i(\bar{z})^T dy = 0, \quad d\lambda_i \text{ free}, \quad i \in \mathcal{I}_+,$$

$$d\lambda_i = 0, \quad i \in \mathcal{I}(\bar{x})^c,$$

where

$$\mathcal{I}(\bar{x}) \equiv \{i \ : \ g_i(\bar{z}) = 0\},$$

$$\mathcal{I}_0 \equiv \{i \ : \ g_i(\bar{z}) = 0 = \bar{\lambda}_i\}, \quad \mathcal{I}_+ \equiv \{i \ : \ g_i(\bar{z}) = 0 < \bar{\lambda}_i\},$$

and $\mathcal{I}(\bar{x})^c$ is the complement of $\mathcal{I}(\bar{x})$ in $\{1, \ldots, \ell\}$; i.e.,

$$\mathcal{I}(\bar{x})^c \equiv \{i \ : \ g_i(\bar{z}) < 0\}.$$

Under this setting, we will present a set of second-order necessary and second-order sufficient conditions for the triple $(\bar{x}, \bar{y}, \bar{\lambda})$ by extending the approach in the previous section. Specifically, we consider the KKT constraints in (1) as a parametric, mixed NCP defined by the function:

$$\begin{pmatrix} x \\ y \\ \lambda \end{pmatrix} \mapsto \begin{pmatrix} L(x, y, \lambda) \\ -g(x, y) \end{pmatrix},$$

with (y, λ) as the primary variables and x as the parameter. Although the results in Section **5.4** pertain to the standard NCP, their extensions to the mixed NCP are straightforward. In the following, all the results are stated without proof.

It would be useful to consider the index sets $\{1, \ldots, m\} \cup \mathcal{I}_+$ and $\mathcal{I}_0$ playing the role of $\alpha(\bar{z})$ and $\beta(\bar{z})$ in the last section, respectively. We include $\{1, \ldots, m\}$ in $\alpha(\bar{z})$ because the Lagrangean equation $L(x, y, \lambda) = 0$ always holds as an equation. (This is the mixed nature of the KKT constraint system.) The assumptions set forth above are enough for us to establish a result similar to Proposition **5.4.1**. To explain this result, we introduce some notation. A generic vector triple $(x, y, \lambda) \in \Re^{n+m+\ell}$ will be denoted w and the particular triple $(\bar{x}, \bar{y}, \bar{\lambda})$ by $\bar{w}$; similarly, the directional triple $(dx, y'(\bar{x}; dx), \lambda'(\bar{x}; dx))$ will be denoted dw. We continue to use z for (x, y) and dz for $(dx, y'(\bar{x}; dx))$. Also for ease of notation, we write

$$dw^T \nabla^2 L(\bar{w}) dw \equiv \left(dw^T \nabla^2 L_i(\bar{w}) dw \right)_{i=1}^m \in \Re^m.$$

Similarly to the three index sets $d\beta_a(\bar{x}; dx)$, $d\beta_b(\bar{x}; dx)$, and $d\beta_g(\bar{x}; dx)$, we define

$$d\mathcal{I}_a \equiv \{i \in \mathcal{I}_0 : \nabla_x g_i(\bar{z})^T dx + \nabla_y g_i(\bar{z})^T y'(\bar{x}; dx) = 0 < \lambda_i'(\bar{x}; dx)\},$$

$$d\mathcal{I}_b \equiv \{i \in \mathcal{I}_0 : \nabla_x g_i(\bar{z})^T dx + \nabla_y g_i(\bar{z})^T y'(\bar{x}; dx) = 0 = \lambda_i'(\bar{x}; dx)\},$$

$$d\mathcal{I}_g \equiv \{i \in \mathcal{I}_0 : \nabla_x g_i(\bar{z})^T dx + \nabla_y g_i(\bar{z})^T y'(\bar{x}; dx) < 0 = \lambda_i'(\bar{x}; dx)\}.$$

It can then be shown that for all vectors $dx \in \Re^n$, the limits

$$\lim_{\tau \to 0+} \frac{y(\bar{x} + \tau \, dx) - \bar{y} - \tau \, y'(\bar{x}; dx)}{\tau^2} \equiv \tfrac{1}{2} y^{(2)}(\bar{x}; dx)$$

and

$$\lim_{\tau \to 0+} \frac{\lambda(\bar{x} + \tau \, dx) - \bar{\lambda} - \tau \, \lambda'(\bar{x}; dx)}{\tau^2} \equiv \tfrac{1}{2} \lambda^{(2)}(\bar{x}; dx)$$

exist and they are the unique pair $(d^{(2)}y, d^{(2)}\lambda)$ satisfying the system:

$$dw^T \nabla^2 L(\bar{w}) dw + \nabla_y L(\bar{w}) d^{(2)} y + \nabla_y g_{\mathcal{I}(\bar{x})}(\bar{z})^T d^{(2)} \lambda_{\mathcal{I}(\bar{x})} = 0,$$

$$\left. \begin{array}{c} dz^T \nabla^2 g_i(\bar{z}) dz + \nabla_y g_i(\bar{z})^T d^{(2)} y \leq 0, \quad d^{(2)} \lambda_i \geq 0 \\[2mm] d^{(2)} \lambda_i \left(dz^T \nabla^2 g_i(\bar{z}) dz + \nabla_y g_i(\bar{z})^T d^{(2)} y \right) = 0 \end{array} \right\} \quad i \in d\mathcal{I}_b,$$

$$dz^T \nabla^2 g_i(\bar{z}) dz + \nabla_y g_i(\bar{z})^T d^{(2)} y = 0, \quad d^{(2)} \lambda_i \text{ free}, \quad i \in \mathcal{I}_+ \cup d\mathcal{I}_a$$

$$d^{(2)} \lambda_i = 0, \quad i \in d\mathcal{I}_g \cup \mathcal{I}(\bar{x})^c.$$

Similar to the index set $d^{(2)}\alpha(\bar{x}; dx)$ and $d^{(2)}\gamma(\bar{x}; dx)$, we define

$$d^{(2)}\mathcal{I}_+ \equiv \{i \in d\mathcal{I}_b : \lambda_i^{(2)}(\bar{x}; dx) > 0\}$$

$$d^{(2)}\mathcal{I}_- \equiv \{i \in d\mathcal{I}_b : 0 > dz^T \nabla^2 g_i(\bar{z}) dz + \nabla_y g_i(\bar{z})^T y^{(2)}(\bar{x}; dx)\}.$$

We recall the MPEC Lagrangean function (3.3.9) introduced in Subsection **3.3.1**. Specifically, for $(x, y, \lambda, \zeta, \pi, \eta) \in \Re^{n+m+\ell+s+m+\ell}$, let

$$\mathcal{L}^{\mathrm{MPEC}}(x, y, \lambda, \zeta, \pi, \eta) \equiv$$
$$f(x, y) + \zeta^T (Gx + a) - \pi^T L(x, y, \lambda) + \eta^T g(x, y). \tag{2}$$

We also recall the first-order stationarity conditions for the problem (1) as stated in Corollary **4.2.32**, with $\lambda = \bar{\lambda}$. For ease of reference, we restate these conditions as follows: for each index subset α of $\mathcal{I}_0$ with complement $\bar{\alpha}$ in $\mathcal{I}_0$, there exist multipliers $\zeta \in \Re^s$, $\eta \in \Re^\ell$ and $\pi \in \Re^m$ such that

$$\nabla_x f(\bar{z}) + G^T \zeta + \nabla_x g(\bar{z})^T \eta = \nabla_x L(\bar{w})^T \pi$$

$$\nabla_y f(\bar{z}) + \nabla_y g(\bar{z})^T \eta = \nabla_y L(\bar{w})^T \pi$$

$$\pi^T \nabla_y g_i(\bar{z}) \leq 0, \quad i \in \alpha,$$

$$\bar{\lambda}_i \left(\pi^T \nabla_y g_i(\bar{z}) \right) = 0, \quad i \in \mathcal{I}(\bar{x}), \tag{3}$$

$$\eta_i \geq 0, \quad i \in \bar{\alpha},$$

$$\eta_i = 0, \quad i \notin \mathcal{I}(\bar{x}),$$

$$\zeta \geq 0, \quad \zeta^T (G\bar{x} + a) = 0.$$

Finally, we let

$$\mathcal{C}(\bar{w}; \mathcal{F}^{\mathrm{KKT}}) \equiv \mathcal{T}(\bar{w}; \mathcal{F}^{\mathrm{KKT}}) \cap \nabla f(\bar{z})^\perp \subseteq \Re^{n+m+\ell}$$

be the critical cone of (1) at $\bar{w}$.

In what follows, we extend Corollary **5.4.4** to the context of the VI constrained MP (1). Like this corollary, the result below is obtained by specifying one choice of the index set α, which corresponds to the choice $\beta_0 = d^{(2)}\alpha(\bar{x}; dx)$ that led to (5.4.11). In part (a) of the result, we write $dz^T \nabla^2 g(\bar{z})dz$ to denote the m-vector with components $dz^T \nabla^2 g_i(\bar{z})dz$, for $i = 1, \ldots, m$.

5.5.1 Theorem. Assume the setting described above.

(a) If $\bar{w}$ is a local minimum of (1), then for each critical vector dw in $\mathcal{C}(\bar{w}; \mathcal{F}^{\mathrm{KKT}})$ and each pair (π, η) satisfying (3) corresponding to the index set $\alpha \subseteq \mathcal{I}_0$, where

$$\alpha \equiv d\mathcal{I}_a \cup d^{(2)}\mathcal{I}_+,$$

it holds that

$$dw^T \nabla^2_{ww} \mathcal{L}^{\text{MPEC}}(\bar{w}, \zeta, \pi, \eta)dw - (\nabla_y g(\bar{z})\pi)^T \lambda^{(2)}(\bar{x}; dx)$$

$$- \eta^T \left(dz^T \nabla^2 g(\bar{z})dz + \nabla_y g(\bar{z})y^{(2)}(\bar{x}; dx) \right) \geq 0.$$

(b) Conversely, if for every nonzero critical vector $dw \in \mathcal{C}(\bar{w}; \mathcal{F}^{\text{KKT}})$ there exists a pair (π, η) satisfying (3) corresponding to $\alpha \subseteq \mathcal{I}_0$, where α is as defined above, such that

$$dw^T \nabla^2_{ww} \mathcal{L}^{\text{MPEC}}(\bar{w}, \zeta, \pi, \eta)dw > 0,$$

then there exist a neighborhood $V \subseteq \Re^{n+m+\ell}$ of $\bar{w}$ and a constant $\gamma > 0$ such that for all triples $w \equiv (x, y, \lambda) \in \mathcal{F}^{\text{KKT}} \cap V$,

$$f(z) \geq f(\bar{z}) + \gamma \|w - \bar{w}\|^2. \qquad (4)$$

An interesting consequence of the conclusion of part (b) should be mentioned. The inequality (4) implies the uniqueness of the KKT multiplier $\bar{\lambda}$ associated with $\bar{z}$ as a solution of the VI $(F(\bar{x}, \cdot), C(\bar{x}))$; i.e., $M(\bar{x}) = \{\bar{\lambda}\}$. To see this, assume that there is $\lambda \in M(\bar{x}) \setminus \{\bar{\lambda}\}$. Since $M(\bar{x})$ is convex, we may assume without loss of generality that λ is sufficiently close to $\bar{\lambda}$ so that $(\bar{z}, \lambda) \in V$. By (4), we have

$$f(\bar{z}) \geq f(\bar{z}) + \gamma \|\lambda - \bar{\lambda}\|,$$

which implies $\lambda = \bar{\lambda}$, a contradiction.

The uniqueness of $\bar{\lambda}$ is related to a special structure of the MPEC Lagrangean function $\mathcal{L}^{\text{MPEC}}$. Namely, $\mathcal{L}^{\text{MPEC}}$ contains no quadratic term in λ; this implies that $\nabla^2_{\lambda\lambda} \mathcal{L}^{\text{MPEC}}$ is identically equal to zero. See Subsection **5.6.1**, particularly Proposition **5.6.8**, for more discussion on this property of $\mathcal{L}^{\text{MPEC}}$.

5.6 A Piecewise Programming Approach

In this section, we use the PCP approach to derive some second-order optimality conditions. Such an approach was used in Section **4.3** for the verification of the MPEC hypotheses. Theorem **5.1.1** plays a central role in the following analysis.

Given twice continuously differentiable mappings $f : \Re^{n+m} \to \Re$ and $F : \Re^{n+m} \to \Re^m$, and the set $Z = \{(x, y) \in \Re^{n+m} : Gx + Hy \le a\}$ where $G \in \Re^{s \times n}$, $H \in \Re^{s \times m}$ and $a \in \Re^s$, consider the following NCP constrained MP:

$$\text{minimize} \quad f(x, y)$$

$$\text{subject to} \quad (x, y) \in Z$$

$$y^T F(x, y) = 0, \tag{1}$$

$$F(x, y) \ge 0, \quad y \ge 0.$$

Let $\mathcal{F}$ denote its feasible region. We mention that in the following development, the constraint $Gx + Hy \le a$ can be replaced by a nonlinear constraint $h(x, y) \le 0$ for a twice continuously function $h : \Re^{n+m} \to \Re^s$; we use the former merely to take advantage of previous discussion. Also, although the tangent cone to Z at $\bar{z} \in Z$ can be given explicitly as

$$\{(dx, dy) : G_i dx + H_i dy \le 0, \text{ for } i \text{ with } G_i \bar{x} + H_i \bar{y} = a_i\},$$

we will use the notation $\mathcal{T}(\bar{z}; Z)$ for brevity.

For a given feasible vector $\bar{z} \in \mathcal{F}$, we recall three index sets defined in (3.2.7):

$$\alpha(\bar{z}) = \{i : F_i(\bar{z}) = 0 < \bar{y}_i\},$$

$$\beta(\bar{z}) = \{i : F_i(\bar{z}) = 0 = \bar{y}_i\},$$

$$\gamma(\bar{z}) = \{i : F_i(\bar{z}) > 0 = \bar{y}_i\}.$$

The disjunctive approach to the complementarity condition in (1) is to decompose it into inequalities and equations corresponding to each subset $\beta \subseteq \beta(\bar{z})$:

$$\text{minimize} \quad f(x, y)$$

$$\text{subject to} \quad (x, y) \in Z$$

$$\left.\begin{array}{r} F_i(x, y) = 0 \\ y_i \ge 0 \end{array}\right\} \quad i \in \beta \cup \alpha(\bar{z}) \tag{2}$$

$$\left.\begin{array}{r} F_i(x, y) \ge 0 \\ y_i = 0 \end{array}\right\} \quad i \in \bar{\beta} \cup \gamma(\bar{z}),$$

where $\bar{\beta}$ denotes the complement of β in $\beta(\bar{z})$. Let $\mathcal{F}_\beta$ denote the feasible region of (2). Clearly, $\mathcal{F}_\beta \subseteq \mathcal{F}$. We can apply Theorem **5.1.1** to (2) to obtain some PCP second-order necessary/sufficient optimality conditions for the NCP constrained MP (1). For this purpose, we introduce the critical cone of (2) at $\bar{z}$:

$$\mathcal{C}(\bar{z}; \mathcal{F}_\beta) \equiv \{(dx, dy) \in \mathcal{T}(\bar{z}; Z) \; : \; \nabla_x f(\bar{z})dx + \nabla_y f(\bar{z})dy = 0,$$

$$\nabla_x F_i(\bar{z})^T dx + \nabla_y F_i(\bar{z})^T dy = 0, \text{ for } i \in \beta \cup \alpha(\bar{z}),$$

$$\nabla_x F_i(\bar{z})^T dx + \nabla_y F_i(\bar{z})^T dy \geq 0, \text{ for } i \in \bar{\beta}, \tag{3}$$

$$dy_i = 0, \text{ for } i \in \bar{\beta} \cup \gamma(\bar{z}),$$

$$dy_i \geq 0, \text{ for } i \in \beta\}.$$

5.6.1 Theorem. Suppose $\bar{z} \in \mathcal{F}$ is a local minimum of (1). Let $\beta \subseteq \beta(\bar{z})$ be an index set such that the MFCQ holds at $\bar{z}$ for (2). Then for each $dz = (dx, dy) \in \mathcal{C}(\bar{z}; \mathcal{F}_\beta)$, there exist multipliers $\zeta \in \Re^s$, $\pi \in \Re^{|\alpha(\bar{z})|+|\beta(\bar{z})|}$, and $\eta \in \Re^{|\beta|}$ satisfying the first-order optimality condition (5.3.4) and

$$dz^T \left(\nabla^2 f(\bar{z}) - \sum_{i \in \alpha(\bar{z}) \cup \beta(\bar{z})} \pi_i \nabla^2 F_i(\bar{z}) \right) dz \geq 0. \tag{4}$$

Proof. Since $\bar{z}$ is a local minimum of (1), $\bar{z}$ must be a local minimum of (2). Moreover, the KKT conditions for (2) at $\bar{z}$ are analogous to the system (5.3.4). Thus the desired conclusion is a direct consequence of part (a) of Theorem **5.1.1**.   **Q.E.D.**

The second-order sufficient optimality condition is stated below.

5.6.2 Theorem. Suppose $\bar{z} \in \mathcal{F}$ is a feasible vector of (1). Suppose further that for each $\beta \subseteq \beta(\bar{z})$, MFCQ holds at $\bar{z}$ for (2); moreover for each $dz = (dx, dy)$ in the critical cone $\mathcal{C}(\bar{z}; \mathcal{F}_\beta)$, there exist multipliers $\zeta \in \Re^s$, $\pi \in \Re^{|\alpha(\bar{z})|+|\beta(\bar{z})|}$, and $\eta \in \Re^{|\beta|}$ satisfying the KKT conditions (5.3.4) and

$$dz^T \left(\nabla^2 f(\bar{z}) - \sum_{i \in \alpha(\bar{z}) \cup \beta(\bar{z})} \pi_i \nabla^2 F_i(\bar{z}) \right) dz > 0. \tag{5}$$

There exist then a neighborhood $V \subseteq \Re^{n+m}$ of $\bar{z}$ and a constant $\gamma > 0$,

$$f(z) \geq f(\bar{z}) + \gamma \|z - \bar{z}\|^2$$

for each $z \equiv (x, y) \in \mathcal{F} \cap V$; hence $\bar{z}$ is a strict local minimizer of (1).

Proof. We first fix any $\beta \subseteq \beta(\bar{z})$. By Theorem **5.1.1**(b), it follows that for some neighborhood N_β of $\bar{z}$ and constant $\gamma_\beta > 0$, $f(z) \geq f(\bar{z}) + \gamma_\beta \|z - \bar{z}\|^2$ for each $z \in N_\beta \cap \mathcal{F}_\beta$.

By continuity, there exists some neighborhood N_0 of $\bar{z}$ such that for all $z \in N_0 \cap \mathcal{F}$, $\beta(z) \subseteq \beta(\bar{z})$. Thus, each vector z in $N_0 \cap \mathcal{F}$ must belong to $\mathcal{F}_\beta$ for some $\beta \subseteq \beta(\bar{z})$ (e.g., take $\beta = \beta(z)$). Define a neighborhood of $\bar{z}$,

$$N \equiv N_0 \cap \left(\bigcap_\beta N_\beta \right),$$

where β ranges over all possible subsets of $\beta(\bar{z})$. Then, for each $z \in N \cap \mathcal{F}$, we have $z \in \mathcal{F}_\beta$ for some $\beta \subseteq \beta(\bar{z})$ and $z \in N_\beta$. It then follows that for the positive quantity $\gamma \equiv \min_\beta \gamma_\beta$, that $f(z) \geq f(\bar{z}) + \gamma \|z - \bar{z}\|^2$ if $z \in N \cap \mathcal{F}$. **Q.E.D.**

We pause for a moment to compare Theorems **5.6.1** and **5.6.2** with the results in Theorem **5.4.3**. We first examine the necessary optimality conditions. The conclusions of Theorem **5.6.1** and Theorem **5.4.3**(a) are different in two major respects. First, in both theorems the MPEC multipliers (π, η_β) are dependent on the choice of dz; but in Theorem **5.4.3**(a) this dependence is stated more explicitly. Second, the condition (4) is weaker than the corresponding condition (5.4.6). Indeed, the latter condition contains four terms, with the first identical to the quadratic form (4) and the remaining three terms nonnegative; thus the condition (5.4.6) is a refinement of (4). This refinement is made possible by the use of second-order information of the implicit solution function $y(x)$; in contrast no such information is exploited in the classical derivations of second-order optimality conditions.

The assumptions in Theorems **5.6.1** and **5.4.3**(a) are also quite different; we illustrated this difference in Example **4.3.4** whereby the MFCQ for (2) is shown to be not comparable with SCOC, which is equivalent to SRC in NCP constraints. If the LICQ holds for (2) at $\bar{z}$, then there is a unique (π, η_β) satisfying (5.3.4); in this case, the condition (4) holds for all $dz \in \mathcal{C}(\bar{z}; \mathcal{F}_\beta)$ with this pair (π, η_β). The following example illustrates that in general this LICQ for (2) is also not comparable with SRC for the NCP constrained MP (1).

5.6.3 Example. Consider the NCP constrained MP in the joint variables $(x, y) \in \Re^2 \times \Re^2$ whereby the inner NCP is defined by

$$
F(x, y) = \begin{pmatrix} x_1 - y_1 \\ x_2 - y_2 \end{pmatrix}.
$$

Clearly, $\bar{z} = ((0, 0), (0, 0))$ is a feasible solution and $\beta(\bar{z})$ is equal to $\{1, 2\}$. It can easily be checked that the LICQ is satisfied for (2) with each $\beta \subseteq \beta(\bar{z})$. However, since the inner NCP has multiple solutions around $\bar{z}$, the BIF condition is violated, which further implies that the SRC condition is not satisfied at $\bar{z}$. This shows that LICQ does not imply SRC in general. To see if the reverse implication holds, consider the case where the inner NCP is defined by

$$
F(x, y) = \begin{pmatrix} x_1^2 + y_1 \\ x_2^2 + y_2 \end{pmatrix}.
$$

Consider the feasible solution $\bar{z} = ((0, 0), (0, 0))$. Clearly, $\beta(\bar{z}) = \{1, 2\}$. Since $F(x, \cdot)$ is strongly monotone, the SRC holds. However, it can be checked that LICQ does not hold for (2) at $\bar{z}$ with $\beta = \beta(\bar{z})$. Thus, SRC cannot imply LICQ either.

Theorem **5.6.2** can be similarly compared with Theorem **5.4.3**(b). In particular, since the last three terms in (5.4.6) are nonnegative, the condition (5) in Theorem **5.6.2** implies the corresponding condition imposed in Theorem **5.4.3**(b). It should be noted that there is no regularity assumption (other than the existence of the multipliers) in Theorem **5.6.2**. Thus, the weakening of (5) to (5.4.6) is achieved with the aid of the additional SRC condition.

We can also extend the results of Theorems **5.6.1** and **5.6.2** to the general VI constrained MP by rewriting its constraints in KKT form. In particular, consider the KKT constrained MP (1.3.8):

$$\text{minimize} \quad f(x,y)$$

$$\text{subject to} \quad (x,y) \in Z,$$

$$F(x,y) + \sum_{i=1}^{\ell} \lambda_i \nabla_y g_i(x,y) = 0, \tag{6}$$

$$\lambda \geq 0, \quad g(x,y) \leq 0, \quad \lambda^T g(x,y) = 0,$$

where in addition to the functions f and F and the set Z defined above, we use the twice continuously differentiable function $g : \Re^{n+m} \to \Re^{\ell}$. Throughout, we continue to use the notation set in Subsection **3.3.1** and Section **4.3**. Fix a vector $\bar{z} \in \mathcal{F}$ and assume that SBCQ holds at $\bar{z}$. If $\bar{z}$ is a local minimizer of the MPEC in general form, (1.1.1), then by Proposition **3.4.1**, $(\bar{z}, \bar{\lambda})$ is a local minimizer of (6) for all $\lambda \in M(\bar{z})$. Thus for each pair of index sets $(\alpha, \bar{\alpha})$ partitioning $\mathcal{I}_0(\bar{z}, \bar{\lambda})$, $(\bar{z}, \bar{\lambda})$ is a local minimizer of the following subproblem of (6):

$$\text{minimize} \quad f(x,y)$$

$$\text{subject to} \quad Gx + Hy + a \leq 0,$$

$$L(x,y,\lambda) = 0,$$

$$\left.\begin{array}{l} \lambda_i = 0 \\ g_i(x,y) \leq 0 \end{array}\right\} \forall i \in \mathcal{J}_1, \tag{7}$$

$$\left.\begin{array}{l} \lambda_i \geq 0 \\ g_i(x,y) = 0 \end{array}\right\} \forall i \in \mathcal{J}_2,$$

where $(\mathcal{J}_1, \mathcal{J}_2)$ is defined (cf. (3.3.11)) by

$$\mathcal{J}_1 \equiv \bar{\alpha} \cup \{i : g_i(\bar{z}) < 0\}, \quad \text{and} \quad \mathcal{J}_2 \equiv \alpha \cup \mathcal{I}_+(\bar{z}, \bar{\lambda}). \tag{8}$$

Under some standard constraint qualifications, Theorem **5.1.1** implies that $(\bar{z}, \bar{\lambda})$ satisfies some second-order necessary optimality conditions specialized to (7).

For simplicity, we take Z to be the polyhedron (3.3.3). We recall the MPEC Lagrangean function; see (3.3.9):

$$\mathcal{L}^{\text{MPEC}}(x,y,\lambda,\zeta,\pi,\eta) \equiv$$

$$f(x,y) + \zeta^T(Gx + Hy + a) - \pi^T L(x,y,\lambda) + \eta^T g(x,y).$$

264 Chapter 5. Second-Order Optimality Conditions

In what follows, we establish second-order necessary/sufficient optimality conditions for (6). Continuing the use of the generic notation $w = (x, y, \lambda)$, we introduce the critical cone of (7) at $\bar{w}$:

$$\mathcal{C}_\alpha(\bar{w}) \equiv$$

$$\{(dx, dy, d\lambda) \in \mathcal{T}(\bar{z}; Z) \times \Re^\ell : \nabla_x f(\bar{z})dx + \nabla_y f(\bar{z})dy = 0,$$

$$\nabla_x L(\bar{w})dx + \nabla_y L(\bar{w})dy + \sum_{i \in \mathcal{J}_2} d\lambda_i \nabla g_i(\bar{z}) = 0,$$

$$\nabla_x g_i(\bar{z})^T dx + \nabla_y g_i(\bar{z})^T dy = 0, \text{ for } i \in \mathcal{J}_2, \qquad (9)$$

$$\nabla_x g_i(\bar{z})^T dx + \nabla_y g_i(\bar{z})^T dy \leq 0, \text{ for } i \in \bar{\alpha},$$

$$d\lambda_i = 0, \text{ for } i \in \mathcal{J}_1,$$

$$d\lambda_i \geq 0, \text{ for } i \in \alpha\}.$$

5.6.4 Theorem. Let $\bar{z} = (\bar{x}, \bar{y}) \in \mathcal{F}$ and the multiplier $\bar{\lambda} \in M(\bar{z})$ be given; let $\bar{w} = (\bar{z}, \bar{\lambda})$. Let α and $\bar{\alpha}$ be two arbitrary index sets partitioning $\mathcal{I}_0(\bar{w})$; define $\mathcal{J}_1$ and $\mathcal{J}_2$ as in (8). Suppose the MFCQ holds at $\bar{w}$ for (7). If $\bar{w}$ is a local minimizer of (6), then for each $dw \equiv (dz, d\lambda)$ belonging to the cone $\mathcal{C}_\alpha(\bar{w})$, there exist multipliers ζ, π, and η satisfying the system (3.3.12) such that

$$dw^T \nabla^2_{ww} \mathcal{L}^{\text{MPEC}}(\bar{w}, \zeta, \pi, \eta)dw \geq 0.$$

Proof. As in Theorem **5.6.1**, the conclusion follows from part (a) of Theorem **5.1.1**. **Q.E.D.**

We can also derive the following second-order sufficient optimality condition for the KKT constrained MP (6).

5.6.5 Theorem. Let $\bar{w} = (\bar{z}, \bar{\lambda}) \in \mathcal{F}^{\text{KKT}}$ be a feasible vector of (6). Suppose for each $\alpha \subseteq \mathcal{I}_0(\bar{w})$, the MFCQ holds at $\bar{w}$ for (7), and for each nonzero $dw \equiv (dz, d\lambda) \in \mathcal{C}_\alpha(\bar{w})$, there exist multipliers ζ, π and η satisfying the first-order optimality condition (3.3.12) and

$$dw^T \nabla^2_{ww} \mathcal{L}^{\text{MPEC}}(\bar{w}, \zeta, \pi, \eta)dw > 0. \qquad (10)$$

Then there exists a neighborhood $V \subseteq \Re^{n+m+\ell}$ of $\bar{w}$ and a constant $\gamma > 0$ such that for each $w \equiv (z, \lambda) \in \mathcal{F}^{\mathrm{KKT}} \cap V$,

$$f(z) \geq f(\bar{z}) + \gamma \|w - \bar{w}\|^2 \geq f(\bar{z}) + \gamma \|z - \bar{z}\|^2.$$

Proof. The proof is similar to that of Theorem **5.6.2**. We first fix any $\alpha \subseteq \mathcal{I}_0(\bar{w})$. Let $\mathcal{F}_\alpha^{\mathrm{KKT}}$ denote the feasible set of (7), where $\mathcal{J}_1$, $\mathcal{J}_2$ are defined by (8). By Theorem **5.1.1**, it follows that for some neighborhood N_α of $\bar{w}$ and constant $\gamma_\alpha > 0$,

$$f(z) \geq f(\bar{z}) + \gamma_\alpha \|w - \bar{w}\|^2 \geq f(\bar{z}) + \gamma_\alpha \|z - \bar{z}\|^2$$

for all $(z, \lambda) \in N_\alpha \cap \mathcal{F}_\alpha^{\mathrm{KKT}}$.

By continuity, there exists a neighborhood N_0 of $\bar{w}$ such that for all $w \in N_0 \cap \mathcal{F}^{\mathrm{KKT}}$, $\mathcal{I}_0(w) \subseteq \mathcal{I}_0(\bar{w})$. Thus, each vector w in $N_0 \cap \mathcal{F}^{\mathrm{KKT}}$ must belong to $\mathcal{F}_\alpha^{\mathrm{KKT}}$ for some $\alpha \subseteq \mathcal{I}_0(\bar{w})$ (e.g., take $\alpha = \mathcal{I}_0(w)$). Define a neighborhood of $\bar{w}$ as

$$N \equiv N_0 \cap \left(\bigcap_\alpha N_\alpha \right),$$

where α ranges over all possible subsets of $\mathcal{I}_0(\bar{w})$. Similarly, define a positive constant $\gamma = \min_\alpha \gamma_\alpha$. Then, for each $w \in N \cap \mathcal{F}^{\mathrm{KKT}}$, we have $w \in \mathcal{F}_\alpha^{\mathrm{KKT}}$ for some $\alpha \in \mathcal{I}_0(\bar{w})$ and $w \in N_\alpha$. It then follows that $f(z) \geq f(\bar{z}) + \gamma \|w - \bar{w}\|^2 \geq f(\bar{z}) + \gamma \|z - \bar{z}\|^2$ for all $(z, \lambda) \in N \cap \mathcal{F}^{\mathrm{KKT}}$. **Q.E.D.**

As in the discussion following Theorem **5.5.1**, we can see that the conclusion of the previous result implies uniqueness of the KKT multiplier $\bar{\lambda}$ associated with $\bar{z}$, because for any such multiplier λ we can assume without loss of generality that λ is close enough to $\bar{\lambda}$ to give $(\bar{z}, \lambda) \in \mathcal{F}^{\mathrm{KKT}} \cap V$.

5.6.1 *Sufficiency based on the relaxed NLP*

At the end of Subsection **4.3.1**, we introduced the relaxed NLP (4.3.11) associated with a feasible solution $(\bar{z}, \bar{\lambda})$ of the KKT constrained MP (6). Proposition **4.3.8** shows that these relaxed NLPs for various $\bar{\lambda} \in M(\bar{z})$ offer a sufficient condition for $\bar{z}$ to be a local minimizer of the original MPEC (3.1.1). In what follows, we discuss the application of Theorem **5.1.1** to the relaxed NLP.

For convenience, we repeat the relaxed NLP associated with the given $\bar{w} \equiv (\bar{z}, \bar{\lambda}) \in \mathcal{F}^{\text{KKT}}$:

$$\text{minimize} \quad f(x, y)$$

$$\text{subject to} \quad Gx + Hy + a \leq 0,$$

$$L(x, y, \lambda) = 0,$$

$$\left. \begin{array}{l} \lambda_i = 0 \\ g_i(x, y) \leq 0 \end{array} \right\} \forall i \notin \mathcal{I}(\bar{z}), \tag{11}$$

$$\left. \begin{array}{l} \lambda_i \geq 0 \\ g_i(x, y) \leq 0 \end{array} \right\} \forall i \in \mathcal{I}_0(\bar{z}, \bar{\lambda}),$$

$$\left. \begin{array}{l} \lambda_i \geq 0 \\ g_i(x, y) = 0 \end{array} \right\} \forall i \in \mathcal{I}_+(\bar{z}, \bar{\lambda}),$$

where for consistency with the treatment in this section, we have taken the function $h(x, y)$ in (4.3.11) to be $Gx + Hy + a$. Without the constraint set $Z \equiv \{(x, y) : Gx + Hy + a \leq 0\}$, the tangent cone of (11) at $\bar{w}$ is exactly the cone $\mathcal{L}^{\text{NLP}}(\bar{w}, \text{Gr}(M))$ defined in (3.4.6), under appropriate CQs. Hence the critical cone of this problem at $\bar{w}$ is given by

$$\mathcal{C}_{\text{rl}}(\bar{w}) \equiv \{(dx, dy, d\lambda) \in \mathcal{T}(\bar{z}; Z) \times \Re^\ell : \nabla_x f(\bar{z}) dx + \nabla_y f(\bar{z}) dy = 0,$$

$$\nabla_x L(\bar{w}) dx + \nabla_y L(\bar{w}) dy + \sum_{i \in \mathcal{I}(\bar{z})} d\lambda_i \nabla g_i(\bar{z}) = 0,$$

$$\nabla_x g_i(\bar{z})^T dx + \nabla_y g_i(\bar{z})^T dy \leq 0, \text{ for } i \in \mathcal{I}_0(\bar{w}),$$

$$\nabla_x g_i(\bar{z})^T dx + \nabla_y g_i(\bar{z})^T dy = 0, \text{ for } i \in \mathcal{I}_+(\bar{w}),$$

$$d\lambda_i = 0, \text{ for } i \notin \mathcal{I}(\bar{z}),$$

$$d\lambda_i \geq 0, \text{ for } i \in \mathcal{I}_0(\bar{w})\},$$

where the subscript "rl" indicates "relaxed". Recalling the discussion in Section **3.2** where the set-valued map $\mathcal{LS}_{(\bar{z}, \lambda)}$ was defined for any λ in $M(\bar{z})$, we note that the canonical projection of $\mathcal{C}_{\text{rl}}(\bar{w})$ onto $\Re^{n+m}$ is a superset of $\text{Gr}(\mathcal{LS}_{\bar{w}}) \cap \mathcal{T}(\bar{z}; Z) \cap \nabla f(\bar{z})^\perp$; i.e.,

$$(dx, dy) \in \text{Gr}(\mathcal{LS}_{\bar{w}}) \cap \mathcal{T}(\bar{z}; Z) \cap \nabla f(\bar{z})^\perp \implies$$

$$\exists d\lambda \in \Re^\ell \text{ such that } (dx, dy, d\lambda) \in \mathcal{C}_{\text{rl}}(\bar{w}).$$

Moreover the reverse implication holds if $\bar{w}$ is strictly complementary, that is, if $\bar{\lambda} - g(\bar{z}) > 0$; see the discussion at the end of Chapter **3**.

For each subset α of $\mathcal{I}_0(\bar{w})$, the constraints in the nonlinear program (11) are relaxations of the constraints in (7). Consequently, we have

$$\bigcup_{\alpha \subseteq \mathcal{I}_0(\bar{w})} \mathcal{C}_\alpha(\bar{w}) \subseteq \mathcal{C}_{rl}(\bar{w}), \tag{12}$$

where $\mathcal{C}_\alpha(\bar{w})$ is given by (9). The KKT system of the nonlinear program (11) is as follows:

$$\begin{aligned}
\nabla_x f(\bar{z}) + G^T \zeta + \nabla_x g(\bar{z})^T \eta &= \nabla_x L(\bar{w})^T \pi \\
\nabla_y f(\bar{z}) + H^T \zeta + \nabla_y g(\bar{z})^T \eta &= \nabla_y L(\bar{w})^T \pi \\
\pi^T \nabla_y g_i(\bar{z}) &\leq 0, \quad i \in \mathcal{I}_0(\bar{w}) \\
\pi^T \nabla_y g_i(\bar{z}) &= 0, \quad i \in \mathcal{I}_+(\bar{w}) \tag{13} \\
\eta_i &\geq 0, \quad i \in \mathcal{I}_0(\bar{w}) \\
\eta_i &= 0, \quad i \notin \mathcal{I}(\bar{z}) \\
\zeta \geq 0, \quad \zeta^T(G\bar{x} + a) &= 0.
\end{aligned}$$

5.6.6 Proposition. If the MFCQ holds for (11) at $\bar{w} \in \mathcal{F}^{KKT}$ and for each nonzero $dw \equiv (dz, d\lambda) \in \mathcal{C}_{rl}(\bar{w})$, there exist multipliers ζ, π and η satisfying (13) and

$$dw^T \nabla^2_{ww} \mathcal{L}^{MPEC}(\bar{w}, \zeta, \pi, \eta) dw > 0, \tag{14}$$

then $\bar{w}$ is a strict local minimizer of (6).

Proof. By part (b) for Theorem **5.1.1**, it follows that $\bar{w}$ is a strict local minimizer of the NLP (11). As in the proof of Proposition **4.3.8**, we can deduce that $\bar{w}$ is a strict local minimizer of (6). **Q.E.D.**

The difference between Theorem **5.6.5** and Proposition **5.6.6** is that the former result assumes conditions on the nonlinear programs (7) corresponding to various subsets α of $\mathcal{I}_0(\bar{w})$, whereas the latter result assumes conditions on a single nonlinear program, namely (11). Due to the inclusion (12), it follows that if (14) holds for all $dw \in \mathcal{C}_{rl}(\bar{w}) \setminus \{0\}$, then for each

$\alpha \subseteq \mathcal{I}_0(\bar{w})$, (14) holds for all $dw \in \mathcal{C}_\alpha(\bar{w}) \setminus \{0\}$. Nevertheless, as Proposition **5.6.7** below shows, if the MFCQ holds for the nonlinear program (7) at $\bar{w}$ corresponding to any one subset α of $\mathcal{I}_0(\bar{w})$, then the MFCQ holds for (11) at $\bar{w}$. Consequently, Proposition **5.6.6** can be thought of, roughly speaking, as obtained from Theorem **5.6.5** by relaxing the MFCQ at $\bar{w}$ for the NLPs (7) for all subsets α of $\mathcal{I}(\bar{w})$ (to any one subset α) but requiring the strict copositivity of the matrix $\nabla^2_{ww}\mathcal{L}^{\mathrm{MPEC}}(\bar{w}, \zeta, \pi, \eta)$ on the larger cone $\mathcal{C}_{\mathrm{rl}}(\bar{w})$.

5.6.7 Proposition. Let $\Phi : \Re^n \to \Re^p$ and $\Psi : \Re^n \to \Re^q$ be C^1 functions in a neighborhood of a vector $u \in \Re^n$ that satisfies the system:

$$\Phi(x) = 0, \quad \Psi(x) \leq 0. \tag{15}$$

For any subset $\alpha \subseteq \{i : \Psi_i(u) = 0\}$, if the MFCQ holds at u for the restricted system:

$$\Phi(x) = 0, \quad \Psi_{\bar{\alpha}}(x) \leq 0, \quad \Psi_\alpha(x) = 0, \tag{16}$$

where $\bar{\alpha} \equiv \{1, \ldots, q\} \setminus \alpha$, then the MFCQ holds at u for the system (15).

Proof. As in the discussion in Subsection **1.3.2**, the MFCQ at u for (16) is equivalent to the following implication; see (1.3.21):

$$\left. \begin{array}{c} \sum_{i=1}^p \eta_i \nabla \Phi_i(u) + \sum_{j=1}^q \delta_j \nabla \Psi_j(u) = 0 \\ \delta_i \geq 0, \ i \in \bar{\alpha} \end{array} \right\} \Rightarrow \left\{ \begin{array}{c} \eta_i = 0, \ \forall i \\ \delta_j = 0, \ \forall j. \end{array} \right.$$

Writing down a similar implication that is equivalent to the MFCQ at u for (15), we easily deduce that the MFCQ for (16) implies the MFCQ for (15). **Q.E.D.**

We conclude this chapter with an observation about the second-order sufficient condition for the MPEC where the function g is affine, say

$$g(x, y) \equiv Dx + Ey + b, \quad (x, y) \in \Re^{n+m},$$

for some matrices D and E and vector b of appropriate dimensions (the function F is not necessarily affine, however). In this case, we have for any vector $w \equiv (z, \lambda)$ and $v \equiv (\zeta, \pi, \eta)$, with $z \equiv (x, y)$,

$$dw^T \nabla^2_{ww}\mathcal{L}^{\mathrm{MPEC}}(w, v)dw = dz^T \left(\nabla^2 f(z) - \sum_{i=1}^m \pi_i \nabla^2 F_i(z) \right) dz, \tag{17}$$

where $dw \equiv (dx, dy, d\lambda)$ and $dz \equiv (dx, dy)$. Incidentally, the matrix

$$\nabla^2 f(\bar{z}) - \sum_{i=1}^{m} \pi_i \nabla^2 F_i(\bar{z}) \in \Re^{(n+m) \times (n+m)} \tag{18}$$

has appeared several times before in the context of the NCP constrained MP; see Sections **5.3** and **5.4**. Exploiting the reduction in (17), we have the following result.

5.6.8 Proposition. Let $g(x, y)$ be affine and $\bar{w} \in \mathcal{F}^{\text{KKT}}$. If the matrix $\nabla^2_{ww} \mathcal{L}^{\text{MPEC}}(\bar{w}, \zeta, \pi, \eta) \in \Re^{(n+m+\ell) \times (n+m+\ell)}$ is strictly copositive on the cone $\mathcal{C}_{\text{rl}}(\bar{w}) \subseteq \Re^{n+m+\ell}$, then $M(\bar{z}) = \{\bar{\lambda}\}$ and the matrix (18) is strictly copositive on the critical cone for the KKT constrained MP at $\bar{w}$ (using the notation in Section **5.2**):

$$\mathcal{C}(\bar{w}) \equiv \text{Gr}(\mathcal{L}\mathcal{S}_{\bar{w}}) \cap \mathcal{T}(\bar{z}; Z) \cap \nabla f(\bar{z})^{\perp} \subseteq \Re^{n+m}.$$

Moreover, the converse holds if $\bar{w}$ is strictly complementary.

Proof. To prove the necessity part, assume that $\nabla^2_{ww} \mathcal{L}^{\text{MPEC}}(\bar{w}, \zeta, \pi, \eta)$ is strictly copositive on $\mathcal{C}_{\text{rl}}(\bar{w})$. We claim that $M(\bar{z}) = \{\bar{\lambda}\}$. Since this is equivalent to the SMFCQ holding for the VI $(F(\bar{x}, \cdot), C(\bar{x}))$ at $\bar{w}$, it suffices to show the implication:

$$\left. \begin{array}{l} \sum_{i \in \mathcal{I}(\bar{z})} d\lambda_i \nabla g_i(\bar{z}) = 0 \\[2mm] d\lambda_i \geq 0, \ \forall i \in \mathcal{I}_0(\bar{z}) \end{array} \right\} \implies d\lambda_i = 0, \ \forall i \in \mathcal{I}(\bar{z}).$$

Let $d\lambda_i$, $i \in \mathcal{I}(\bar{z})$ satisfy the left-hand side of the above implication. With $d\lambda_i \equiv 0$ for $i \notin \mathcal{I}(\bar{z})$, the vector $dw \equiv (0, 0, d\lambda) \in \mathcal{C}_{\text{rl}}(\bar{w})$. By the equation (17), we deduce $dw^T \nabla^2_{ww} \mathcal{L}^{\text{MPEC}}(\bar{w}, \zeta, \pi, \eta) dw = 0$. Thus it follows that $d\lambda = 0$. The strict copositivity of the matrix (18) on $\mathcal{C}(\bar{w})$ follows from the same equation (17) and the aforementioned canonical relation between the two cones $\mathcal{C}_{\text{rl}}(\bar{w})$ and $\mathcal{C}(\bar{w})$.

Conversely, suppose that $\bar{w}$ is strictly complementary, $M(\bar{z}) = \{\bar{\lambda}\}$, and the matrix (18) is strictly copositive on $\mathcal{C}(\bar{w})$. Let $dw \equiv (dx, dy, d\lambda)$ be in $\mathcal{C}_{\text{rl}}(\bar{w})$. Then $dz \equiv (dx, dy) \in \mathcal{C}(\bar{w})$. If $dz \neq 0$, then we are done. If $dz = 0$, then $d\lambda = 0$ by the SMFCQ for the VI $(F(\bar{x}, \cdot), C(\bar{x}))$ at $\bar{w}$. Consequently, if $dw \neq 0$, then we must have $dz \neq 0$, and hence $dw^T \nabla^2_{ww} \mathcal{L}^{\text{MPEC}}(\bar{w}, \zeta, \pi, \eta) dw > 0$, as desired.   **Q.E.D.**

6

Algorithms for MPEC

In the previous three chapters, we studied the necessary optimality conditions and sufficient optimality conditions of the MPEC. This chapter discusses three general classes of iterative algorithms for computing a stationary point of an MPEC. The first class of algorithms is based on a penalty interior point approach (PIPA); the second class of algorithms is based on an implicit programming approach. The third class of algorithms is of the Newton type for solving MPEC, based on the piecewise programming formulation. For the latter algorithms, some conditions pertaining to the relaxed nonlinear program introduced at the end of Subsection **4.3.1** turn out to provide useful conditions for their convergence. In essence, the interior point approach is applicable to MPECs where the lower-level problems possess certain generalized monotonicity properties; the second approach relies on the implicit program formulation of the MPEC; in particular, the SCOC assumption introduced in Subsection **4.2.7** will play an important role in this approach. In both approaches, we establish that any accumulation point of the iterates produced by the algorithms must be a stationary point of MPEC provided that the point satisfies a strict complementarity assumption. The third class of algorithms is an extension of some

271

locally convergent Newton methods for solving smooth nonlinear programs extended to a piecewise smooth setting; thus these algorithms are locally convergent in the sense that a closeness assumption is required on the initial iterate in order to guarantee convergence. Like their smooth counterpart, the piecewise smooth Newton algorithms have superlinear (and even quadratic) rates of convergence under mild assumptions. The chapter ends with a section that reports some preliminary computational results with a MATLAB implementation of PIPA; see Section **6.5**.

6.1 A Penalty Interior Point Algorithm

Motivated by the formulation (1.3.11), we consider the MPEC formulated with (parametric) mixed complementarity constraints:

$$
\begin{aligned}
\text{minimize} \quad & f(x, y, w, z) \\
\text{subject to} \quad & x \in X \\
& F(x, y, w, z) = 0 \\
& (y, w) \geq 0, \quad w^T y = 0,
\end{aligned}
\tag{1}
$$

where $f : \Re^{n+2m+\ell} \to \Re$ is assumed to be continuously differentiable on an open set containing $X \times \Re^{2m}_+ \times \Re^\ell$, X is a nonempty polyhedron in $\Re^n$, and $F : \Re^{n+2m+\ell} \to \Re^{m+\ell}$ is assumed to be twice continuously differentiable on an open set containing $X \times \Re^{2m}_+ \times \Re^\ell$. Subsequently, we shall specialize the problem (1) to an MPEC where the inner problem is defined by the set of KKT conditions of a VI or by a mixed nonlinear complementarity problem. Throughout this section, we shall write

$$
X \equiv \{ x \in \Re^n : Gx \leq a \}
\tag{2}
$$

where $G \in \Re^{p \times n}$ and $a \in \Re^p$ are given. Notice that neither PIPA nor the implicit programming approach to be described in a later section permits a joint feasible constraint like $(x, y, w, z) \in Z \subseteq \Re^{n+2m+\ell}$ in (1). Let $\mathcal{F}$ denote the feasible region of (1). We shall assume that $\mathcal{F} \neq \emptyset$.

In the formulation (1), x is the upper-level variable and the triple (y, w, z) is the lower-level variable. For each given x, the lower-level problem is characterized by the mixed NCP in the variables (y, w, z):

$$F(x, y, w, z) = 0,$$
$$(y, w) \geq 0, \quad y^T w = 0;$$

the mixed nature of this NCP is due to the presence of the free variable z. The functional relationship of the variables (y, w, z) and their dependence on the upper-level variable x are completely described by the function F which, in the general analysis, is not assumed to have any special structure. Subsequently, we consider various special forms of F.

6.1.1 *Optimality conditions*

Before explaining PIPA for solving the problem (1), we find it useful to state the first-order optimality conditions for this problem. For this purpose, let

$$u^* \equiv (x^*, y^*, w^*, z^*) \in \mathcal{F}$$

be given. Then u^* is a stationary point of (1) if and only if

$$du \in \mathcal{T}(u^*; \mathcal{F}) \implies \nabla f(u^*)^T du \geq 0.$$

As in Section **3.2**, we give an equivalent primal-dual formulation of this implication to facilitate the convergence analysis of the algorithms. Specifically, we need to identify a basic CQ to equate the tangent cone $\mathcal{T}(u^*; \mathcal{F})$ with the linearized cone of $\mathcal{F}$ at u^*. With the latter cone, we can then obtain the primal-dual formulation of the stationary conditions, as in Section **3.3**.

For this purpose, we introduce some notation used heavily throughout this and subsequent sections. Define the following index sets that are derived from the complementarity between y^* and w^*:

$$\alpha \equiv \text{supp}(y^*) = \{i : y_i^* > 0 = w_i^*\},$$
$$\beta \equiv \{i : y_i^* = 0 = w_i^*\},$$
$$\gamma \equiv \text{supp}(w^*) = \{i : y_i^* = 0 < w_i^*\};$$

also define the notation:

$$DF_x \equiv \nabla_x F(u^*), \qquad DF_z \equiv \nabla_z F(u^*),$$

$$DF_{y_\alpha} \equiv \nabla_{y_\alpha} F(u^*), \qquad DF_{y_\beta} \equiv \nabla_{y_\beta} F(u^*),$$

$$DF_{w_\gamma} \equiv \nabla_{w_\gamma} F(u^*), \qquad DF_{w_\beta} \equiv \nabla_{w_\beta} F(u^*),$$

$$df_x^* \equiv \nabla_x f(u^*), \qquad df_z^* \equiv \nabla_z f(u^*),$$

$$df_{y_\alpha}^* \equiv \nabla_{y_\alpha} f(u^*), \qquad df_{y_\beta}^* \equiv \nabla_{y_\beta} f(u^*),$$

$$df_{w_\gamma}^* \equiv \nabla_{w_\gamma} f(u^*), \qquad df_{w_\beta}^* \equiv \nabla_{w_\beta} f(u^*).$$

Define the mapping $H : \Re^{n+2m+\ell} \to \Re^{2m+\ell}$ by

$$H(x,y,w,z) \equiv \left(\begin{array}{c} F(x,y,w,z) \\ \min(y,w) \end{array} \right)$$

and the piecewise linear mapping $LH^* : \Re^{m+\ell+|\beta|} \to \Re^{m+\ell+|\beta|}$ by

$$LH^*(dy_\alpha, dw_\gamma, dz, dy_\beta, dw_\beta)$$

$$\equiv \left(\begin{array}{c} DF_{y_\alpha} dy_\alpha + DF_{w_\gamma} dw_\gamma + DF_z dz + DF_{y_\beta} dy_\beta + DF_{w_\beta} dw_\beta \\ \min(dy_\beta, dw_\beta) \end{array} \right).$$

This map LH^* is obtained from H as follows: fix the variables (x, y_γ, w_α) at $(x^*, y_\gamma^*, w_\alpha^*) = (x^*, 0, 0)$, compute the directional derivative with respect to the remaining variables $(y_\alpha, w_\gamma, z, y_\beta, w_\beta)$ at the point $(y_\alpha^*, w_\gamma^*, z^*, y_\beta^*, w_\beta^*)$ (see Lemma **2.4.8** for the directional derivative of the min function); the resulting derivative as a function of the direction is LH^*. As such LH^* is intimately related to the directional derivative of H at (x^*, y^*, w^*, z^*). Recalling the $\mathcal{W}$ property defined for pairs of matrices in Subsection **4.2.7**, we establish the following result.

6.1.1 Lemma. The following statements are equivalent.

(a) LH^* is a global Lipschitzian homeomorphism.

(b) LH^* is coherently oriented.

(c) The pair (W_0, W_1) has the $\mathcal{W}$ property, where

$$W_0 \equiv [DF_{y_\alpha} \quad DF_{w_\gamma} \quad DF_z \quad DF_{y_\beta}],$$

$$W_1 \equiv [DF_{y_\alpha} \quad DF_{w_\gamma} \quad DF_z \quad -DF_{w_\beta}].$$

Proof. Define two matrices

$$
D_0 \equiv \left[\begin{array}{ccccc} DF_{y_\alpha} & DF_{w_\gamma} & DF_z & DF_{y_\beta} & DF_{w_\beta} \\ 0 & 0 & 0 & I_{|\beta|} & 0 \end{array} \right],
$$

$$
D_1 \equiv \left[\begin{array}{ccccc} DF_{y_\alpha} & DF_{w_\gamma} & DF_z & DF_{y_\beta} & DF_{w_\beta} \\ 0 & 0 & 0 & 0 & I_{|\beta|} \end{array} \right].
$$

The linear pieces of LH^* are defined by the column representatives of the pair (D_0, D_1). Any such column representative D induces a corresponding column representative W of (W_0, W_1) with $\det D = \det W$. Therefore (b) and (c) are equivalent. That (a) and (b) are equivalent is a consequence of results from [145]. **Q.E.D.**

Similar to the stationarity results derived in the last chapter, we can establish a set of necessary and sufficient conditions for (x^*, y^*, w^*, z^*) to be a stationary point of (1). The proof of this result consists of applying Lemma **6.1.1** to show that Theorem **4.2.15** is applicable to the parametric system

$$
H(x, y, w, z) = 0
$$

at u^*; the details are omitted.

6.1.2 Theorem. Let $(x^*, y^*, w^*, z^*) \in \mathcal{F}$ be such that the map LH^* is a global homeomorphism. Then (x^*, y^*, w^*, z^*) is a stationary point of (1) if and only if for every partition (β_1, β_2) of β, there exist $\zeta \in \Re^p$ and $\pi \in \Re^{m+\ell}$ such that

$$
\begin{aligned}
& df_x^* - (DF_x)^T \pi + G^T \zeta = 0, \\
& df_{y_\alpha}^* - (DF_{y_\alpha})^T \pi = 0, \\
& df_{w_\gamma}^* - (DF_{w_\gamma})^T \pi = 0, \\
& df_{y_{\beta_1}}^* - (DF_{y_{\beta_1}})^T \pi \geq 0, \\
& df_{w_{\beta_2}}^* - (DF_{w_{\beta_2}})^T \pi \geq 0, \\
& \zeta \geq 0, \quad \zeta^T (a - Gx^*) = 0.
\end{aligned}
\tag{3}
$$

A special case of the above result occurs when (x^*, y^*, w^*, z^*) is non-degenerate, that is, when $y^* + w^* > 0$. In this case the set $\beta = \emptyset$ and the map LH^* becomes linear:

$$LH^*(dy_\alpha, dw_\gamma, dz) = \begin{bmatrix} DF_{y_\alpha} & DF_{w_\gamma} & DF_z \end{bmatrix} \begin{pmatrix} dy_\alpha \\ dw_\gamma \\ dz \end{pmatrix}.$$

Thus LH^* is a global homeomorphism if and only if the following matrix

$$\begin{bmatrix} DF_{y_\alpha} & DF_{w_\gamma} & DF_z \end{bmatrix} \in \Re^{(m+\ell) \times (m+\ell)} \tag{4}$$

is nonsingular. Furthermore, since β is empty, the stationarity conditions
(3) in Theorem **6.1.2** simplify significantly. In the following corollary, we
write

$$W_* \equiv \text{diag}(w^*), \quad Y_* \equiv \text{diag}(y^*).$$

6.1.3 Corollary. Let $u^* \equiv (x^*, y^*, w^*, z^*) \in \mathcal{F}$ be nondegenerate. Suppose that the matrix (4) is nonsingular. Then (x^*, y^*, w^*, z^*) is a stationary
point of (1) if and only if there exist $\zeta \in \Re^p$, $\pi \in \Re^{m+\ell}$ and $\xi \in \Re^m$ such
that

$$df_x^* - (DF_x)^T \pi + G^T \zeta = 0,$$

$$df_y^* - (DF_y)^T \pi - W_* \xi = 0,$$

$$df_w^* - (DF_w)^T \pi - Y_* \xi = 0, \tag{5}$$

$$\zeta \geq 0, \quad \zeta^T (a - Gx^*) = 0,$$

where $DF_y \equiv \nabla_y F(u^*)$ and $DF_w \equiv \nabla_w F(u^*)$.

Proof. If (3) holds, then it suffices to define

$$\xi \equiv (W_* + Y_*)^{-1}(df_y^* + df_w^* - (DF_y + DF_w)^T \pi).$$

Conversely, if (5) holds, then (3) holds by the definition of the index sets
α and γ.   **Q.E.D.**

   More generally, if $\beta \neq \emptyset$, then the condition that the map LH^* is
a global homeomorphism is equivalent to this map being a bijection. In
turn, this bijective condition can be further characterized based on the
relationship between LH^* and certain mixed LCPs. Since the rest of this

section deals only with the nondegenerate case, we do not discuss this matter further.

6.1.2 *Preliminaries for PIPA*

The following definition plays a central role in the interior point based algorithm for computing a stationary point of (1).

6.1.4 Definition. A partitioned matrix

$$Q = [\ A \quad B \quad C \ ]$$

where $A, B \in \Re^{(m+\ell) \times m}$ and $C \in \Re^{(m+\ell) \times \ell}$ is said to have the *mixed P_0 property* if C has full column rank and the implication below holds:

$$\left. \begin{array}{l} Ar + Bs + Ct = 0 \\ (r, s) \neq 0 \end{array} \right\} \implies r_i s_i \geq 0 \ \text{ for some } i \text{ with } |r_i| + |s_i| > 0.$$

The same partitioned matrix Q is said to have the *mixed P property* if C has full column rank and the stronger implication holds:

$$\left. \begin{array}{l} Ar + Bs + Ct = 0 \\ (r, s) \neq 0 \end{array} \right\} \implies r_i s_i > 0 \ \text{ for some } i.$$

When $\ell = 0$, that is, C is vacuous, we drop the word "mixed" in these concepts.

The above definition originates from the classes of P_0 and P matrices in the LCP theory [54]. Indeed a square matrix M is said to be a P_0 matrix if

$$x \neq 0 \implies \max_{i : x_i \neq 0} x_i (Mx)_i \geq 0;$$

and M is a P matrix if (see Proposition **1.3.12**)

$$x \neq 0 \implies \max_i x_i (Mx)_i > 0.$$

It is easy to see that if M is a P_0 matrix (P matrix), then $Q \equiv [M \ -I]$ (with C vacuous) has the P_0 property (P property, respectively).

The mixed P property implies that certain "complementary" columns of the matrices A and B must be linearly independent. In the next result, we denote by M_γ the columns of a matrix M indexed by γ.

6.1.5 Proposition. Suppose that the matrix $Q = [\,A\ B\ C\,]$ as given in Definition **6.1.4** has the mixed P property. Then for every subset α of $\{1,\ldots,m\}$ with complement $\bar{\alpha}$, the submatrix

$$[\,A_\alpha\quad B_{\bar\alpha}\quad C\,]$$

has linearly independent columns.

Proof. Indeed, if there exists $(r_\alpha, s_{\bar\alpha}, t) \neq 0$ such that

$$A_\alpha r_\alpha + B_{\bar\alpha} s_{\bar\alpha} + Ct = 0,$$

then by defining

$$r \equiv (r_\alpha, 0), \quad s \equiv (0, s_{\bar\alpha}),$$

we have

$$Ar + Bs + Ct = 0, \quad \text{and} \quad r \circ s = 0.$$

By assumption, we deduce that $r = s = 0$. Since C has linearly independent columns, we also have $t = 0$, which is a contradiction.   **Q.E.D.**

Throughout the rest of this section, we assume that the partitioned matrix

$$\left[\ \nabla_y F(x,y,w,z)\quad \nabla_w F(x,y,w,z)\quad \nabla_z F(x,y,w,z)\ \right] \quad (6)$$

has the mixed P_0 property for all $(x,y,w,z) \in X \times \Re_+^{2m} \times \Re^\ell$. A consequence of this assumption is the following nonsingularity property which is central to the interior point based algorithm.

6.1.6 Proposition. Assume that the partitioned matrix (6) has the mixed P_0 property at a vector $(x,y,w,z) \in X \times \Re_{++}^{2m} \times \Re^\ell$. Then the matrix

$$\left[\begin{array}{ccc} \nabla_y F(x,y,w,z) & \nabla_w F(x,y,w,z) & \nabla_z F(x,y,w,z) \\ W & Y & 0 \end{array}\right], \quad (7)$$

where $W \equiv \operatorname{diag}(w)$ and $Y \equiv \operatorname{diag}(y)$, is nonsingular. The same conclusion holds if (6) has the mixed P property at $(x,y,w,z) \in X \times \Re_+^{2m} \times \Re^\ell$ satisfying $y + w > 0$.

Proof. We prove only the second assertion, as the first is similar. Assume that the matrix (7) is singular for some $(x, y, w, z) \in X \times \Re_+^{2m} \times \Re^\ell$ satisfying $y + w > 0$. Then there exists a nonzero vector $(r, s, t) \in \Re^{2m+\ell}$ such that

$$\nabla_y F(x, y, w, z)r + \nabla_w F(x, y, w, z)s + \nabla_z F(x, y, w, z)t = 0,$$

$$Wr + Ys = 0.$$

Since $(y, w) \geq 0$ and $y + w > 0$, we have for each $i = 1, \ldots, m$, either $y_i > 0$ or $w_i > 0$. If $y_i > 0$, then

$$r_i s_i = -w_i r_i^2 / y_i \leq 0.$$

Similarly, we also have $r_i s_i \leq 0$ if $w_i > 0$. Consequently, it follows that $r \circ s \leq 0$. Thus by the mixed P property, we must have $(r, s) = 0$. Since the matrix $\nabla_z F(x, y, w, z)$ has full column rank, we obtain $t = 0$. This contradicts the assumption that (r, s, t) is nonzero. Consequently, the desired matrix (7) is nonsingular. **Q.E.D.**

We next introduce two real-valued functions that play an important role in the algorithm to be described later. First is the nonnegative function $\phi : X \times \Re_+^{2m} \times \Re^\ell \to R_+$ which measures the constraint violation of a vector $(x, y, w, z) \in X \times \Re_+^{2m} \times \Re^\ell$:

$$\phi(x, y, w, z) \equiv F(x, y, w, z)^T F(x, y, w, z) + w^T y.$$

Clearly, with $(x, y, w, z) \in X \times \Re_+^{2m} \times \Re^\ell$, $\phi(x, y, w, z) = 0$ if and only if (x, y, w, z) is feasible to the MPEC (1). Note also that $\phi(x, y, w, z) > 0$ if $(y, w) > 0$. A level set of this function ϕ is denoted by

$$\mathcal{L}(\delta) \equiv \left\{ (x, y, w, z) \in X \times \Re_{++}^{2m} \times \Re^\ell : \phi(x, y, w, z) \leq \delta \right\}, \qquad (8)$$

where $\delta > 0$ is an arbitrary scalar. Next is the penalized objective function of the problem (1) defined as follows: for $(x, y, w, z) \in X \times \Re_+^{2m} \times \Re^\ell$

$$P_\alpha(x, y, w, z) \equiv f(x, y, w, z) + \alpha\phi(x, y, w, z), \qquad (9)$$

where $\alpha > 0$ is a penalty parameter that will be adjusted during the algorithm.

As general references on interior point methods for solving linear programs and complementarity problems, we refer to [142, 276, 287, 296]. The key idea of the penalty interior point method is as follows. Given an iterate $(x^\nu, y^\nu, w^\nu, z^\nu) \in X \times \Re^{2m}_{++} \times \Re^\ell$, we solve a quadratic program whose solution yields a descent direction for the function P_{α_ν} at $(x^\nu, y^\nu, w^\nu, z^\nu)$ for some suitably defined α_ν. A one-dimensional search is carried out along this direction so as to decrease P_{α_ν} while maintaining the positivity of the (y, w) components and the feasibility of the x component (i.e., $x \in X$). In order to prevent any of the (y, w) components reaching zero prematurely, a centrality condition is imposed on these components. An added benefit of this condition is that large steps can be taken to reach the next iterate.

We begin by explaining the quadratic program for the search direction and the centrality of the (y, w) components. Specifically let $\bar\rho \in (0, 1)$ be a given scalar and let $(x^\nu, y^\nu, w^\nu, z^\nu) \in X \times \Re^{2m}_{++} \times \Re^\ell$ be a given iterate that satisfies the condition:

$$y^\nu \circ w^\nu \geq \bar\rho \mu_\nu\, e, \tag{10}$$

where $\mu_\nu \equiv (y^\nu)^T w^\nu / m$ and e is the vector of all ones. This is the centrality condition mentioned above. Note that equality holds in (10) if and only if all products $y^\nu_i w^\nu_i$, $i = 1, \ldots, m$, are equal. Roughly speaking, the condition (10) disallows any particular product $y^\nu_i w^\nu_i$ to be too small unless all the products are small. We wish to maintain a similar inequality for all subsequent iterates.

The constraints of the quadratic program are defined by the set X, the linearization of the equations $F(x, y, w, z) = 0$ and $y \circ w = 0$ at the current iterate, and a bound on the magnitude of the search direction in the x component. Whereas the linearization of the equation $F(x, y, w, z) = 0$ is a standard routine borrowed from Newton's method for solving systems of nonlinear equations [211], the linearization of the complementarity equation $y \circ w = 0$ is less standard and is a key feature in the family of interior point methods for solving linear programs and LCPs [142, 296]. Specifically, a perturbation term is added to the standard linearization of this complementarity equation; that is, we consider the perturbed complementarity equation:

$$y \circ w = \sigma_\nu \mu_\nu e,$$

where $\sigma_\nu \in (0, 1)$ is a given scalar, and apply the standard Newton linearization to the latter equation at (y^ν, w^ν). The scalar σ_ν is a factor balancing the relative importance of taking a pure Newton direction (corresponding to $\sigma_\nu = 0$) versus a purely centralizing direction (corresponding to $\sigma_\nu = 1$). The pure Newton step is useful for fast convergence; whereas the centralizing direction is essential for maintaining the positivity of (y, w) and taking large steps. As we shall see, the above perturbation bears a close connection to the centrality condition (10) and helps to maintain it. Denoting the variables of the quadratic program by (dx, dy, dw, dz), we impose a bound on $\|dx\|$ in order to control the magnitude of the remaining components (dy, dw, dz). The necessity for such a restriction will become clear in the convergence analysis of the method. A scalar $c > 0$ is fixed for this purpose.

We now give the details of the quadratic program for finding the search direction. Let $Q_\nu \in \Re^{n \times n}$ be a given symmetric positive semidefinite matrix. Presumably a practical choice for such a matrix Q_ν would involve some second-order information of the functions f and F in (1). Throughout the discussion herein, we do not specify the matrix Q_ν; see also condition (29). We denote by QP_ν the following convex quadratic program in the variables $(dx, dy, dw, dz) \in \Re^{n+2m+\ell}$:

$$\text{minimize} \quad \sum_{s=w,x,y,z} (df_s^\nu)^T ds + \tfrac{1}{2}(dx)^T Q_\nu dx$$

$$\text{subject to} \quad x^\nu + dx \in X,$$

$$(dx)^T dx \leq c \left(\|F^\nu\| + (w^\nu)^T y^\nu \right), \tag{11}$$

$$\nabla_x F^\nu dx + \nabla_y F^\nu dy + \nabla_w F^\nu dw + \nabla_z F^\nu dz = -F^\nu,$$

$$W_\nu dy + Y_\nu dw = -Y_\nu w^\nu + \sigma_\nu \mu_\nu e,$$

where for $s = w, x, y, z$,

$$df_s^\nu \equiv \nabla_s f(x^\nu, y^\nu, w^\nu, z^\nu), \quad \nabla_s F^\nu \equiv \nabla_s F(x^\nu, y^\nu, w^\nu, z^\nu),$$

$$F^\nu \equiv F(x^\nu, y^\nu, w^\nu, z^\nu), \quad W_\nu \equiv \mathrm{diag}(w^\nu), \quad Y_\nu \equiv \mathrm{diag}(y^\nu). \tag{12}$$

Under the assumed mixed P_0 property of the partitioned matrix (6), this quadratic program must have an optimal solution.

6.1.7 Lemma. Let $(x^\nu, y^\nu, w^\nu, z^\nu) \in X \times \Re_{++}^{2m} \times \Re^\ell$ and $Q_\nu \in \Re^{n \times n}$ be a symmetric positive definite matrix. Suppose that the partitioned matrix

$$\left[\begin{array}{ccc} \nabla_y F^\nu & \nabla_w F^\nu & \nabla_z F^\nu \end{array} \right]$$

has the mixed P_0 property. Then the QP_ν has a unique optimal solution.

Proof. By Proposition **6.1.6**, the matrix

$$
\begin{bmatrix}
\nabla_y F^\nu & \nabla_w F^\nu & \nabla_z F^\nu \\
W_\nu & Y_\nu & 0
\end{bmatrix}
\tag{13}
$$

is nonsingular. Thus we can use the last two constraint equations in QP_ν to eliminate the variables (dy, dw, dz) by expressing them in terms of dx; this elimination results in a strictly convex quadratic program in the variable dx alone. Moreover, the resulting quadratic program has $dx = 0$ as a feasible solution; consequently, this program has a unique optimal solution that in turn yields a unique optimal solution to QP_ν.   **Q.E.D.**

The fact that there is no joint constraint on (x, y, w, z) except in the function F and the complementarity condition $y^T w = 0$ in the problem (1) is important for the validity of the above proof. With the presence of a constraint like $(x, y, w, z) \in Z \subseteq \Re^{n+2m+\ell}$, the quadratic subproblem would require a corresponding constraint of the form

$$
(x^\nu, y^\nu, w^\nu, z^\nu) + (dx, dy, dw, dz) \in Z,
$$

which could render the quadratic subproblem infeasible.

The mixed P_0 property of the iterate $(x^\nu, y^\nu, w^\nu, z^\nu)$ ensures the non-singularity of the matrix (13) which in turn is key to the solvability of the QP_ν. It turns out this property is needed in PIPA only for this purpose. More specifically, if the matrix (13) is nonsingular for all ν, then the algorithm to be described later is well defined and the convergence analysis to be given remains valid.

In what follows, we denote the unique optimal solution of QP_ν by

$$
(dx^\nu, dy^\nu, dw^\nu, dz^\nu).
$$

This solution, along with some multipliers $\zeta^\nu, \theta_\nu, \pi^\nu$, and ξ^ν, satisfies the following conditions (we recall the representation (2) of the set X):

$$Q_\nu dx^\nu + \theta_\nu dx^\nu + df_x^\nu - (\nabla_x F^\nu)^T \pi^\nu + G^T \zeta^\nu = 0,$$

$$df_y^\nu - (\nabla_y F^\nu)^T \pi^\nu - W_\nu \xi^\nu = 0,$$

$$df_w^\nu - (\nabla_w F^\nu)^T \pi^\nu - Y_\nu \xi^\nu = 0,$$

$$df_z^\nu - (\nabla_z F^\nu)^T \pi^\nu = 0,$$

$$\theta_\nu \geq 0, \quad \theta_\nu \left((dx^\nu)^T dx^\nu - c(\|F^\nu\| + (y^\nu)^T w^\nu)\right) = 0,$$

$$\zeta^\nu \geq 0, \quad (\zeta^\nu)^T (G(x^\nu + dx^\nu) - a) = 0. \tag{14}$$

Since $Gx^\nu \leq a$, we have

$$(\zeta^\nu)^T G dx^\nu = (\zeta^\nu)^T (a - Gx^\nu) \geq 0.$$

Multiplying the first four equations of (14) by dx^ν, dy^ν, dw^ν, and dz^ν respectively, adding the results and simplifying, we obtain

$$(dx^\nu)^T (Q_\nu + \theta_\nu I) dx^\nu + (df_x^\nu)^T dx^\nu + (df_y^\nu)^T dy^\nu + (df_w^\nu)^T dw^\nu$$

$$+ (df_z^\nu)^T dz^\nu + (F^\nu)^T \pi^\nu + (\xi^\nu)^T (W_\nu y^\nu - \sigma_\nu \mu_\nu e) \leq 0; \tag{15}$$

moreover, we can solve for the multipliers π^ν and ξ^ν from the second, third, and fourth equations in (14):

$$\begin{pmatrix} \pi^\nu \\ \xi^\nu \end{pmatrix} = \begin{bmatrix} \nabla_y F^\nu & \nabla_w F^\nu & \nabla_z F^\nu \\ W_\nu & Y_\nu & 0 \end{bmatrix}^{-T} \begin{pmatrix} df_y^\nu \\ df_w^\nu \\ df_z^\nu \end{pmatrix}, \tag{16}$$

where the superscript $^{-T}$ denotes matrix inverse followed by transpose. The equations (14)–(16) are used later in the convergence analysis.

As we have indicated, the vector $(dx^\nu, dy^\nu, dw^\nu, dz^\nu)$ provides the direction along which we search for the next iterate. For this purpose, we define for each scalar $\tau \in [0, 1]$,

$$\begin{pmatrix} x^\nu(\tau) \\ y^\nu(\tau) \\ w^\nu(\tau) \\ z^\nu(\tau) \end{pmatrix} \equiv \begin{pmatrix} x^\nu \\ y^\nu \\ w^\nu \\ z^\nu \end{pmatrix} + \tau \begin{pmatrix} dx^\nu \\ dy^\nu \\ dw^\nu \\ dz^\nu \end{pmatrix}. \tag{17}$$

Note that we have $x^\nu(\tau) \in X$. Our goal is to determine an appropriate step size $\tau_\nu \in (0, 1]$ such that the next iterate

$$(x^{\nu+1}, y^{\nu+1}, w^{\nu+1}, z^{\nu+1}) \equiv (x^\nu(\tau_\nu), y^\nu(\tau_\nu), w^\nu(\tau_\nu), z^\nu(\tau_\nu)) \qquad (18)$$

will remain positive and satisfy the centrality condition

$$y^{\nu+1} \circ w^{\nu+1} \geq \bar\rho \mu_{\nu+1} \, e$$

with the same constant $\bar\rho \in (0, 1)$ as in (10). To this end we define the scalar τ_ν' to be the supremum of scalars $\tau' \in (0, 1]$ such that for all $\tau \in [0, \tau')$,

$$(y^\nu(\tau), w^\nu(\tau)) > 0; \qquad (19)$$

this scalar τ_ν' must be positive and can easily be determined by a ratio test. Since for all $i = 1, \ldots, m$,

$$w_i^\nu dy_i^\nu + y_i^\nu dw_i^\nu = -y_i^\nu w_i^\nu + \sigma_\nu \mu_\nu,$$

we deduce

$$y_i^\nu(\tau) w_i^\nu(\tau) - \bar\rho \mu_\nu(\tau) =$$

$$(1 - \tau)(y_i^\nu w_i^\nu - \bar\rho \mu_\nu) + \tau(1 - \bar\rho)\sigma_\nu \mu_\nu + \tau^2 (dy_i^\nu dw_i^\nu - \bar\rho (dy^\nu)^T dw^\nu / m),$$

where $\mu_\nu(\tau) \equiv (y^\nu(\tau))^T w^\tau(\tau)/m$. Since we want to ensure that the above difference is nonnegative for all $\tau \geq 0$ sufficiently small, we consider the following real-valued function in the step size $\tau \in [0, 1]$:

$$g_\nu(\tau) \equiv (1 - \bar\rho)\sigma_\nu \mu_\nu + \tau \left(\min_{1 \leq i \leq m} dy_i^\nu dw_i^\nu - \bar\rho (dy^\nu)^T dw^\nu / m \right). \qquad (20)$$

Clearly $g_\nu(0) > 0$; moreover, for all $i = 1, \ldots, m$ and all $\tau \in [0, 1]$,

$$y_i^\nu(\tau) w_i^\nu(\tau) - \bar\rho \mu_\nu(\tau) \geq (1 - \tau)(y_i^\nu w_i^\nu - \bar\rho \mu_\nu) + \tau g_\nu(\tau). \qquad (21)$$

Let $\bar\tau_\nu$ be the unique root of $g_\nu(\tau)$ in the interval $(0, 1]$ if such a root exists; otherwise let $\bar\tau_\nu$ be equal to 1. Since g_ν is linear in τ, $\bar\tau_\nu$ is uniquely defined; moreover, we have $g_\nu(\tau) > 0$ for all $\tau \in [0, \bar\tau_\nu)$.

6.1.8 Lemma. The following statements are valid:

(a) $(y^\nu(\tau), w^\nu(\tau)) \geq 0$ for all $\tau \in [0, \bar\tau_\nu]$;

(b) for all $\tau \in [0, \bar{\tau}_\nu]$ and all i,

$$y_i^\nu(\tau)w_i^\nu(\tau) - \bar{\rho}\mu_\nu(\tau) \geq 0; \tag{22}$$

(c) either $\bar{\tau}_\nu = \tau_\nu' = 1$, or $(y^\nu(\bar{\tau}_\nu), w^\nu(\bar{\tau}_\nu)) > 0$.

Proof. For each $\tau \in [0, \bar{\tau}_\nu]$, (21) implies

$$y_i^\nu(\tau)w_i^\nu(\tau) \geq \bar{\rho}\mu_\nu(\tau), \quad \text{for all } i = 1, \ldots, m.$$

Summing up these m inequalities and using the fact that $\bar{\rho} < 1$, we deduce that

$$(y^\nu(\tau))^T w^\nu(\tau) \geq 0.$$

Assume that $\bar{\tau}_\nu > \tau_\nu'$. Then for some $\tau \in (\tau_\nu', \bar{\tau}_\nu)$ and some index j we have

$$y_j^\nu(\tau)w_j^\nu(\tau) = 0.$$

For this index j, (21) implies

$$0 \geq -\bar{\rho}\mu_\nu(\tau) \geq \tau g_\nu(\tau) > 0,$$

which is a contradiction. Hence, for all $\tau \in [0, \bar{\tau}_\nu)$, (19) holds. By continuity (a) follows. Part (b) is obvious.

To prove (c), we note that one of the following three situations must hold: (i) $\bar{\tau}_\nu < \tau_\nu'$, (ii) $\bar{\tau}_\nu = \tau_\nu' < 1$, or (iii) $\bar{\tau}_\nu = \tau_\nu' = 1$. Clearly $(y^\nu(\bar{\tau}_\nu), w^\nu(\bar{\tau}_\nu)) > 0$ in case (i). It remains to show that in case (ii), $(y^\nu(\bar{\tau}_\nu), w^\nu(\bar{\tau}_\nu)) > 0$. Suppose that for some index j, either $y_j^\nu(\bar{\tau}_\nu)$ or $w_j^\nu(\bar{\tau}_\nu)$ is equal to zero. Since (22) holds at $\bar{\tau}_\nu$ for $i = j$, it follows that $(y^\nu(\bar{\tau}_\nu))^T w^\nu(\bar{\tau}_\nu) = 0$. Since $(y(\bar{\tau}_\nu), w(\bar{\tau}_\nu)) \geq 0$, it follows that for all $i = 1, \ldots, m$,

$$y_i^\nu(\bar{\tau}_\nu)w_i^\nu(\bar{\tau}_\nu) = 0.$$

Since $g_\nu(\tau) \geq 0$ for all $\tau \in [0, \bar{\tau}_\nu]$ and (10) holds, we deduce from (21) that for all $i = 1, \ldots, m$,

$$(1 - \bar{\tau}_\nu)(y_i^\nu w_i^\nu - \bar{\rho}\mu_\nu) = 0.$$

Since $\bar{\tau}_\nu < 1$, we obtain

$$y_i^\nu w_i^\nu - \bar{\rho}\mu_\nu = 0$$

for all i. This is impossible since $(y^\nu, w^\nu) > 0$ and $\bar{\rho} < 1$. Consequently, we must have $(y^\nu(\bar{\tau}_\nu), w^\nu(\bar{\tau}_\nu)) > 0$ as desired.   **Q.E.D.**

The scalar $\bar{\tau}_\nu$ provides an upper bound for the desired step size τ_ν which will be determined ultimately with the aid of the function ϕ and the penalized merit function P_α. The next result establishes an important identity of the former function.

6.1.9 Lemma. For $s = x, y, w, z$, let $d\phi_s^\nu \equiv \nabla_s \phi(x^\nu, y^\nu, w^\nu, z^\nu)$. It holds that

$$
(d\phi_x^\nu)^T dx^\nu + (d\phi_y^\nu)^T dy^\nu + (d\phi_w^\nu)^T dw^\nu + (d\phi_z^\nu)^T dz^\nu
$$
$$
= -2\|F^\nu\|^2 - (1 - \sigma_\nu)(y^\nu)^T w^\nu. \tag{23}
$$

Proof. We drop the iteration counter ν from the notation in the following proof. For each $j = 1, \ldots, n$ we have

$$
\frac{\partial \phi(x, y, w, z)}{\partial x_j} = 2 \sum_{i=1}^{m+\ell} F_i(x, y, w, z) \frac{\partial F_i(x, y, w, z)}{\partial x_j};
$$

for each $j = 1, \ldots, m$,

$$
\frac{\partial \phi(x, y, w, z)}{\partial y_j} = 2 \sum_{i=1}^{m+\ell} F_i(x, y, w, z) \frac{\partial F_i(x, y, w, z)}{\partial y_j} + w_j,
$$

$$
\frac{\partial \phi(x, y, w, z)}{\partial w_j} = 2 \sum_{i=1}^{m+\ell} F_i(x, y, w, z) \frac{\partial F_i(x, y, w, z)}{\partial w_j} + y_j,
$$

and for each $j = 1, \ldots, \ell$,

$$
\frac{\partial \phi(x, y, w, z)}{\partial z_j} = 2 \sum_{i=1}^{m+\ell} F_i(x, y, w, z) \frac{\partial F_i(x, y, w, z)}{\partial z_j}.
$$

Using the above expressions for the partial derivatives of ϕ and the following constraints in the QP_ν:

$$
\nabla_x F(x, y, w, z)^T dx + \nabla_y F(x, y, w, z)^T dy + \nabla_w F(x, y, w, z)^T dw
$$
$$
+ \nabla_z F(x, y, w, z)^T dz = -F(x, y, w, z),
$$

and for $j = 1, \ldots, m$,

$$
w_j dy_j + y_j dw_j = -w_j y_j + \sigma \frac{y^T w}{m},
$$

we easily obtain the desired inequality (23).   **Q.E.D.**

In view of the expression (23), we present an update rule for the penalty parameter α in the penalized merit function P_α; see (9). Specifically, for a given iterate $(x^\nu, y^\nu, w^\nu, z^\nu) \in X \times \Re_{++}^{2m} \times \Re^\ell$ and the associated search vector $(dx^\nu, dy^\nu, dw^\nu, dz^\nu)$ and for a given value $\alpha_{\nu-1} > 1$ of the penalty parameter, we propose the following rule:

(Penalty update rule): Let $p_\nu \geq 1$ be the smallest integer $p \geq 1$ such that

$$
\begin{aligned}
(df_x^\nu)^T dx^\nu &+ (df_y^\nu)^T dy^\nu + (df_w^\nu)^T dw^\nu + (df_z^\nu)^T dz^\nu \\
&- \alpha_{\nu-1}^p \left(2\|F^\nu\|^2 + (1 - \sigma_\nu)(y^\nu)^T w^\nu \right) < -\phi_\nu,
\end{aligned}
\tag{24}
$$

where

$$
\phi_\nu \equiv \phi(x^\nu, y^\nu, w^\nu, z^\nu).
$$

Let $\alpha_\nu = \alpha_{\nu-1}^{p_\nu}$. Note that since (y^ν, w^ν) is positive, ϕ_ν is positive, hence p_ν is finite and can be determined by testing successive values of p, starting with $p = 1$.

Based on the inequality (24), the following lemma is easily proved.

6.1.10 Lemma. Let α_ν be determined by the above penalty update rule. Then for any constant $\gamma' \in (0, 1)$, there exists a scalar $\tau' > 0$ such that for all $\tau \in (0, \tau']$,

$$
\phi(x^\nu(\tau), y^\nu(\tau), w^\nu(\tau), z^\nu(\tau)) \leq \phi_\nu,
$$

and

$$
\begin{aligned}
P_{\alpha_\nu}&(x^\nu(\tau), y^\nu(\tau), w^\nu(\tau), z^\nu(\tau)) - P_{\alpha_\nu}(x^\nu, y^\nu, w^\nu, z^\nu) \\
&\leq \gamma'\tau \left((df_x^\nu)^T dx^\nu + (df_y^\nu)^T dy^\nu + (df_w^\nu)^T dw^\nu + (df_z^\nu)^T dz^\nu \right. \\
&\quad \left. - \alpha_\nu(2\|F^\nu\|^2 + (1 - \sigma_\nu)(y^\nu)^T w^\nu) \right) < -\gamma'\tau\phi_\nu.
\end{aligned}
$$

Proof. We prove only the assertion about the function P_{α_ν}. Dropping ν from the notation, we note that

$$
\begin{aligned}
(dP_\alpha)_x^T dx &+ (dP_\alpha)_y^T dy + (dP_\alpha)_w^T dw + (dP_\alpha)_z^T dz = \\
(df_x)^T dx &+ (df_y)^T dy + (df_w)^T dw + (df_z)^T dz - \alpha(2\|F\|^2 + (1 - \sigma)y^T w),
\end{aligned}
$$

where $(dP_\alpha)_s$ denotes $\nabla_s P_\alpha(x, y, w, z)$ for $s = x, y, w, z$. Since

$$
\lim_{\tau \to 0+} \tau^{-1} [P_\alpha(x(\tau), y(\tau), w(\tau), z(\tau)) - P_\alpha(x, y, w, z)]
$$

$$
= (dP_\alpha)_x^T dx + (dP_\alpha)_y^T dy + (dP_\alpha)_w^T dw + (dP_\alpha)_z^T dz < 0,
$$

the desired conclusion follows by a standard argument for an iterative descent algorithm. Specifically, we proceed by way of contradiction and assume that no such scalars τ' exist for a certain γ'. This will easily lead to a contradiction that invalidates the above limit. The details are omitted. **Q.E.D.**

6.1.3 Algorithm description and convergence analysis

With the development in the previous subsection, we are now ready to state the penalty interior point algorithm for solving the problem (1). The algorithm is initialized with some given scalars and an initial iterate; the two main steps of the algorithm are (i) the solution of a quadratic program and (ii) a one-dimensional line search of the Armijo type. The feasibility of the second step is justified by Lemma **6.1.10**.

The Penalty Interior Point Algorithm

Step 0. (Initialization) Let $c > 0$, $\bar{\sigma}$, ρ_1, γ, γ', $\varepsilon \in (0,1)$, and $\alpha_{-1} > 1$. Also let $(x^0, y^0, w^0, z^0) \in X \times \Re^{2m}_{++} \times \Re^{\ell}$ be given. Let $\bar{\rho} \in (0,1)$ and $\sigma_0 \in [0,1)$ satisfy

$$\bar{\rho}\,\frac{(y^0)^T w^0}{m} \leq \min_{1 \leq i \leq m} y_i^0 w_i^0 \quad \text{and} \quad \sigma_0 \leq \min(\bar{\sigma}, \bar{\rho}).$$

Let Q_0 be a given symmetric positive semidefinite matrix. Set $\nu = 0$.

Step 1. (Direction generation) Let $(dx^\nu, dy^\nu, dw^\nu, dz^\nu)$ be the unique optimal solution of the QP_ν. Let α_ν be determined by the penalty update rule.

Step 2. (Step size determination) Let $g_\nu(\tau)$ be the linear function defined in (20).

Let $\bar{\tau}_\nu$ be the (unique) root of the function $g_\nu(\tau)$ for $\tau \in (0,1]$ if this root exists; let $\bar{\tau}_\nu$ be equal to 1 if g_ν has no root in $(0,1]$. If $\bar{\tau}_\nu = 1$, redefine $\bar{\tau}_\nu$ to be $1 - \varepsilon$.

Let $\tau_\nu \equiv \bar{\tau}_\nu \rho_1^{r_\nu}$, where r_ν is the smallest nonnegative integer r such that with $\tau \equiv \bar{\tau}_\nu \rho_1^r$,

$$\phi(x^\nu(\tau), y^\nu(\tau), w^\nu(\tau), z^\nu(\tau)) \leq \phi(x^\nu, y^\nu, w^\nu, z^\nu), \tag{25}$$

$$P_{\alpha_\nu}(x^\nu(\tau), y^\nu(\tau), w^\nu(\tau), z^\nu(\tau)) - P_{\alpha_\nu}(x^\nu, y^\nu, w^\nu, z^\nu)$$

$$\leq \gamma'\tau \left((df_x^\nu)^T dx^\nu + (df_y^\nu)^T dy^\nu + (df_w^\nu)^T dw^\nu + (df_z^\nu)^T dz^\nu \right) \qquad (26)$$

$$-\alpha_\nu(2\|F^\nu\|^2 + (1-\sigma_\nu)(y^\nu)^T w^\nu)) \leq -\gamma'\tau\phi_\nu.$$

Step 3. (Termination check) Define $(x^{\nu+1}, y^{\nu+1}, w^{\nu+1}, z^{\nu+1})$ according to (18). Test this new iterate for termination. Specifically, if a prescribed stopping rule is satisfied by $(x^{\nu+1}, y^{\nu+1}, w^{\nu+1}, z^{\nu+1})$, then terminate; otherwise choose $Q_{\nu+1}$ to be a symmetric positive semidefinite matrix and $\sigma_{\nu+1}$ to be a scalar satisfying $0 < \sigma_{\nu+1} \leq \sigma_\nu$ and return to Step 1 with ν replaced by $\nu + 1$.

A possible stopping rule to be used in Step 3 above is

$$\|(dx^\nu, dy^\nu, dw^\nu, dz^\nu)\| \leq \text{tolerance}.$$

As we shall see from the following analysis, this is a reasonable way to terminate the algorithm because under this rule, the final iterate $(x^\nu, y^\nu, w^\nu, z^\nu)$ is an acceptable stationary point of (1) satisfying a set of approximate stationarity conditions, provided that a certain assumption is satisfied.

In what follows, we perform a limit analysis of the algorithm. Specifically, assuming that the algorithm does not terminate in a finite number of iterations, we demonstrate under two key assumptions that the algorithm generates a well-defined, bounded, infinite sequence of iterates

$$\{(x^\nu, y^\nu, w^\nu, z^\nu)\} \subset X \in \Re_{++}^{2m} \times \Re^\ell \qquad (27)$$

such that (10) holds for all ν. The two assumptions are:

(A1) the partitioned matrix (6) has the mixed P_0 property for all vectors $(x, y, w, z) \in X \times \Re_{++}^{2m} \times \Re^\ell$;

(A2) the level set $\mathcal{L}(\phi_0)$ defined in (8) is bounded.

Consider the initial iterate (x^0, y^0, w^0, z^0) which clearly belongs to the set $\mathcal{L}(\phi_0)$. Inductively, we assume that for some $\nu \geq 0$, $(x^\nu, y^\nu, w^\nu, z^\nu)$ is well defined and belongs to $\mathcal{L}(\phi_0)$, the pair (y^ν, w^ν) is positive, and satisfies the centrality condition (10). By Lemma **6.1.7**, the search direction $(dx^\nu, dy^\nu, dw^\nu, dz^\nu)$ is well defined; moreover, since $\phi_\nu > 0$, the penalty value α_ν is well defined. Since the scalar $\bar{\tau}_\nu$ is always less than 1 (possibly by the redefinition), Lemma **6.1.8** implies that for all $\tau \in [0, \bar{\tau}_\nu]$, (19) and (22) must hold. Finally Lemma **6.1.10** establishes that the integer r_ν can be computed in a finite number of trials, starting with the initial value $r = 0$ and increasing r by one if either (25) or (26) fails to hold; consequently, the step size τ_ν is well defined and so is the next iterate $(x^{\nu+1}, y^{\nu+1}, w^{\nu+1}, z^{\nu+1})$. Moreover by (25) and the inductive hypothesis, we deduce

$$\phi(x^{\nu+1}, y^{\nu+1}, w^{\nu+1}, z^{\nu+1}) \le \phi(x^0, y^0, w^0, z^0).$$

Thus $(x^{\nu+1}, y^{\nu+1}, w^{\nu+1}, z^{\nu+1}) \in \mathcal{L}(\phi_0)$. This completes the inductive argument that establishes the well-definedness and boundedness of the sequence in (27). We summarize this discussion in the following result.

6.1.11 Proposition. Under assumptions (A1) and (A2), the penalty interior point algorithm generates a well-defined, bounded sequence (27). In addition the sequence $\{dx^\nu\}$ is also bounded.

Proof. It suffices to prove the additional boundedness statement. But this follows easily from the constraint:

$$\| dx^\nu \|^2 \le c \left(\| F^\nu \| + (w^\nu)^T y^\nu \right), \tag{28}$$

in the QP_ν.   **Q.E.D.**

The above proposition does not assert the boundedness of the sequence $\{(dy^\nu, dw^\nu, dz^\nu)\}$. This issue is dealt with in Lemma **6.1.12** below.

Next, we analyze the accumulation points (which must exist) of the sequence (27). For this analysis, we need to assume that there exists a scalar $c_2 > 0$ such that

$$0 \le x^T Q_\nu x \le c_2 \| x \|^2, \quad \text{for all } x \in \Re^n \text{ and all } \nu. \tag{29}$$

(The constant choice $Q_\nu = I$ obviously satisfies this requirement.) Our ultimate goal is to show that a certain limit point (x^*, y^*, w^*, z^*) of the sequence (27) must be a stationary point of the MPEC (1), provided that the following strict complementarity and nonsingularity assumptions are satisfied:

(SC) $y^* + w^* > 0$,

(NS) the matrix (7) is nonsingular at the vector (x^*, y^*, w^*, z^*).

To this end we note that any limit point (x^*, y^*, w^*, z^*) of the sequence (27) must belong to the set $X \times \Re_+^{2m} \times \Re^\ell$. Extending the notation in (12), we replace ν by $*$ to denote the various vectors and matrices evaluated at the vector (x^*, y^*, w^*, z^*). The following result establishes several consequences of the two assumptions SC and NS.

6.1.12 Lemma. Suppose $\{(x^\nu, y^\nu, w^\nu, z^\nu) : \nu \in \kappa\}$ is a subsequence converging to (x^*, y^*, w^*, z^*) which satisfies the assumptions SC and NS. Then

(a) the sequences $\{(dy^\nu, dw^\nu, dz^\nu) : \nu \in \kappa\}$ and $\{(\pi^\nu, \xi^\nu) : \nu \in \kappa\}$, where each pair (π^ν, ξ^ν) is given by (16), are bounded; indeed, there exists a scalar $c' > 0$ such that for all $\nu \in \kappa$,

$$\|(dy^\nu, dw^\nu, dz^\nu)\| \leq c' \left(\|F^\nu\| + \|dx^\nu\| + (y^\nu)^T w^\nu \right); \qquad (30)$$

(b) there exists a scalar $\beta > 0$ such that for all $\nu \in \kappa$ and all i,

$$|dy_i^\nu| \leq \beta y_i^\nu, \quad \text{and} \quad |dw_i^\nu| \leq \beta w_i^\nu;$$

hence $\|dy^\nu \circ dw^\nu\|_1 \leq \beta^2 (y^\nu)^T w^\nu$.

Proof. We have for each ν,

$$\begin{pmatrix} dy^\nu \\ dw^\nu \\ dz^\nu \end{pmatrix} = - \begin{bmatrix} \nabla_y F^\nu & \nabla_w F^\nu & \nabla_z F^\nu \\ W_\nu & Y_\nu & 0 \end{bmatrix}^{-1} \begin{pmatrix} F^\nu + \nabla_x F^\nu dx^\nu \\ Y_\nu w^\nu - \sigma_\nu \mu_\nu e \end{pmatrix}.$$

The sequence of matrices

$$\left\{ \begin{bmatrix} \nabla_y F^\nu & \nabla_w F^\nu & \nabla_z F^\nu \\ W_\nu & Y_\nu & 0 \end{bmatrix} : \nu \in \kappa \right\}$$

converges to the limiting matrix

$$\begin{bmatrix} \nabla_y F^* & \nabla_w F^* & \nabla_z F^* \\ W_* & Y_* & 0 \end{bmatrix}$$

which is nonsingular by NS. Hence the sequence of inverses:

$$\left\{ \left[\begin{array}{ccc} \nabla_y F^\nu & \nabla_w F^\nu & \nabla_z F^\nu \\ W_\nu & Y_\nu & 0 \end{array} \right]^{-1} : \nu \in \kappa \right\}$$

is bounded. Since $\{dx^\nu\}$ is bounded by Lemma **6.1.11**, so is

$$\{(dy^\nu, dw^\nu, dz^\nu) : \nu \in \kappa\}.$$

The existence of the scalar c' is obvious. Similarly, by (16), the sequence $\{(\pi^\nu, \xi^\nu) : \nu \in \kappa\}$ is also bounded. This establishes (a).

The proof of (b) relies on the equation (which holds for all $i = 1, \dots, m$)

$$w_i^\nu dy_i^\nu + y_i^\nu dw_i^\nu = -y_i^\nu w_i^\nu + \sigma_\nu \mu_\nu, \tag{31}$$

and the boundedness of the sequences $\{(y^\nu, w^\nu)\}$ and $\{(dy^\nu, dw^\nu) : \nu \in \kappa\}$. Consider an arbitrary index i and suppose $y_i^* > 0$. Then the sequence $\{(y_i^\nu)^{-1} : \nu \in \kappa\}$ is bounded above; hence the ratios $|dy_i^\nu|/y_i^\nu$, $\nu \in \kappa$ are bounded. Dividing (31) by $w_i^\nu y_i^\nu$, we obtain

$$\frac{dw_i^\nu}{w_i^\nu} = -\frac{dy_i^\nu}{y_i^\nu} - 1 + \sigma_\nu \frac{\mu_\nu}{y_i^\nu w_i^\nu}.$$

Since $y_i^\nu w_i^\nu \geq \bar{\rho}\mu_\nu$, it follows that the ratios $|dw_i^\nu|/w_i^\nu$, $\nu \in \kappa$ are also bounded for such index i. Similarly, we can establish that both $|dw_i^\nu|/w_i^\nu$ and $|dy_i^\nu|/y_i^\nu$ are bounded for an index i such that $w_i^* > 0$. By SC, the existence of the constant β follows. The last assertion of the lemma is obvious.   **Q.E.D.**

We are now ready to establish the main convergence results of this section. The first of these results deals with the case where the sequence of penalty values $\{\alpha_\nu\}$ is unbounded.

6.1.13 Theorem. Suppose that the sequence $\{Q_\nu\}$ is bounded according to (29). If $\{\alpha_\nu\}$ is unbounded, then

$$\lim_{\nu(\in\kappa)\to\infty} dx^\nu = 0, \tag{32}$$

where $\kappa \equiv \{\nu : \alpha_{\nu-1} < \alpha_\nu\}$. Furthermore, if (x^*, y^*, w^*, z^*) is any limit point of the subsequence $\{(x^\nu, y^\nu, w^\nu, z^\nu) : \nu \in \kappa\}$ satisfying the assumptions SC and NS, then (x^*, y^*, w^*, z^*) is a stationary point of the problem (1).

Proof. Without loss of generality, we may assume that $\{(x^*, y^*, w^*, z^*)\}$ is the limit of $\{(x^\nu, y^\nu, w^\nu, z^\nu) : \nu \in \kappa\}$. By Lemma **6.1.12**, we may also assume that $\{(\pi^\nu, \xi^\nu) : \nu \in \kappa\}$ converges to (π^*, ξ^*).

By definition, $\alpha_\nu = \alpha_{\nu-1}^{p_\nu}$ where p_ν is the smallest positive integer p for which (24) is valid. If $\nu \in \kappa$, then $p_\nu \geq 2$ and (24) is violated at $p = p_\nu - 1$. Thus we have for each $\nu \in \kappa$,

$$
(df_x^\nu)^T dx^\nu + (df_y^\nu)^T dy^\nu + (df_w^\nu)^T dw^\nu + (df_z^\nu)^T dz^\nu
$$
$$
- \alpha_{\nu-1}^{p_\nu - 1} \left(2\|F^\nu\|^2 + (1 - \sigma_\nu)(y^\nu)^T w^\nu \right) \geq -\phi_\nu.
\tag{33}
$$

Since $\{\alpha_\nu\}$ is nondecreasing, unboundedness of this sequence means the sequence $\{\alpha_\nu\} \to \infty$, so $\alpha_{\nu-1}^{p_\nu - 1} \geq \alpha_{\nu-1} \to \infty$. Since both sequences $\{(x^\nu, y^\nu, w^\nu, z^\nu)\}$ and $\{(dx^\nu, dy^\nu, dw^\nu, dz^\nu) : \nu \in \kappa\}$ are bounded, it follows that

$$
\lim_{\nu(\in\kappa) \to \infty} \left(2\|F^\nu\|^2 + (1 - \sigma_\nu)(y^\nu)^T w^\nu \right) = 0.
$$

Since $\sigma_\nu \leq \bar{\sigma} < 1$, we obtain $F(x^*, y^*, w^*, z^*) = 0$ and $(y^*)^T w^* = 0$; thus (x^*, y^*, w^*, z^*) is feasible to the MPEC (1). Let

$$
\alpha \equiv \mathrm{supp}(y^*) \quad \text{and} \quad \gamma \equiv \mathrm{supp}(w^*).
$$

Then α and γ partition $\{1, \ldots, m\}$, by the condition SC. It is easy to see that the condition NS implies that the matrix

$$
[\, \nabla_{y_\alpha} F^* \quad \nabla_{w_\gamma} F^* \quad \nabla_z F^* \,]
$$

is nonsingular; this nonsingularity property is exactly the one required in Corollary **6.1.3**.

Since for all ν,

$$
(dx^\nu)^T dx^\nu \leq c \left(\|F^\nu\| + (w^\nu)^T y^\nu \right),
$$

it follows that (32) holds. From (33), we obtain for all $\nu \in \kappa$ sufficiently large

$$
(df_x^\nu)^T dx^\nu + (df_y^\nu)^T dy^\nu + (df_w^\nu)^T dw^\nu + (df_z^\nu)^T dz^\nu \geq
$$
$$
\alpha_{\nu-1}^{p_\nu - 1} \left(2\|F^\nu\|^2 + (1 - \sigma_\nu)(y^\nu)^T w^\nu \right) - \left(\|F^\nu\|^2 + (y^\nu)^T w^\nu \right).
$$

Since $\alpha_{\nu-1}^{p_\nu - 1} \to \infty$ and $\sigma_\nu \leq \bar{\sigma} < 1$, it follows that

$$
(df_x^\nu)^T dx^\nu + (df_y^\nu)^T dy^\nu + (df_w^\nu)^T dw^\nu + (df_z^\nu)^T dz^\nu \geq 0
$$

for all $\nu \in \kappa$ sufficiently large. By this inequality, (15), and (29), we deduce

$$\theta_\nu \|dx^\nu\|^2 \leq -(F^\nu)^T \pi^\nu - (\xi^\nu)^T (W_\nu y^\nu - \sigma_\nu \mu_\nu e).$$

Dividing by $\|dx^\nu\|^2$ and using the Cauchy-Schwartz inequality and also the boundedness of the sequence $\{(\pi^\nu, \xi^\nu) : \nu \in \kappa\}$, we deduce

$$\theta_\nu \leq \frac{\text{constant}}{\|dx^\nu\|^2} (\|F^\nu\| + (w^\nu)^T y^\nu).$$

We claim that $\{\theta_\nu : \nu \in \kappa\}$ is bounded. This is certainly true if $\theta_\nu = 0$. If $\theta_\nu > 0$, then we have

$$\|dx^\nu\|^2 = c(\|F^\nu\| + (y^\nu)^T w^\nu);$$

which yields

$$\theta_\nu \leq c^{-1} \text{ constant.}$$

Thus $\{\theta_\nu : \nu \in \kappa\}$ is bounded.

For each ν, let $\mathcal{I}_\nu$ be the set of all indices i for which

$$(G(x^\nu + dx^\nu) - a)_i = 0.$$

Then $\zeta_i^\nu = 0$ for all $i \notin \mathcal{I}_\nu$. Since there are only finitely many such index sets $\mathcal{I}_\nu$, we may assume without loss of generality that $\mathcal{I}_\nu$ is the same for all $\nu \in \kappa$. Let $\mathcal{I}$ denote this common index set. Since for each ν,

$$Q_\nu dx^\nu + \theta_\nu dx^\nu + df_x^\nu - (\nabla_x F^\nu)^T \pi^\nu + G^T \zeta^\nu = 0,$$

and the sequence $\{Q_\nu : \nu \in \kappa\}$ is bounded by (29), which implies

$$\lim_{\nu(\in\kappa)\to\infty} \left[Q_\nu dx^\nu + \theta_\nu dx^\nu + df_x^\nu - (\nabla_x F^\nu)^T \pi^\nu \right] = df_x^* - (\nabla_x F^*)\pi^*,$$

it follows that there must exist a vector $\zeta^* \geq 0$ satisfying $\zeta_i^* = 0$ for all $i \notin \mathcal{I}$ such that

$$df_x^* - (\nabla_x F^*)^T \pi^* + G^T \zeta^* = 0.$$

Passing to the limit $\nu(\in \kappa) \to \infty$ in the second, third, and fourth equations in (14), we deduce

$$df_y^* - (\nabla_y F^*)^T \pi^* - W_* \xi^* = 0,$$

$$df_w^* - (\nabla_w F^*)^T \pi^* - Y_* \xi^* = 0,$$

$$df_z^* - (\nabla_z F^*)^T \pi^* = 0.$$

By Corollary **6.1.3**, the last four equations, together with the nonneg-ativity of η^* and the complementarity condition $(\zeta^*)^T(Gx^* - a) = 0$, are exactly the stationarity conditions for the MPEC (1) at the vector (x^*, y^*, w^*, z^*). Consequently, (x^*, y^*, w^*, z^*) is a stationary point of the problem (1).   **Q.E.D.**

It should be noted that the index set κ in Theorem **6.1.13** is quite specific and the limit point (x^*, y^*, w^*, z^*) pertains to the corresponding subsequence $\{(x^\nu, y^\nu, w^\nu, z^\nu) : \nu \in \kappa\}$. Specifically, κ refers to those itera-tions indexed by ν where there is a change in the penalty value α_ν. The fact that Theorem **6.1.13** deals only with this particular subsequence is a limitation of its proof; nevertheless, this result seems reasonable as the set κ has much to do with the assumption that the sequence of penalty values $\{\alpha_\nu\}$ is unbounded in the first place. Moreover, such a conclusion is consis-tent with a general penalty approach for solving constrained optimization problems which in the case of unbounded penalty values provides the only conclusion similar to Theorem **6.1.13**; see [40, 42, 43].

Next we analyze the case where the sequence $\{\alpha_\nu\}$ is bounded. In this case, we may assume without loss of generality that α_ν is a constant for all ν and we will drop the subscript ν from this penalty constant. Unlike the previous case, the subsequent analysis does not require the specification of the subsequence of $\{(x^\nu, y^\nu, z^\nu, w^\nu)\}$. Indeed we consider any limit point of this sequence and establish its stationarity under the assumptions SC and NS.

Let $\kappa \subseteq \{1, 2, \ldots\}$ be arbitrary; assume that

$$\lim_{\nu(\in\kappa)\to\infty} (x^\nu, y^\nu, w^\nu, z^\nu) = (x^*, y^*, w^*, z^*)$$

and that SC and NS hold at this limit vector. Throughout the analysis below, we assume that the sequence of scalars $\{\sigma_\nu : \nu \in \kappa\}$ is bounded away from zero; specifically, we assume that there exists $\underline{\sigma} > 0$ such that

$$\sigma_\nu \geq \underline{\sigma}, \quad \text{for all } \nu \in \kappa. \tag{34}$$

This assumption is clearly satisfied if we fix σ_ν to be a positive constant for all ν. By Lemma **6.1.12**, we obtain the following bound on the sequence $\{\bar{\tau}_\nu : \nu \in \kappa\}$.

6.1.14 Lemma. If (34) holds, then there exists a positive scalar δ_1 such that for all $\nu \in \kappa$, $\bar{\tau}_\nu \geq \delta_1$.

Proof. By (20) and part (b) of Lemma **6.1.12**, we have

$$g_\nu(\tau) \geq \mu_\nu \left((1 - \bar{\rho})\sigma_\nu - \tau\beta(1 + \bar{\rho}) \right).$$

By (34), it follows that $g_\nu(\tau) > 0$ for all

$$\tau < \frac{(1 - \bar{\rho})\underline{\sigma}}{\beta(1 + \bar{\rho})}.$$

Consequently, the unique positive root of $g_\nu(\tau)$ in the interval $[0, 1]$, if it exists, is no less than the right-hand ratio.   **Q.E.D.**

There is also a positive lower bound on the step size τ for which (25) holds for all $\nu \in \kappa$. The proof of this requires us to bound the quantities $\phi^\nu(\tau)$ and $\|F^\nu(\tau)\|$, where

$$\phi^\nu(\tau) \equiv \phi(x^\nu(\tau), y^\nu(\tau), w^\nu(\tau), z^\nu(\tau)),$$

$$F^\nu(\tau) \equiv F(x^\nu(\tau), y^\nu(\tau), w^\nu(\tau), z^\nu(\tau)).$$

We write

$$\nabla F^\nu \equiv \nabla F(x^\nu(\tau), y^\nu(\tau), w^\nu(\tau), z^\nu(\tau)).$$

By (28), (30), and the boundedness of the sequence $\{(x^\nu, y^\nu, w^\nu, z^\nu)\}$, we deduce the existence of a constant $c'' > 0$ such that for all $\nu \in \kappa$,

$$\|(dx^\nu, dy^\nu, dw^\nu, dz^\nu)\| \leq c'' \left(\|F^\nu\| + (w^\nu)^T y^\nu \right)^{1/2}. \tag{35}$$

6.1.15 Lemma. There exists a constant $\tilde{c} > 0$ such that for all $\nu \in \kappa$ and all $\tau \in [0, 1]$,

$$y^\nu(\tau)^T w^\nu(\tau) \leq (y^\nu)^T w^\nu \left[1 - (1 - \sigma_\nu)\tau + \tilde{c}\tau^2 \right],$$

$$\|F^\nu(\tau)\| \leq (1 - \tau)\|F^\nu\| + \tau^2 \tilde{c} \left[\|F^\nu\| + (y^\nu)^T w^\nu \right]. \tag{36}$$

Proof. Let $\nu \in \kappa$ and $\tau \in [0, 1]$ be arbitrary. By direct substitution, we have

$$y^\nu(\tau)^T w^\nu(\tau) = (y^\nu)^T w^\nu \left[1 - (1 - \sigma_\nu)\tau \right] + \tau^2 (dy^\nu)^T dw^\nu.$$

By part (b) of Lemma **6.1.12**, the first inequality in (36) follows.

Since F is C^2, by Taylor expansion, for each index i there exists a scalar $\tau' \in [0, 1]$ such that

$$F_i^\nu(\tau) = F_i^\nu + \tau \delta F_i^\nu + \frac{\tau^2}{2} \delta^2 F_i^\nu(\tau'),$$

where

$$\delta F_i^\nu \equiv (\nabla_x F_i^\nu)^T dx^\nu + (\nabla_y F_i^\nu)^T dy^\nu + (\nabla_w F_i^\nu)^T dw^\nu + (\nabla_z F_i^\nu)^T dz^\nu,$$

$$\delta^2 F_i^\nu(\tau') \equiv \begin{bmatrix} dx^\nu \\ dy^\nu \\ dw^\nu \\ dz^\nu \end{bmatrix}^T \nabla^2 F_i(x^\nu(\tau'), y^\nu(\tau'), w^\nu(\tau'), z^\nu(\tau')) \begin{bmatrix} dx^\nu \\ dy^\nu \\ dw^\nu \\ dz^\nu \end{bmatrix}.$$

Since $\{(x^\nu, y^\nu, w^\nu, z^\nu)\}$ is bounded, the C^2 property of F implies that

$$\nabla^2 F_i(x^\nu(\tau'), y^\nu(\tau'), w^\nu(\tau'), z^\nu(\tau'))$$

is bounded by a constant which is independent of $\nu \in \kappa$, i, and $\tau' \in [0, 1]$. Moreover, since $\delta F_i^\nu = -F_i^\nu$, we deduce

$$|F_i^\nu(\tau)| \leq (1 - \tau)|F_i^\nu| + \frac{\tau^2}{2}(\sup \|\nabla^2 F_i\|)\|(dx^\nu, dy^\nu, dw^\nu, dz^\nu)\|^2.$$

Applying (35), we obtain the second inequality in (36).   **Q.E.D.**

6.1.16 Corollary. There exists a scalar $\delta_3 > 0$ such that for all $\tau \in [0, \delta_3]$ and all $\nu \in \kappa$, $\phi^\nu(\tau) \leq \phi_\nu$.

Proof. From the inequalities in (36), we obtain

$$\phi^\nu(\tau) \leq (1 - 2\tau)\|F^\nu\|^2 + (1 - (1 - \sigma_\nu)\tau)(y^\nu)^T w^\nu$$

$$+ \tau^2 \left[(\|F^\nu\|^2 + \tilde{c}(y^\nu)^T w^\nu) + 2\tilde{c}\|F^\nu\| (\|F^\nu\| + (y^\nu)^T w^\nu) \right]$$

$$- 2\tau^3 \tilde{c}\|F^\nu\| \left[\|F^\nu\| + (y^\nu)^T w^\nu \right] + \tau^4 \tilde{c}^2 \left[\|F^\nu\| + (y^\nu)^T w^\nu \right]^2.$$

Since $\{\|F^\nu\| : \nu \in \kappa\}$ and $\{(y^\nu)^T w^\nu : \nu \in \kappa\}$ are bounded, we deduce the existence of a constant $c_0 > 0$, independent of ν, such that

$$\phi^\nu(\tau) \leq \left[\left(1 - \frac{1 - \sigma_\nu}{2}\tau \right) - \tau \left(\frac{1 - \sigma_\nu}{2} - \tau c_0 \right) \right] \phi_\nu.$$

Consequently with

$$\delta_3 \equiv \frac{1 - \sigma_\nu}{2c_0},$$

we have

$$\phi^\nu(\tau) \le \phi_\nu$$

for all $\tau \in [0, \delta_3]$.   **Q.E.D.**

We may now state the main convergence result for the interior point algorithm when the sequence of penalty parameters $\{\alpha_\nu\}$ is bounded.

6.1.17 Theorem. Suppose that the sequence $\{Q_\nu\}$ is bounded according to (29) and that the sequence $\{\sigma_\nu\}$ is bounded away from zero. If $\{\alpha_\nu\}$ is bounded, then every limit point of the sequence $\{(x^\nu, y^\nu, w^\nu, z^\nu)\}$ that satisfies the assumptions SC and NS is a stationary point of the problem (1).

Proof. Let $\{(x^\nu, y^\nu, w^\nu, z^\nu) : \nu \in \kappa\}$ denote an arbitrary subsequence converging to a limit $(x^*, y^*, w^*, z^*) \in X \times \Re_+^{2m} \times \Re^\ell$ satisfying SC and NS. Without loss of generality, we may assume that the sequences $\{\sigma_\nu : \nu \in \kappa\}$ and $\{(dx^\nu, dy^\nu, dw^\nu, dz^\nu) : \nu \in \kappa\}$ converge, respectively, to $\sigma_* \in (0, 1)$ and (dx^*, dy^*, dw^*, dz^*).

The first step is to show that the limit (32) is valid for this infinite index set κ. Since $\{\alpha_\nu\}$ is bounded, we may assume without loss of generality that α_ν is equal to a constant α for all ν. The sequence

$$\{P_\alpha(x^\nu, y^\nu, w^\nu, z^\nu)\} \tag{37}$$

is decreasing. Since $\{(x^\nu, y^\nu, w^\nu, z^\nu)\}$ is bounded, the sequence (37) converges. Hence, it follows from (26) that

$$\lim_{\nu \to \infty} \tau_\nu \left((df_x^\nu)^T dx^\nu + (df_y^\nu)^T dy^\nu + (df_w^\nu)^T dw^\nu + (df_z^\nu)^T dz^\nu \right.$$
$$\left. -\alpha(2\|F^\nu\|^2 + (1 - \sigma_\nu)(y^\nu)^T w^\nu) \right) = 0 \tag{38}$$

and

$$\lim_{\nu \to \infty} \tau_\nu \phi_\nu = 0. \tag{39}$$

We divide the remaining analysis into two cases:

$$\text{(i) } \liminf_{\nu(\in\kappa) \to \infty} \tau_\nu > 0, \quad \text{or (ii) } \liminf_{\nu(\in\kappa) \to \infty} \tau_\nu = 0.$$

In case (i), it follows from (39) that

$$\lim_{\nu(\in\kappa)\to\infty} \phi_\nu = \phi(x^*,y^*,w^*,z^*) = 0.$$

Thus (x^*,y^*,w^*,z^*) is feasible to (1). From (38), we have

$$\lim_{\nu(\in\kappa)\to\infty} \left((df_x^\nu)^T dx^\nu + (df_y^\nu)^T dy^\nu + (df_w^\nu)^T dw^\nu + (df_z^\nu)^T dz^\nu\right) = 0.$$

Henceforth the remaining proof is similar to that of Theorem **6.1.13**.

Consider case (ii). By Lemma **6.1.14**, the sequence $\{\bar{\tau}_\nu : \nu \in \kappa\}$ is bounded away from zero. Thus

$$\lim_{\nu(\in\kappa)\to\infty} \rho_1^{r_\nu} = 0,$$

or equivalently,

$$\lim_{\nu(\in\kappa)\to\infty} r_\nu = \infty.$$

By Corollary **6.1.16**, it follows that (25) holds for

$$\tilde{\tau}_\nu \equiv \bar{\tau}_\nu \rho_1^{r_\nu-1}$$

for all $\nu \in \kappa$. Thus we have

$$P_\alpha(x^\nu(\tilde{\tau}_\nu), y^\nu(\tilde{\tau}_\nu), w^\nu(\tilde{\tau}_\nu), z^\nu(\tilde{\tau}_\nu)) - P_\alpha(x^\nu, y^\nu, w^\nu, z^\nu)$$

$$> -\gamma'\tilde{\tau}_\nu \left((df_x^\nu)^T dx^\nu + (df_y^\nu)^T dy^\nu + (df_w^\nu)^T dw^\nu + (df_z^\nu)^T dz^\nu\right)$$

$$-\alpha(2\|F^\nu\|^2 + (1-\sigma_\nu)(y^\nu)^T w^\nu)).$$

Dividing by $\tilde{\tau}_\nu$ and letting $\nu \to \infty$ along the subsequence κ, we obtain

$$(dP_\alpha^*)_x^T dx^* + (dP_\alpha^*)_y^T dy^* + (dP_\alpha^*)_w^T dw^* + (dP_\alpha^*)_z^T dz^* \geq$$

$$- \gamma' \left((df_x^*)^T dx^* + (df_y^*)^T dy^* + (df_w^*)^T dw^* + (df_z^*)^T dz^*\right)$$

$$- \alpha(2\|F^*\| + (1-\sigma_*)(y^*)^T w^*)),$$

where $(dP_\alpha^*)_s$ and df_s^* denote, respectively,

$$\nabla_s P_\alpha(x^*,y^*,w^*,z^*) \quad \text{and} \quad \nabla_s f(x^*,y^*,w^*,z^*)$$

for $s = x,y,w,z$. Since $\gamma' < 1$, it follows that

$$(df_x^*)^T dx^* + (df_y^*)^T dy^* + (df_w^*)^T dw^* + (df_z^*)^T dz^*$$

$$- \alpha\left(2\|F^*\| + (1-\sigma_*)(y^*)^T w^*\right) = 0,$$

which implies, by the penalty update rule, that

$$\phi(x^*, y^*, w^*, z^*) = 0.$$

In turn, this yields

$$(df_x^*)^T dx^* + (df_y^*)^T dy^* + (df_w^*)^T dw^* + (df_z^*)^T dz^* = 0.$$

We are now back to the previous case.   **Q.E.D.**

6.1.4 *Special cases*

In what follows, we discuss some special cases of (1). In each case, we present sufficient conditions for the satisfaction of the two assumptions (A1) and (A2), i.e., the mixed P_0 property of the partitioned Jacobian matrix (6) and the boundedness of the level set $\mathcal{L}(\delta)$. We are also concerned with the nonsingularity condition NS.

First, consider the case where

$$F(x, y, w, z) \equiv \begin{pmatrix} g(x, y, z) - w \\ h(x, y, z) \end{pmatrix}, \tag{40}$$

where $g : \Re^{n+m+\ell} \to \Re^m$ and $h : \Re^{n+m+\ell} \to \Re^\ell$ are twice continuously differentiable on an open set containing $X \times \Re^m_+ \times \Re^\ell$. This case corresponds to the MPEC where each inner VI is a mixed NCP given by

$$w = g(x, y, z) \geq 0, \quad y \geq 0, \quad w^T y = 0$$
$$0 = h(x, y, z).$$

For the function F given above, the partitioned matrix (6) is equal to

$$\begin{bmatrix} \nabla_y g(x, y, z) & -I & \nabla_z g(x, y, z) \\ \nabla_y h(x, y, z) & 0 & \nabla_z h(x, y, z) \end{bmatrix}. \tag{41}$$

6.1.18 Proposition. Let $(x, y, w, z) \in \Re^{n+2m+\ell}$ be an arbitrary vector. The following statements are valid.

(a) If the matrix

$$\begin{bmatrix} \nabla_z g(x, y, z) \\ \nabla_z h(x, y, z) \end{bmatrix} \tag{42}$$

has full column rank and the matrix

$$
\begin{bmatrix}
\nabla_y g(x,y,z) & \nabla_z g(x,y,z) \\
\nabla_y h(x,y,z) & \nabla_z h(x,y,z)
\end{bmatrix}
\tag{43}
$$

is positive semidefinite, or alternatively, if $\nabla_z h(x,y,z)$ is nonsingular and

$$
\nabla_y g(x,y,z) - \nabla_z g(x,y,z)\nabla_z h(x,y,z)^{-1}\nabla_y h(x,y,z)
\tag{44}
$$

is a P_0 matrix, then the matrix (41) has the mixed P_0 property.

(b) If the matrix (43) is P, then the matrix (41) has the mixed P property.

(c) If $(y,w) \geq 0$, $y + w > 0$, the assumptions of (a) hold, and the matrix

$$
\begin{bmatrix}
\nabla_{y_\alpha} g_\alpha(x,y,z) & \nabla_z g_\alpha(x,y,z) \\
\nabla_{y_\alpha} h(x,y,z) & \nabla_z h(x,y,z)
\end{bmatrix}
\tag{45}
$$

where $\alpha \equiv \mathrm{supp}(y)$, is nonsingular, then the matrix (7) is nonsingular.

Proof. To prove (a) and (b), assume that

$$
\begin{aligned}
s &= \nabla_y g(x,y,z)r + \nabla_z g(x,y,z)t \\
0 &= \nabla_y h(x,y,z)r + \nabla_z h(x,y,z)t
\end{aligned}
\tag{46}
$$

for some $(r,s) \neq 0$. Then $(r,t) \neq 0$; moreover $r^T s \geq 0$ if the matrix (43) is positive semidefinite. Thus $r_i s_i \geq 0$ for at least one i with $(r_i, s_i) \neq 0$. This establishes the first assertion in (a). If $\nabla_z h(x,y,z)$ is nonsingular, then $r \neq 0$; moreover, we can eliminate the variable t in (46), resulting in

$$
s = Mr
$$

where M is the matrix in (44). Since M is a P_0 matrix and $r \neq 0$, it follows that there exists an index i for which $r_i \neq 0$ and $r_i s_i \geq 0$. Thus (a) is now proved. If the matrix (43) is P, then we must have $r_i s_i > 0$ for some index i. Thus (41) has the mixed P property and (b) holds. To prove (c), suppose that in addition to (46) we have

$$
Wr + Ys = 0.
$$

Since $y + w > 0$, we have $w_j > 0$ for all $j \notin \text{supp}(y)$ which implies $r_j = 0$ for all these j. For $i \in \text{supp}(y)$, we have

$$r_i s_i = -\frac{w_i r_i^2}{y_i} \leq 0.$$

Thus $r^T s \leq 0$. Since the matrix (43) is positive semidefinite, it follows that $r_i s_i = 0$ for all i. Hence we must have $s_i = 0$ for all $i \in \text{supp}(y)$. Thus

$$
\begin{aligned}
0 &= \nabla_{y_\alpha} g_\alpha(x, y, z) r_\alpha + \nabla_z g_\alpha(x, y, z) t, \\
0 &= \nabla_{y_\alpha} h(x, y, z) r_\alpha + \nabla_z h(x, y, z) t.
\end{aligned}
\tag{47}
$$

Hence if the matrix (45) is nonsingular, then $r_\alpha = 0$ and $t = 0$ which, along with the fact that $r_j = 0$ for all $j \notin \alpha$, yields $s = 0$. Thus the matrix (7) is nonsingular.   **Q.E.D.**

6.1.19 Remark. Recalling Proposition **6.1.6**, we note that the matrix (7) is also nonsingular if (43) is a P matrix and $(y, w) \geq 0$ satisfies $y + w > 0$.

Next we give a sufficient condition for the level set $\mathcal{L}(\delta)$ to be bounded when F is given by (40).

6.1.20 Proposition. Suppose that the set

$$\left\{ (x, y, z) \in X \times \Re^m_{++} \times \Re^\ell : g(x, y, z) \geq u, \, \|h(x, y, z)\| \leq c \right\}$$

is bounded for every vector $u \in \Re^m$ and scalar $c \in \Re_{++}$. Then the set $\mathcal{L}(\delta)$ is bounded for every $\delta > 0$.

Proof. Let $\{(x^k, y^k, w^k, z^k)\}$ be an arbitrary sequence in $\mathcal{L}(\delta)$ for some $\delta > 0$. Write

$$u^k \equiv g(x^k, y^k, z^k) - w^k, \quad v^k \equiv h(x^k, y^k, z^k).$$

Then $\{(u^k, v^k)\}$ is a bounded sequence. Without loss of generality, we may assume that $\{(u^k, v^k)\}$ converges to $(\bar{u}, \bar{v})$. Then for some finite $\varepsilon > 0$, we deduce that $\{(x^k, y^k, z^k)\}$ is a subset of

$$\left\{ (x, y, z) \in X \times \Re^m_{++} \times \Re^\ell : g(x, y, z) \geq \bar{u} - \varepsilon e, \, \|h(x, y, z)\| \leq \|\bar{v}\| + \varepsilon \right\},$$

where e is the vector of all ones. Hence $\{(x^k, y^k, z^k)\}$ is bounded and so is $\{w^k\} = \{g(x^k, y^k, z^k) - u^k\}$.   **Q.E.D.**

Combining Propositions **6.1.18** and **6.1.20** with the convergence results of the interior point algorithm, we obtain the following consequence pertaining to the convergence of this algorithm for solving the MPEC with monotone mixed NCPs as its inner problems:

$$\text{minimize} \quad f(x, y, w, z)$$

$$\text{subject to} \quad x \in X$$

$$w = g(x, y, z), \quad 0 = h(x, y, z) \tag{48}$$

$$(y, w) \geq 0, \quad w^T y = 0.$$

In the next and all subsequent convergence results, we will let

$$\kappa \equiv \begin{cases} \{1, 2, \ldots\} & \text{if } \{\alpha_\nu\} \text{ is bounded} \\ \{\nu : \alpha_{\nu-1} < \alpha_\nu\} & \text{if } \{\alpha_\nu\} \text{ is unbounded.} \end{cases} \tag{49}$$

6.1.21 Corollary. Let $f : \Re^{n+2m+\ell} \to \Re$ be continuously differentiable, and $g : \Re^{n+m+\ell} \to \Re^m$ and $h : \Re^{n+m+\ell} \to \Re^\ell$ be twice continuously differentiable on an open set in $\Re^{n+m+\ell}$ containing $X \times \Re_+^m \times \Re^\ell$, where X is a nonempty polyhedral subset of $\Re^n$. Assume that for every $x \in X$, the mapping

$$\begin{pmatrix} y \\ z \end{pmatrix} \in \Re^{m+\ell} \mapsto \begin{pmatrix} g(x, y, z) \\ h(x, y, z) \end{pmatrix} \in \Re^{m+\ell} \tag{50}$$

is monotone. Assume also that the matrix (42) has full column rank for all $(x, y, z) \in X \times \Re_+^m \times \Re^\ell$ and that the set

$$\{(x, y, z) \in X \times \Re_{++}^m \times \Re^\ell : g(x, y, z) \geq u, \|h(x, y, z)\| \leq c\}$$

is bounded for every vector $u \in \Re^m$ and scalar $c \in R_{++}$. Then for any initial $(x^0, y^0, w^0, z^0) \in X \times \Re_{++}^{2m} \times \Re^\ell$, the PIPA applied to (48) generates a well-defined, bounded sequence of iterates $\{(x^\nu, y^\nu, w^\nu, z^\nu)\}$; moreover, if (x^*, y^*, w^*, z^*) is a limit point of the subsequence

$$\{(x^\nu, y^\nu, w^\nu, z^\nu) : \nu \in \kappa\} \tag{51}$$

such that $y^* + w^* > 0$ and the matrix

$$\begin{bmatrix} \nabla_{y_\alpha} g_\alpha(x^*, y^*, z^*) & \nabla_z g_\alpha(x^*, y^*, z^*) \\ \nabla_{y_\alpha} h(x^*, y^*, z^*) & \nabla_z h(x^*, y^*, z^*) \end{bmatrix}$$

is nonsingular, then (x^*, y^*, w^*, z^*) is a stationary point of (48).

Proof. The monotonicity of the mapping (50) implies that the matrix (43) is positive semidefinite. The desired conclusion requires no further proof. **Q.E.D.**

We next establish the boundedness of the level sets $\mathcal{L}(\delta)$ by restricting the class of mappings g and h. We introduce some definitions. For a function $G : X \times U \subseteq X \times \Re^N \to \Re^N$, we say that $G(x, \cdot)$ is *uniformly coercive* in U for all $x \in X$ if

$$\|G(x, u)\| \to \infty \quad \text{uniformly in } x \in X \quad \text{as } \|u\| \to \infty.$$

We note that if $G(x, \cdot)$ is a global homeomorphism on U with a uniformly Lipschitzian inverse so that

$$\|G(x, u) - G(x, v)\| \geq \gamma \|u - v\|, \quad \forall u, v \in U, \ x \in X,$$

where γ is a positive constant independent of $x \in X$, then G must be uniformly coercive in U for all $x \in X$. In particular if G is a uniform P function on U with a modulus that is independent of $x \in X$, such that

$$\max_{1 \leq i \leq N} (u_i - v_i)(G_i(x, u) - G_i(x, v)) \geq c \|u - v\|^2, \quad \forall u, v \in U,$$

then $G(x, \cdot)$ is uniformly coercive in U for all $x \in X$. Hence if $G(x, \cdot)$ is strongly monotone on U uniformly for $x \in X$, then $G(x, \cdot)$ is uniformly coercive in U for all $x \in X$.

Let $G : \prod_{i=1}^{N} U_i \subseteq \Re^N \to \Re^N$ be given. For each subset $\alpha \subseteq \{1, \ldots, N\}$ with complement $\bar{\alpha}$ and each vector $a \in \prod_{i=1}^{N} U_i$, define $H : \prod_{i \in \bar{\alpha}} U_i \to \Re^k$, where $k \equiv |\bar{\alpha}|$, by

$$H(u_{\bar{\alpha}}) \equiv G_{\bar{\alpha}}(a_\alpha, u_{\bar{\alpha}}), \quad \text{for all } u_{\bar{\alpha}} \in \prod_{i \in \bar{\alpha}} U_i.$$

This function H is a *principal subfunction* of G obtained by fixing the u_α components at the constant a_α and removing the corresponding α components from the function G.

Returning to the function F given by (40), we let

$$G(x, y, z) \equiv \begin{pmatrix} g(x, y, z) \\ h(x, y, z) \end{pmatrix}, \quad \text{for } (x, y, z) \in X \times \Re_+^m \times \Re^\ell. \tag{52}$$

Let $\mathcal{G}$ be the collection of principal subfunctions of G obtained by fixing the y_α components at some nonnegative vector a_α and removing the corresponding g_α functions, for all index subsets α of $\{1, \ldots, m\}$ and nonnegative vectors a_α.

6.1.22 Proposition. Let X be a bounded subset of $\mathfrak{R}^n$. Suppose that the function G is Hölder continuous on $X \times \mathfrak{R}^m_+ \times \mathfrak{R}^\ell$. If each function in the family $\mathcal{G}$ is uniformly coercive in its domain for all $x \in X$, then for any scalar $\delta > 0$, the set $\mathcal{L}(\delta)$ is bounded.

Proof. Let $\{(x^k, y^k, w^k, z^k)\}$ be an arbitrary sequence in $\mathcal{L}(\delta)$ for some $\delta > 0$. Since X is bounded, it follows that $\{x^k\}$ is bounded. For each k, write

$$g^k \equiv g(x^k, y^k, z^k), \quad h^k \equiv h(x^k, y^k, z^k).$$

By working with an appropriate subsequence if necessary, we may assume that there is a subset $\alpha \subseteq \{1, \ldots, m\}$ such that for each $i \in \alpha$, $\{w_i^k\} \to \infty$ and for each $i \notin \alpha$, $\{w_i^k\}$ is bounded. This index set α may be empty or equal to the entire set $\{1, \ldots, m\}$. Let $\bar{\alpha}$ be the complement of α in $\{1, \ldots, m\}$. Since $\{y_i^k w_i^k\}$ is bounded for each i, it follows that $\{y_i^k\} \to 0$ for each $i \in \alpha$. For each k, let G^k be the principal subfunction of G obtained by fixing the y_α components at zero and removing the α components from g. We have

$$G^k(x^k, y_{\bar{\alpha}}^k, z^k) = \begin{pmatrix} g_{\bar{\alpha}}(x^k, y_{\bar{\alpha}}^k, 0, z^k) \\ h(x^k, y_{\bar{\alpha}}^k, 0, z^k) \end{pmatrix}$$

$$= \begin{pmatrix} g_{\bar{\alpha}}^k \\ h^k \end{pmatrix} + \begin{pmatrix} g_{\bar{\alpha}}(x^k, y_{\bar{\alpha}}^k, 0, z^k) - g_{\bar{\alpha}}^k \\ h(x^k, y_{\bar{\alpha}}^k, 0, z^k) - h^k \end{pmatrix}.$$

By the definition of the index set α and the Hölderian property of the function G, it follows that the sequence $\{G^k(x^k, y_{\bar{\alpha}}^k, z^k)\}$ is bounded. Hence by the uniform coercive assumption of the function G^k, it follows that $\{(y_{\bar{\alpha}}^k, z^k)\}$ is bounded. Hence the sequence $\{(x^k, y^k, z^k)\}$ is bounded; thus so is $\{g^k\}$. Since $\{g^k - w^k\}$ is bounded, it follows that $\{w^k\}$ is also bounded.

Consequently, the set $\mathcal{L}(\delta)$ must be bounded, as desired.   **Q.E.D.**

Combining Propositions **6.1.18** and **6.1.22** with the convergence results of the interior point algorithm, we obtain the following consequence
pertaining to the convergence of this algorithm for solving the MPEC (48).

6.1.23 Corollary. Let $f : \Re^{n+2m+\ell} \to \Re$ be continuously differentiable,
and $g : \Re^{n+m+\ell} \to \Re^m$ and $h : \Re^{n+m+\ell} \to \Re^\ell$ be twice continuously
differentiable on an open set in $\Re^{n+m+\ell}$ containing $X \times \Re_+^m \times \Re^\ell$, where
X is a nonempty bounded polyhedron in $\Re^n$. Assume that the function G
defined by (52) is Hölder continuous on $X \times \Re_+^m \times \Re^\ell$ and that $G(x, \cdot, \cdot)$ is
a uniform P function on $\Re_+^m \times \Re^\ell$ with a P modulus that is independent of
$x \in X$. For any initial $(x^0, y^0, w^0, z^0) \in X \times \Re_{++}^{2m} \times \Re^\ell$, the interior point
algorithm applied to (48) generates a well-defined, bounded sequence of
iterates $\{(x^\nu, y^\nu, w^\nu, z^\nu)\}$; moreover, any limit point (x^*, y^*, w^*, z^*) of the
subsequence (51), where κ is defined by (49), that satisfies $y^* + w^* > 0$, is
a stationary point of (48).

Proof. The uniform P assumption implies that the matrix (43) is a P ma-
trix for all $(x, y, z) \in X \times \Re_+^m \times \Re^\ell$. Thus the sequence $\{(x^\nu, y^\nu, w^\nu, z^\nu)\}$ is
well defined. Moreover, the same P condition implies that the uniform co-
ercive assumption of Proposition **6.1.22** is satisfied. The desired conclusion
now follows from Theorems **6.1.13** and **6.1.17**.   **Q.E.D.**

Next we consider an MPEC where each lower-level problem is defined
by a mixed LCP:

$$\begin{aligned}
\text{minimize} \quad & f(x, y, w, z) \\
\text{subject to} \quad & x \in X, \\
& 0 = q + Ex + Ay + Bw + Cz, \\
& (y, w) \geq 0, \quad w^T y = 0.
\end{aligned} \tag{53}$$

In this linear case, we show that the level set $\mathcal{L}(\delta)$ is bounded if a certain
homogenized constraint system has a unique solution. In what follows,
$0^+ X$ denotes the recession cone of the set X. It is well known from convex
analysis that any closed convex set X is bounded if and only if $0^+ X = \{0\}$.
In the following result, we do not assume that X is bounded.

6.1.24 Corollary. Let X be a polyhedron in $\Re^n$ and $f : \Re^{n+2m+\ell} \to \Re$
be continuously differentiable. Let $E \in \Re^{(m+\ell) \times n}$, A, $B \in \Re^{(m+\ell) \times m}$,

$C \in \Re^{(m+\ell) \times \ell}$, and $q \in \Re^{m+\ell}$ be given. Suppose that matrix $Q = [A\ B\ C]$ has the mixed P property and that the homogeneous system

$$x \in 0^+ X,$$

$$0 = Ex + Ay + Bw + Cz, \qquad (54)$$

$$(y, w) \geq 0, \quad y^T w = 0,$$

has $(x, y, w, z) = (0, 0, 0, 0)$ as the unique solution. Then the conclusion of Corollary **6.1.23** holds for the problem (53).

Proof. It suffices to show that the set $\mathcal{L}(\delta)$ is bounded for any $\delta > 0$. Assume for contradiction that $\{(x^k, y^k, w^k, z^k)\} \subset \mathcal{L}(\delta)$ is an unbounded sequence. Without loss of generality, we may assume that

$$\lim_{k \to \infty} \|(x^k, y^k, w^k, z^k)\| = \infty,$$

$$\lim_{k \to \infty} \frac{(x^k, y^k, w^k, z^k)}{\|(x^k, y^k, w^k, z^k)\|} = (x^\infty, y^\infty, w^\infty, z^\infty) \neq 0,$$

where $x^\infty \in 0^+ X$ and $(y^\infty, w^\infty) \geq 0$. Consider the two sequences $\{r^k\}$ and $\{s^k\}$ defined for every $k \geq 0$ by

$$r^k \equiv q + Ex^k + Ay^k + Bw^k + Cz^k, \quad s^k = y^k \circ w^k.$$

Since $\|r^k\|^2 + (y^k)^T w^k \leq \delta$ for all k, it can easily be shown that

$$(x^\infty, y^\infty, w^\infty, z^\infty)$$

must be a solution of the homogeneous system (54). But this is a contradiction. Hence the sequence $\{(x^k, y^k, w^k, z^k)\}$ must be bounded. **Q.E.D.**

6.1.25 Remark. If X is bounded, then the mixed P property of Q implies that the system (54) has a unique solution. Thus this uniqueness assumption becomes redundant when X is bounded.

Finally we consider an MPEC where each lower-level problem is a monotone VI. Specifically, this MPEC is

$$\text{minimize} \quad \theta(x, z)$$

$$\text{subject to} \quad x \in X, \qquad (55)$$

$$\text{and} \quad z \in \text{SOL}(F(x, \cdot), C(x)),$$

where $\theta : \Re^{n+\ell} \to \Re$ is continuously differentiable; X is a nonempty poly-hedron in $\Re^n$, $F : \Re^{n+\ell} \to \Re^\ell$ is twice continuously differentiable, and for each $x \in X$, $F(x, \cdot)$ is a monotone mapping on $\Re^\ell$, and

$$C(x) \equiv \{z \in \Re^\ell : g(x, z) \leq 0\},$$

where $g : \Re^{n+\ell} \to \Re^m$ is three times continuously differentiable, and for each $x \in X$ and each $i = 1, \ldots, m$, $g_i(x, \cdot)$ is a convex function. We now cast (55) in the form of (1). Let $L(x, z, y)$ denote the vector-valued Lagrangean function of the VI $(F(x, \cdot), C(x))$; i.e.,

$$L(x, z, y) \equiv F(x, z) + \sum_{i=1}^{m} y_i \nabla_z g_i(x, z).$$

Define the function $H : X \times \Re_+^{2m} \times \Re^\ell \to \Re^{m+\ell}$ by

$$H(x, y, w, z) \equiv \begin{pmatrix} w + g(x, z) \\ L(x, z, y) \end{pmatrix}, \quad \text{for } (x, y, w, z) \in X \times \Re_+^{2m} \times \Re^\ell.$$

The problem (55) is equivalent to

$$\begin{aligned} \text{minimize} \quad & \theta(x, z) \\ \text{subject to} \quad & x \in X, \\ & H(x, y, w, z) = 0, \\ & (y, w) \geq 0, \quad w^T y = 0, \end{aligned} \tag{56}$$

which is essentially the formulation (1.3.11). Here the function H plays the role of F in (1). We have

$$\left[\nabla_y H(x, y, w, z) \quad \nabla_w H(x, y, w, z) \quad \nabla_z H(x, y, w, z) \right] =$$
$$\begin{bmatrix} 0 & I & \nabla_z g(x, z) \\ \nabla_z g(x, z)^T & 0 & \nabla_z L(x, z, w) \end{bmatrix}. \tag{57}$$

The following result identifies a sufficient condition for this partitioned matrix to have the mixed $\mathbf{P_0}$ property.

6.1.26 Proposition. Let $(x, y, w, z) \in X \times \Re^{2m}_+ \times \Re^\ell$ be arbitrary. If the matrix $\nabla_z L(x, z, y)$ is positive semidefinite on $\Re^\ell$ and positive definite on the null space of the gradient vectors:

$$\{\nabla_z g_i(x, z) : i = 1, \ldots, m\}, \tag{58}$$

that is

$$\left. \begin{array}{l} \nabla_z g_i(x, z)^T u = 0, \text{ for } i = 1, \ldots, m \\[2mm] 0 \neq u \in \Re^\ell \end{array} \right\} \implies u^T \nabla_z L(x, z, y) u > 0,$$

then the partitioned matrix in (57) has the mixed P_0 property. If, in addition, $y + w > 0$, and the matrix $\nabla_z L(x, z, y)$ is positive definite on the null space of the gradient vectors:

$$\{\nabla_z g_i(x, z) : i \in \mathrm{supp}(y)\}, \tag{59}$$

then the matrix

$$\begin{bmatrix} \nabla_y H(x, y, w, z) & \nabla_w H(x, y, w, z) & \nabla_z H(x, y, w, z) \\[2mm] W & Y & 0 \end{bmatrix}$$

is nonsingular.

Proof. We first show that

$$\begin{bmatrix} \nabla_z g(x, z) \\[2mm] \nabla_z L(x, z, y) \end{bmatrix}$$

has full column rank. Let $u \in \Re^\ell$ such that

$$\nabla_z g_i(x, z)^T u = 0, \quad i = 1, \ldots, m$$

$$\nabla_z L(x, z, y) u = 0.$$

Then $u^T \nabla_z L(x, z, y) u = 0$ which implies $u = 0$. Next let (r, s, t) satisfy

$$0 = s + \nabla_z g(x, z) t$$

$$0 = \nabla_z g(x, z)^T r + \nabla_z L(x, z, y) t.$$

We have

$$r^T s = t^T \nabla_z L(x, z, y) t \geq 0.$$

This establishes the first assertion of the proposition. To prove the second assertion, we note that as in the proof of part (c) of Proposition **6.1.18**, it suffices to show that the matrix

$$
\begin{bmatrix}
0 & \nabla_z g_\alpha(x,z) \\
\nabla_z g_\alpha(x,z)^T & \nabla_z L(x,z,y)
\end{bmatrix}
$$

is nonsingular, where $\alpha \equiv \mathrm{supp}(y)$. Let (v_α, u) be such that

$$
\nabla_z g_i(x,z)^T u = 0, \quad i \in \alpha,
$$

$$
\sum_{i \in \alpha} v_i \nabla_z g_i(x,z) + \nabla_z L(x,z,y) u = 0.
$$

As above, we must have $u = 0$. The linear independence of the vectors (59) implies $v_\alpha = 0$.   **Q.E.D.**

Combining the above proposition with the convergence results of the interior point algorithm, we obtain the following consequence pertaining to the convergence of this algorithm for solving the MPEC (55).

6.1.27 Corollary. Let $F : \Re^{n+\ell} \to \Re^\ell$ be twice continuously differentiable, $\theta : \Re^{n+\ell} \to \Re$ be continuously differentiable, and $g : \Re^{n+\ell} \to \Re^m$ be thrice continuously differentiable on an open set in $\Re^{n+\ell}$ containing $X \times \Re^\ell$, where X is a nonempty polyhedron in $\Re^n$. Assume that the set

$$
\{(x,z) \in X \times \Re^\ell : g(x,z) \leq u\}
$$

is bounded for each vector $u \in \Re^m$ and that for each $x \in X$, $F(x, \cdot)$ is monotone and $g_i(x, \cdot)$ is convex for each $i = 1, \ldots, m$. Assume further that for each $(x,z) \in X \times \Re^\ell$, $\nabla_z F(x,z)$ is positive definite on the null space of the gradients (59) and the gradients (58) are linearly independent. For any initial $(x^0, y^0, w^0, z^0) \in X \times \Re^{2m}_{++} \times \Re^\ell$, the interior point algorithm applied to (56) generates a well-defined, bounded sequence of iterates $\{(x^\nu, y^\nu, w^\nu, z^\nu)\}$; moreover, if (x^*, y^*, w^*, z^*) is a limit point of the subsequence (51), where κ is defined by (49), such that $y^* + w^* > 0$, then (x^*, y^*, w^*, z^*) is a stationary point of (55).

Proof. By assumptions it follows that for each $(x,y,z) \in X \times \Re^m_+ \times \Re^\ell$, the matrix $\nabla_z L(x,z,y)$ is positive semidefinite on $\Re^\ell$ and positive definite on the null space of the vectors (58). Thus the sequence $\{(x^\nu, y^\nu, w^\nu, z^\nu)\}$

is well defined. It remains to establish that the level set $\mathcal{L}(\delta)$ is bounded for every $\delta > 0$. Let $\{(x^k, y^k, w^k, z^k)\} \subseteq X \times \Re_{++}^{2m} \times \Re^{\ell}$ be a sequence such that

$$\{L(x^k, z^k, y^k)\}, \quad \{w^k + g(x^k, z^k)\}, \quad \text{and} \quad \{(w^k)^T y^k\}$$

are bounded. We first show that $\{(x^k, z^k)\}$ is bounded. Indeed, as in the proof of Proposition **6.1.20**, we can deduce that the latter sequence must belong to the set

$$\{(x, z) \in X \times \Re^{\ell} : g(x, z) \leq u\}$$

for some suitable vector $u \in \Re^m$. Since the latter set is bounded by assumption, so is the sequence $\{(x^k, z^k)\}$. Since $\{w^k + g(x^k, z^k)\}$ is bounded, it follows that $\{w^k\}$ is also bounded. It remains to show that $\{y^k\}$ is bounded.

Assume for contradiction that

$$\lim_{k \to \infty} \|y^k\| = \infty.$$

Without loss of generality, we may further assume that

$$\lim_{k \to \infty} (x^k, w^k, z^k) = (x^*, w^*, z^*), \quad \lim_{k \to \infty} \frac{y^k}{\|y^k\|} = y^{\infty} \neq 0.$$

We have for each k,

$$L(x^k, z^k, y^k) = F(x^k, z^k) + \sum_{i=1}^{m} y_i^k \nabla_z g_i(x^k, z^k).$$

Since $\{L(x^k, z^k, y^k)\}$ is bounded, a standard normalization followed by a limiting argument gives

$$0 = \sum_{i=1}^{m} y_i^{\infty} \nabla_z g_i(x^*, z^*).$$

Since the gradients $\{\nabla_z g_i(x^*, z^*)\}_{i=1}^{m}$ are linearly independent, it follows that $y^{\infty} = 0$ which is a contradiction. This establishes the boundedness of the set $\mathcal{L}(\delta)$.   **Q.E.D.**

6.1.28 Remark. The linear independence assumption of the gradient vectors (58) is quite strong. This assumption is needed for the boundedness of the multiplier sequence $\{y^{\nu}\}$. It is possible to modify the interior point method so that this sequence is bounded under a Slater-type assumption. The modification entails the use of a different merit function in order to maintain the positivity of the sequence $\{w^{\nu} + g(x^{\nu}, z^{\nu})\}$; in particular, the initial iterate is chosen to satisfy $w^0 + g(x^0, z^0) > 0$. Details of this modification are omitted. The interested reader can consult the paper [280].

A deficiency of the convergence results established in this section lies in the assumption of strict complementarity at the limit point (x^*, y^*, w^*, z^*) : $y^* + w^* > 0$. We are presently not able to remove this assumption. In the case of an interior point algorithm applied to a standard complementarity problem (including the special case of a KKT system), the strict complementarity assumption is inessential; see [280]. It remains to be seen whether the convergence of PIPA can be established without this assumption.

Another remark concerns the polyhedron X; throughout this section (and also the next section), the sequence $\{x^k\}$ is required to lie in X. We believe it is possible to relax this requirement and extend PIPA to allow the sequence $\{x^k\}$ to be somewhat more flexible; we leave it to the interested reader to explore this extension.

6.2 An Alternative PIPA for LCP Constrained MP

We describe a variation of PIPA for solving the MPEC whose lower-level problem is defined by a monotone LCP. Specifically, we consider the problem:

$$
\begin{aligned}
\text{minimize} \quad & f(x, y) \\
\text{subject to} \quad & x \in X \equiv \{x \in \Re^n : Gx \le a\} \\
& w = q + Nx + My \\
& (w, y) \ge 0, \quad w^T y = 0,
\end{aligned}
\tag{1}
$$

where $f : \Re^{n+m} \to \Re$ is continuously differentiable, $G \in \Re^{p \times n}$, $a \in \Re^p$, $q \in \Re^m$, $N \in \Re^{m \times n}$, and $M \in \Re^{m \times m}$. Throughout this section, we assume that M is positive semidefinite. For any vector $(x, y, w) \in X \times \Re_+^{2m}$, define

$$
r(x, y, w) \equiv q + Nx + My - w \quad \text{and} \quad \phi(x, y, w) \equiv y^T w + \|r(x, y, w)\|.
$$

Clearly (x, y, w) is feasible to (1) if and only if $\phi(x, y, w) = 0$.

Let $(x^\nu, y^\nu, w^\nu) \in X \in \Re_{++}^{2m}$ be a given iterate satisfying the centrality condition:

$$\bar{\rho}\frac{(y^\nu)^T w^\nu}{m} \leq \min_{1 \leq i \leq m} y_i^\nu w_i^\nu, \quad \forall i = 1, \ldots, m,$$

for some given constant $\bar{\rho} \in (0, 1)$. Write

$$df_s^\nu \equiv \nabla_s f(x^\nu, y^\nu), \quad \text{for } s = x, y,$$

$$\phi_\nu \equiv \phi(x^\nu, y^\nu, w^\nu), \quad r^\nu \equiv r(x^\nu, y^\nu, w^\nu),$$

$$Y_\nu \equiv \text{diag}(y^\nu), \quad W_\nu \equiv \text{diag}(w^\nu), \quad \text{and} \quad \mu_\nu \equiv \frac{(y^\nu)^T w^\nu}{m}.$$

Fix a constant $c > 0$. For a symmetric positive definite matrix $Q_\nu \in \Re^{n \times n}$, let QP_ν denote the quadratic program in the variables (dx, dy, dw):

$$\text{minimize} \quad (df_x^\nu)^T dx + (df_y^\nu)^T dy + \tfrac{1}{2}(dx)^T Q_\nu dx$$

$$\text{subject to} \quad x^\nu + dx \in X,$$

$$(dx)^T dx \leq c\phi_\nu, \tag{2}$$

$$Ndx + Mdy - dw = -r^\nu,$$

$$Y_\nu dw + W_\nu dy = -Y_\nu w^\nu + \sigma_\nu \mu_\nu e.$$

In theory we could drop the step size constraint $(dx)^T dx \leq c\phi_\nu$ from the above QP subproblem and the whole convergence theory would remain valid; we have included this constraint here because it seems to speed up the convergence; see Section **6.5** for more discussion. Since M is positive semidefinite, thus it is a P_0 matrix; the above quadratic program has a unique optimal solution by Proposition **6.1.18** and Lemma **6.1.7**, which we denote (dx^ν, dy^ν, dw^ν). For each $\tau \in [0, 1]$, define

$$\begin{pmatrix} x^\nu(\tau) \\ y^\nu(\tau) \\ w^\nu(\tau) \end{pmatrix} = \begin{pmatrix} x^\nu \\ y^\nu \\ w^\nu \end{pmatrix} + \tau \begin{pmatrix} dx^\nu \\ dy^\nu \\ dw^\nu \end{pmatrix}.$$

Note that we have $x^\nu(\tau) \in X$ and

$$r^\nu(\tau) \equiv r(x^\nu(\tau), y^\nu(\tau), w^\nu(\tau)) = (1 - \tau)r^\nu. \tag{3}$$

We next determine an appropriate step size $\bar{\tau}_\nu \in (0, 1]$ so that for all $\tau \in [0, \bar{\tau}]$ the pair $(w^\nu(\tau), y^\nu(\tau))$ satisfies the following three properties:

(i) positivity: $(y^\nu(\tau), w^\nu(\tau)) > 0$;

(ii) limited complementarity decrease: $(w^\nu(\tau))^T y^\nu(\tau) \geq (1 - \tau)(w^\nu)^T y^\nu$;

(iii) centrality: $w^\nu(\tau) \circ y^\nu(\tau) \geq \bar{\rho}\mu_\nu(\tau)e$, where $\mu_\nu(\tau) = (w^\nu(\tau))^T y^\nu(\tau)/m$.

Property (i) is a natural requirement. Roughly speaking, the rationale for requiring (ii) is to give a higher priority to the satisfaction of the linear constraint $q + Nx + My - w = 0$ than to the complementarity condition $y^T w = 0$; in particular, we do not want complementarity satisfied before feasibility. To see that (ii) accomplishes this objective, note that if

$$\|r^\nu\| \leq c_0(w^\nu)^T y^\nu, \tag{4}$$

for some constant $c_0 > 0$, then in view of (3), we have

$$\|r^\nu(\tau)\| \leq c_0(1 - \tau)(w^\nu)^T y^\nu \leq c_0(w^\nu(\tau))^T y^\nu(\tau).$$

Consequently, the relation (4) can be maintained throughout the algorithm provided we choose the step size τ to satisfy the limited complementarity decrease. A consequence of (4) is that if $(w^\nu)^T y^\nu = 0$ then (x^ν, y^ν, w^ν) is feasible to (1).

In addition to the above three properties, we also want to decrease the residual function ϕ. To induce fast convergence, we attempt to choose a step size τ which will reduce ϕ most rapidly while maintaining the properties (i)–(iii). For this purpose, we introduce the quantity

$$\delta(\tau) \equiv \frac{\phi(x, y, w) - \phi(x(\tau), y(\tau), w(\tau))}{\phi(x, y, w)}, \quad \tau \in [0, 1],$$

where we have dropped the superscript ν from the vectors. Clearly

$$\phi(x(\tau), y(\tau), w(\tau)) = (1 - \delta(\tau))\phi(x, y, w);$$

moreover, if conditions (i), (ii), and (iii) hold, then $\delta(\tau) \in [0, 1]$. Thus we can decrease ϕ by maximizing $\delta(\tau)$ subject to the satisfaction of (i)–(iii). Let

$$g_1(\tau) \equiv \min_{1 \leq i \leq m} w_i(\tau)y_i(\tau) - \bar{\rho}w^T(\tau)y(\tau)/m, \tag{5}$$

and

$$g_2(\tau) \equiv w^T(\tau)y(\tau) - (1 - \tau)w^T y; \tag{6}$$

then we are led to select a $\bar{\tau} \in (0,1]$ which maximizes $\delta(\tau)$ subject to $g_1(\tau) \geq 0$ and $g_2(\tau) \geq 0$, for all $\tau \in [0, \bar{\tau}]$; i.e.,

$$\bar{\tau} = \arg \max_{\tau \in [0,1]} \left\{ \delta(\tau) : g_i(\tau') \geq 0, \text{ for all } \tau' \in [0, \tau], \ i = 1, 2 \right\}. \tag{7}$$

It turns out that the above optimization problem can be solved explicitly; this is done in the following lemma.

6.2.1 Lemma. Let X be a polyhedron in $\Re^n$, M be a positive semidefinite matrix, Q be a symmetric positive definite matrix, and $c > 0$ be a given scalar. Suppose that the triple $(x, y, w) \in X \times \Re^{2m}_{++}$ and scalar $\rho \in (0, 1)$ satisfy

$$y \circ w \geq \rho \mu \, e, \tag{8}$$

where $\mu \equiv y^T w / m$. Fix some $\sigma \in (0, 1)$ and let $(d\bar{x}, d\bar{y}, d\bar{w})$ be the unique optimal solution of the QP. The optimal $\bar{\tau}$ as defined by (7) is positive and is given by

$$\bar{\tau} = \begin{cases} \min(1, \tau_1, \tau_2), & \text{if } (d\bar{w})^T d\bar{y} \leq 0, \\ \min(1, \tau_1, \tau_3), & \text{if } (d\bar{w})^T d\bar{y} > 0, \end{cases} \tag{9}$$

where

$$\tau_1 \equiv \min\{\tau > 0 : g_1(\tau) = 0\},$$

$$\tau_2 \equiv \begin{cases} 1, & \text{if } (d\bar{w})^T d\bar{y} \geq 0, \\ \min\{\tau > 0 : g_2(\tau) = 0\}, & \text{otherwise,} \end{cases}$$

and

$$\tau_3 \equiv \frac{(1 - \sigma) w^T y + r(x, y, w)}{2(d\bar{w})^T d\bar{y}}.$$

Furthermore, if $\|d\bar{y} \circ d\bar{w}\| \leq \gamma y^T w$ for some constant $\gamma > 0$, then $\bar{\tau}$ is bounded below by a positive constant depending on γ, ρ, and σ only.

Proof. First of all, we have

$$w(\tau) \circ y(\tau) = (w + \tau d\bar{w}) \circ (y + \tau d\bar{y})$$

$$= w \circ y + \tau(w \circ d\bar{y} + y \circ d\bar{w}) + \tau^2 d\bar{w} \circ d\bar{y}$$

$$= (1 - \tau) w \circ y + \tau \sigma \frac{w^T y}{m} e + \tau^2 d\bar{w} \circ d\bar{y} \tag{10}$$

where the last equality follows from the perturbed Newton equation in the QP. Using (8) to bound the right-hand side of (10), we obtain

$$w(\tau) \circ y(\tau) \geq (1 - \tau)\rho\mu e + \tau\sigma\mu e + \tau^2 d\bar{w} \circ d\bar{y}. \tag{11}$$

Also, premultiplying both sides of (10) by e^T yields

$$w(\tau)^T y(\tau) = (1 - \tau)w^T y + \sigma\tau w^T y + \tau^2 (d\bar{w})^T d\bar{y}. \tag{12}$$

Since $\rho \in (0, 1)$, it follows that

$$0.5(1 + \rho) < 1 \quad \text{and} \quad 0.5(1 + \rho^{-1}) > 1.$$

Using this and the fact that $\sigma \in (0, 1)$ and $w^T y > 0$, we can apply a simple continuity argument to the previous two relations (11) and (12) and obtain

$$w(\tau) \circ y(\tau) \geq (1 - \tau)\rho\mu e + 0.5(1 + \rho)\sigma\tau\mu e$$

and

$$(1 - \tau)w^T y + 0.5(1 + \rho^{-1})\sigma\tau w^T y \geq w(\tau)^T y(\tau) \geq (1 - \tau)w^T y$$

for all small $\tau > 0$. These further imply

$$w(\tau) \circ y(\tau) \geq \rho\mu(\tau)e, \qquad \text{for all small } \tau,$$

where $\mu(\tau) = w^T(\tau)y(\tau)/m$. Therefore, both τ_1 and τ_2 exist and are positive. Also the relation (12) implies that $w(\tau)^T y(\tau) < w^T y$ for all small τ. As noted above, we have $r(x(\tau), y(\tau), w(\tau)) = (1 - \tau)r(x, y, w)$. This shows that

$$\phi(w(\tau), y(\tau), w(\tau)) < \phi(x, y, w)$$

for all $\tau > 0$ sufficiently small; thus $\delta(\tau)$ is positive for all such τ. Hence, the maximum of (7) cannot be attained at $\bar{\tau} = 0$; i.e., $\bar{\tau} > 0$.

Notice that $\delta(\tau)$ is a quadratic function of τ. If $(d\bar{w})^T d\bar{y} \leq 0$, then this quadratic function is concave and therefore it reaches its maximum at the boundary of the feasible interval, thus establishing the first relation of (9). On the other hand, if $(d\bar{w})^T d\bar{y} > 0$, then $\delta(\tau)$ is convex and it has a unique global maximum which is attained at τ_3. Also, in this case we have $g_2(\tau) \geq 0$ for all $\tau \in (0, 1]$; see (12). So, the feasible interval is

$[0, \min(1, \tau_1)]$. The maximum of $\delta(\tau)$ over this interval is clearly equal to the second relation of (9).

Finally, if $\|d\bar{w} \circ d\bar{y}\| \leq \gamma\mu$, then the above argument shows that both τ_1 and τ_2 are bounded below by a positive constant which depends on γ, ρ, and σ only. In particular, we have from (11) and (12) that

$$w(\tau) \circ y(\tau) \geq \rho\mu(\tau)e, \quad \text{for all } \tau \leq \frac{\sigma(1-\rho)}{\gamma(1+\rho)}$$

and

$$w^T(\tau)y(\tau) \geq (1-\tau)w^T y, \quad \text{for all } \tau \leq \frac{\sigma}{\gamma}.$$

Thus, $\tau_1 \geq \sigma(1-\rho)/(\gamma(1+\rho))$ and $\tau_2 \geq \sigma/\gamma$. Also, when $(d\bar{w})^T d\bar{y} > 0$, we have $\tau_3 \geq (1-\sigma)/(2\gamma)$. Consequently, $\bar{\tau}$, which is given by (9), is bounded from below by a positive constant dependent on γ, ρ, and σ only. This completes the proof of the lemma.   **Q.E.D.**

We remark that in practice we can either compute the constants τ_1 and τ_2 explicitly, which basically involves solving some simple quadratic equations, or estimate these values by using the bounds given in the proof of Lemma **6.2.1**. Due to efficiency considerations, the latter is preferred.

Let $\bar{\tau}$ be given by Lemma **6.2.1**. Clearly the triple $(x(\bar{\tau}), y(\bar{\tau}), w(\bar{\tau}))$ satisfies the properties (ii) and (iii). Since $\bar{\tau} \leq \tau_1$, we have

$$w(\tau) \circ y(\tau) \geq \rho\mu(\tau) \quad \text{for all } \tau \in [0, \bar{\tau}].$$

A simple continuity argument shows that $w(\bar{\tau}) > 0$ and $y(\bar{\tau}) > 0$. Thus, $(x(\bar{\tau}), y(\bar{\tau}), w(\bar{\tau}))$ also satisfies the property (i).

As in (6.1.9), we introduce the merit function:

$$P_\alpha(x, y, w) \equiv f(x, y) + \alpha\phi(x, y, w), \quad \text{for } (x, y, w) \in X \times \Re_+^{2m},$$

where α is the penalty parameter. The update rule of this parameter is similar to the previous one:

(**Penalty update rule**): given $\alpha_{\nu-1} \geq 1$, let p_ν be the smallest integer $p \geq 1$ such that

$$(df_x^\nu)^T dx^\nu + (df_y^\nu)^T dy^\nu - \alpha_{\nu-1}^p [(1-\sigma_\nu)(w^\nu)^T y^\nu + \|r^\nu\|] < -\phi_\nu.$$

Set $\alpha_\nu \equiv \alpha_{\nu-1}^{p_\nu}$.

As in Lemma **6.1.10**, we can establish that the function P_{α_ν} can be decreased by moving along the direction (dx^ν, dy^ν, dw^ν), starting at the vector (x^ν, y^ν, w^ν). Instead of repeating the proof, we give a detailed description of a penalty interior point algorithm for solving the problem (1). The main difference between this algorithm and the previous one described in the last section is that we include the rule of limited complementarity decrease in determining the step sizes.

A Variant of PIPA

Step 0. (Initialization) Let $c > 0, \bar\sigma, \rho_1, \gamma, \gamma', \varepsilon \in (0,1)$, and $\alpha_{-1} \geq 1$. Also let $(x^0, y^0, w^0) \in X \in \Re^{2m}_{++}$ be given. Let $\bar\rho \in (0,1)$ and $\sigma_0 \in [0,1)$ satisfy

$$\bar\rho \, \frac{(y^0)^T w^0}{m} \leq \min_{1 \leq i \leq m} y^0_i w^0_i, \quad \text{and} \quad \sigma_0 \leq \min(\bar\sigma, \bar\rho).$$

Let Q_0 be a given symmetric positive definite matrix. Set $\nu = 0$.

Step 1. (Direction generation) Let (dx^ν, dy^ν, dw^ν) be the unique optimal solution of the QP$_\nu$. Let α_ν be determined by the penalty update rule.

Step 2. (Step size determination) Find the scalar $\bar\tau_\nu \in (0,1]$ that maximizes

$$\delta_\nu(\tau) \equiv \frac{\phi_\nu - \phi(x^\nu(\tau), y^\nu(\tau), w^\nu(\tau))}{\phi_\nu}$$

subject to the conditions that for all $\tau \in (0, \bar\tau_\nu]$,

$$(w^\nu(\tau), y^\nu(\tau)) > 0, \tag{13}$$

$$w^\nu(\tau)^T y^\nu(\tau) \geq (1 - \tau)(w^\nu)^T y^\nu, \tag{14}$$

$$w^\nu(\tau) \circ y^\nu(\tau) \geq \rho w^\nu(\tau)^T y^\nu(\tau)/m. \tag{15}$$

Let ℓ_ν be the smallest nonnegative integer ℓ such that with $\tau = \bar\tau_\nu \rho_1^\ell$,

$$P_{\alpha_\nu}(x^\nu(\tau), y^\nu(\tau), w^\nu(\tau)) - P_{\alpha_\nu}(x^\nu, y^\nu, w^\nu) \leq$$

$$\gamma'\tau \left((df^\nu_x)^T dx^\nu + (df^\nu_y)^T dy^\nu - \alpha_\nu[(1 - \sigma_\nu)(w^\nu)^T y^\nu + \|r^\nu\|] \right).$$

Set $\tau_\nu \equiv \bar\tau_\nu \rho_1^{\ell_\nu}$.

Step 3. (Termination check) Define

$$(x^{\nu+1}, y^{\nu+1}, w^{\nu+1}) \equiv (x^\nu(\tau_\nu), y^\nu(\tau_\nu), w^\nu(\tau_\nu)).$$

Test this new iterate for termination. Specifically, if a prescribed stopping rule is satisfied by $(x^{\nu+1}, y^{\nu+1}, w^{\nu+1})$, then terminate; otherwise choose $Q_{\nu+1}$ to be a symmetric positive definite matrix and $\sigma_{\nu+1}$ to be a scalar satisfying $0 < \sigma_{\nu+1} \leq \sigma_\nu$ and return to Step 1 with ν replaced by $\nu + 1$.

As before, a possible stopping rule to be used in Step 3 above is

$$\|(dx^\nu, dy^\nu, dw^\nu)\| \leq \text{tolerance}.$$

In what follows, we give a convergence result for the above algorithm. Let $\{(x^\nu, y^\nu, w^\nu)\}$ be an infinite sequence produced by the algorithm. This sequence must satisfy for each $\nu, (x^\nu, y^\nu, w^\nu) \in X \times \Re^{2m}_{++}$, and

$$\|q + Nx^\nu + My^\nu - w^\nu\| \leq \|q + Nx^0 + My^0 - w^0\|.$$

By using Hoffman's error bound for polyhedra, we can establish the boundedness of the sequence $\{(x^\nu, y^\nu, w^\nu)\}$.

6.2.2 Lemma. If the set

$$G \equiv \{(x,y) \in X \times \Re^m_+ : q + Nx + My \geq 0\} \tag{16}$$

is bounded, then so is $\{(x^\nu, y^\nu, w^\nu)\}$.

Proof. To see this, we consider

$$\begin{aligned}
\|[q + Nx^\nu + My^\nu]_+\| &\leq \|[q + Nx^\nu + My^\nu - w^\nu]_+\| \\
&\leq \|q + Nx^\nu + My^\nu - w^\nu\| \\
&\leq \|q + Nx^0 + My^0 - w^0\|
\end{aligned}$$

where the first inequality is due to $w^\nu > 0$, the second inequality follows from the nonexpansive property of the projection operator $[\cdot]_+$. Thus, by Hoffman's error bound, the distance from (x^ν, y^ν) to G is bounded, which further implies (x^ν, y^ν) is bounded since G is bounded. Moreover, by the relation

$$\|w^\nu\| \leq \|q + Nx^\nu + My^\nu\| + \|q + Nx^0 + My^0 - w^0\|,$$

it follows that w^ν is also bounded. **Q.E.D.**

In what follows, we assume that there exist constants $c_1, c_2 > 0$ such that

$$c_1\|x\|^2 \le x^T Q_\nu x \le c_2\|x\|^2, \quad \text{for all } x \in \Re^n \text{ and all } \nu. \tag{17}$$

Notice that if the step size constraint $(dx)^T dx \le c\phi_\nu$ is included in the quadratic subproblem QP_ν, then we can let $c_1 = 0$ just as in the original version of PIPA (cf. Section 6.1.2). As in Theorems **6.1.13** and **6.1.17**, we can establish the following convergence result whose proof is omitted.

6.2.3 Theorem. Let M be positive semidefinite. Suppose the sequence $\{Q_\nu\}$ is bounded according to (17).

(a) If $\{\alpha_\nu\}$ is unbounded, then

$$\lim_{\nu(\in\kappa)\to\infty} dx^\nu = 0,$$

where $\kappa \equiv \{\nu : \alpha_{\nu-1} < \alpha_\nu\}$. Furthermore, if (x^*, y^*, w^*) is any limit point of the subsequence $\{(x^\nu, y^\nu, w^\nu) : \nu \in \kappa\}$ satisfying (i) $y^* + w^* > 0$, and (ii) M_{JJ} is nonsingular where $J \equiv \{i : y_i^* > 0\}$, then (x^*, y^*, w^*) is a stationary point of the problem (1).

(b) If $\{\sigma_\nu\}$ is bounded away from zero and if $\{\alpha_\nu\}$ is bounded, then any limit point $\{(x^*, y^*, w^*)\}$ of the sequence $\{(x^\nu, y^\nu, w^\nu)\}$ satisfying (i) and (ii) is a stationary point of (1).

6.3 An Implicit Programming Based Algorithm

We consider the MPEC

$$\begin{aligned} \text{minimize} \quad & f(x,y) \\ \text{subject to} \quad & (x,y) \in \mathcal{F} \equiv (X \times \Re^m) \cap \text{Gr}(\mathcal{S}), \end{aligned} \tag{1}$$

where $f : \Re^{n+m} \to \Re$ is a continuously differentiable real-valued function, X is a closed convex set in $\Re^n$, and $\mathcal{S} : X \to \Re^m$ is the set-valued solution map of the parametric VI $(F(x,\cdot), C(x))$ for $x \in X$, with $F : \Re^{n+m} \to \Re^m$ being a continuously differentiable vector-valued mapping and $C(x)$ being a closed convex subset of $\Re^m$.

Based on the implicit programming formulation (4.2.2), we develop a model descent algorithm for computing a stationary point of (1), under a global version of the BIF assumption introduced in Subsection **4.2.2**. The idea of using (4.2.2) as a computational vehicle for solving (1) first appeared in the paper [225]. The paper [215] discusses some related methods based on the computation of "generalized Jacobian matrices" of the implicit function $y(x)$.

For ease of reference, we formally state the blanket condition, which we call *global* BIF, assumed to be in force throughout this section.

(GBIF) For each $(\bar{x}, \bar{y}) \in \mathcal{F}$, there exist a neighborhood $V \times U$ of $(\bar{x}, \bar{y})$ and a Lipschitz continuous function $y : V \cap X \to U$ such that $y(\bar{x}) = \bar{y}$, $y(\cdot)$ is directionally differentiable at $\bar{x}$, and for each $x \in V \cap X$, $y(x)$ is the solution of the VI $(F(x, \cdot), C(x))$ in U. (This implicit function y is thus B-differentiable at $\bar{x}$.)

In Section **4.2**, we discussed extensively the role of the assumption BIF. In particular, we recall Lemma **4.2.5** which characterizes the stationarity of a given feasible pair $(\bar{x}, \bar{y}) \in \mathcal{F}$ in terms of the directional derivative $y'(\bar{x}; dx)$ of the implicit function $y(\cdot)$ in condition GBIF.

6.3.1 *Algorithm description and convergence*

Suppose that an iterate $(x^\nu, y^\nu) \in \mathcal{F}$ is given. We wish to compute a descent direction of f at this iterate. Let V_ν, U_ν, and $\bar{y}^\nu$ be, respectively, the neighborhoods of x^ν, y^ν, and the implicit function asserted by condition GBIF at (x^ν, y^ν). Let $Q_\nu \in \Re^{n \times n}$ be a symmetric positive definite matrix. Write

$$df_s^\nu \equiv \nabla_s f(x^\nu, y^\nu), \quad \text{for } s = x, y.$$

Consider the following constrained minimization problem in the variable $dx \in \Re^n$:

$$\text{minimize} \quad (df_x^\nu)^T dx + (df_y^\nu)^T (\bar{y}^\nu)'(x^\nu; dx) + \tfrac{1}{2}(dx)^T Q_\nu dx \tag{2}$$
$$\text{subject to} \quad x^\nu + dx \in X.$$

Subsequently, we discuss the practical aspects of solving this problem. Here we summarize some important properties of the problem.

6.3.1 Lemma. Let X be a closed convex set and $(x^\nu, y^\nu) \in \mathcal{F}$ be given. Let Q_ν be a symmetric positive definite matrix.

(a) The problem (2) has a globally optimal solution, and the optimum objective value is nonpositive.

(b) If dx^ν is any nonzero optimal solution of (2), then for any scalar $\gamma \in (0,1)$, there exists a scalar $\bar\tau_\nu \in (0,1]$ such that for all $\tau \in [0, \bar\tau_\nu]$, $x^\nu(\tau) \equiv x^\nu + \tau dx^\nu \in X \cap V_\nu$ and

$$f(x^\nu(\tau), \bar y^\nu(x^\nu(\tau))) - f(x^\nu, y^\nu) \le$$
$$\gamma\tau \left((df_x^\nu)^T dx^\nu + (df_y^\nu)^T (\bar y^\nu)'(x^\nu; dx^\nu) \right) < 0.$$

Proof. Clearly $dx = 0$ is feasible to (2). Since the function $\bar y^\nu(x)$ is Lipschitz continuous, there exists a constant $L > 0$ such that

$$\|(\bar y^\nu)'(x^\nu; dx)\| \le L\|dx\|, \quad \text{for all } dx \in \Re^n.$$

Hence the objective function of (2) is coercive; thus an optimal solution to (2) exists. Moreover, the optimal objective value is nonpositive.

If dx^ν is a nonzero optimal solution of (2), then

$$(df_x^\nu)^T dx^\nu + (df_y^\nu)^T (\bar y^\nu)'(x^\nu; dx^\nu) <$$
$$(df_x^\nu)^T dx^\nu + (df_y^\nu)^T (\bar y^\nu)'(x^\nu; dx^\nu) + \tfrac{1}{2}(dx^\nu)^T Q_\nu dx^\nu \le 0.$$

Moreover since

$$\lim_{\tau \downarrow 0} \frac{f(x^\nu(\tau), \bar y^\nu(x^\nu(\tau))) - f(x^\nu, y^\nu)}{\tau} = (df_x^\nu)^T dx^\nu + (df_y^\nu)^T (\bar y^\nu)'(x^\nu; dx^\nu),$$

which holds because of the B-differentiability of the solution function $\bar y^\nu$ at x^ν, the existence of the scalar $\bar\tau_\nu$ follows easily.   **Q.E.D.**

In spite of the positive definiteness of the matrix Q_ν, the objective function of (2) is not necessarily strictly convex (in fact, not even convex in general); this is because of the possible lack of convexity of the function $(\bar y^\nu)'(x^\nu; \cdot)$. Consequently, the assumptions in Lemma **6.3.1** are not enough to imply that the optimal solution to (2) is unique. Nevertheless, if the function y is Fréchet differentiable at $\bar x$, then this directional derivative is linear in its second argument. Hence in the latter case (2) is a strictly convex minimization problem with a quadratic objective function; thus a unique optimal solution exists under the assumptions of Lemma **6.3.1**. If in addition X is polyhedral, then (2) is a strictly convex quadratic program.

Part (b) of Lemma **6.3.1** states that if (2) has a nonzero optimal solution, then such a solution will provide a feasible descent direction for the objective function f at the current iterate x^ν. The following result gives several necessary and sufficient conditions for the case where such an optimal solution does not exist.

6.3.2 Proposition. Let X be a polyhedron and $(x^\nu, y^\nu) \in \mathcal{F}$ be given. Let Q_ν be a symmetric positive definite matrix. the following statements are equivalent.

(a) $dx = 0$ is the only globally optimal solution of (2).

(b) $dx = 0$ is a globally optimal solution of (2).

(c) The optimum objective value of (2) is zero.

(d) (x^ν, y^ν) is a stationary point of (1).

Proof. The implications (a) $\Rightarrow$ (b) $\Rightarrow$ (c) are obvious.

(c) $\Rightarrow$ (d). Since the optimum objective value of (2) is zero, we have for all $dx \in X - x^\nu$ and all scalars $\tau > 0$ sufficiently small,

$$(df_x^\nu)^T(\tau dx) + (df_y^\nu)^T(\bar{y}^\nu)'(x^\nu; \tau dx) + \frac{\tau^2}{2}dx^T Q_\nu dx \geq 0.$$

Dividing for τ and passing to the limit $\tau \downarrow 0$, we deduce

$$(df_x^\nu)^T dx + (df_y^\nu)^T(\bar{y}^\nu)'(x^\nu; dx) \geq 0.$$

Since X is polyhedral, we have $\mathcal{T}(\bar{x}; X) = \bigcup_{\tau > 0} \tau(X - \bar{x})$; consequently, the last inequality holds for all $dx \in \mathcal{T}(\bar{x}; X)$. The stationarity of (x^ν, y^ν) follows from Lemma **4.2.5**.

(d) $\Rightarrow$ (a). Suppose that dx is a nonzero optimal solution of (2). Then as in the proof of Lemma **6.3.1**, we have

$$(df_x^\nu)^T dx + (df_y^\nu)^T(\bar{y}^\nu)'(x^\nu; dx) < 0.$$

Since $dx \in \mathcal{T}(\bar{x}; X)$, we deduce that (x^ν, y^ν) is not a stationary point of (1). This is a contradiction.   **Q.E.D.**

Proposition **6.3.2** implies that if the problem (2) has a zero optimum objective value, then the current pair (x^ν, y^ν) is a desired stationary point of the MPEC (1). If the optimum objective value of (2) is negative, then we can successfully decrease the objective function f from its current value at (x^ν, y^ν) by executing an Armijo-type search on f starting at this pair of iterates and moving along the curve $(x^\nu(\tau), \bar{y}^\nu(x^\nu(\tau)))$ for $\tau \in [0, 1]$, where dx^ν is any optimal solution of (2). Since the function $\bar{y}^\nu$ is generally nonlinear, this search is not along a line segment.

The following is a detailed description of a descent algorithm for computing a stationary point of (1) based on the above ideas.

A Descent Algorithm

Step 0. (Initialization) Let $\rho, \gamma \in (0, 1)$ be given scalars and Q_0 be a given symmetric positive definite matrix. Let $(x^0, y^0) \in \mathcal{F}$ be arbitrary. Set $k = 0$.

Step 1. (Direction generation) Solve the subproblem (2) to obtain a (globally) optimal solution dx^ν. If $dx^\nu = 0$, stop. The current pair (x^ν, y^ν) is a stationary point of (1). Otherwise continue.

Step 2. (Step size determination) Let ℓ_ν be the smallest nonnegative integer ℓ such that with $\tau = \rho^\ell$,

$$
\begin{aligned}
f(x^\nu(\tau), \bar{y}^\nu(x^\nu(\tau))) - f(x^\nu, y^\nu) \le \\
\gamma \tau \left((df_x^\nu)^T dx^\nu + (df_y^\nu)^T (\bar{y}^\nu)'(x^\nu; dx^\nu) \right).
\end{aligned}
\tag{3}
$$

Set $\tau_\nu \equiv \rho^{\ell_\nu}$ and $(x^{\nu+1}, y^{\nu+1}) \equiv (x^\nu(\tau_\nu), \bar{y}^\nu(x^\nu(\tau_\nu)))$. Choose a symmetric positive definite matrix $Q_{\nu+1}$ and return to Step 1 with ν replaced by $\nu+1$.

A practical termination rule to be used in Step 1 is

$$
\|dx^\nu\| \le \text{tolerance}.
$$

This rule is motivated by Proposition **6.3.2**. With the assumption that the algorithm does not terminate finitely in Step 1, the results derived above justify that an infinite sequence of iterates $\{(x^\nu, y^\nu)\} \subset \mathcal{F}$ is generated; moreover, the sequence of objective values $\{f(x^\nu, y^\nu)\}$ is strictly decreasing. Hence if the level set

$$
\{(x, y) \in \mathcal{F} : f(x, y) \le f(x^0, y^0)\}
\tag{4}
$$

is bounded, then so is the sequence $\{(x^\nu, y^\nu)\}$. Let (x^*, y^*) be the limit of a convergent subsequence $\{(x^\nu, y^\nu) : \nu \in \kappa\}$. Our goal is to show that (x^*, y^*) is a stationary point of (1) under certain additional assumptions.

Throughout the following analysis, we assume that X is a polyhedron. Then clearly $x^* \in X$. In order to ensure that (x^*, y^*) is feasible to (1), we impose the following *closed mapping* assumption:

(**CM**) the solution map $\mathcal{S}$ is closed.

As for GBIF, we postpone the discussion of this assumption (CM) until later; see Theorem **1.3.4**. Here we note that (CM) implies that $(x^*, y^*) \in \mathcal{F}$ because

$$y^* = \lim_{\nu(\in \kappa) \to \infty} \bar{y}^\nu(x^\nu)$$

and each $\bar{y}^\nu(x^\nu) \in \mathcal{S}(x^\nu)$. Thus condition GBIF implies that there exists an implicit solution function $\bar{y}^*(x)$ of the parametric VI $(F(x, \cdot), C(x))$ in a neighborhood of (x^*, y^*). As in the case of PIPA, we also need the following *boundedness* assumption on the matrices $\{Q_\nu\}$:

(**BD**) there exist constants $c_1, c_2 > 0$ such that for all ν,

$$c_1 \|x\|^2 \leq x^T Q_\nu x \leq c_2 \|x\|^2, \quad \text{for all } x \in \Re^n.$$

The next result asserts a technical property of the tangent cone associated with a polyhedron.

6.3.3 Lemma. Let X be a polyhedron and let $\bar{x} \in X$. If $d \in \mathcal{T}(\bar{x}; X)$, then there exist positive scalars $\bar{\varepsilon}, \delta$ such that $y + \varepsilon d \in X$ for every $\varepsilon \in [0, \bar{\varepsilon}]$ and every $y \in X$ such that $\|y - \bar{x}\| \leq \delta$.

Proof. Write $X = \{x : Ax \leq b\}$. Then

$$\mathcal{T}(\bar{x}; X) = \bigcap_{i \in \mathcal{I}(\bar{x})} \{d : (Ad)_i \leq 0\},$$

where $\mathcal{I}(\bar{x})$ is the index set of the binding constraints at $\bar{x}$. Using this representation for the tangent cone, we can establish the existence of $\bar{\varepsilon}$ and δ by a simple continuity argument.   **Q.E.D.**

We now state the main convergence result of the descent algorithm for computing a stationary point of (1).

6.3.4 Theorem. Let X be a polyhedron in $\Re^n$ and $f : \Re^{n+m} \to \Re$ be continuously differentiable on an open set containing $\mathcal{F}$. Assume that conditions GBIF, CM, and BD hold. Suppose that the set (4) is bounded.

Let (x^*, y^*) be the limit of a convergent subsequence $\{(x^\nu, y^\nu) : \nu \in \kappa\}$ generated by the above descent algorithm. Let $\bar{y}^*$ be the implicit solution function at (x^*, y^*) asserted by condition GBIF. If $\bar{y}^*$ is strongly Fréchet differentiable at x^*, then (x^*, y^*) is a stationary point of (1).

Proof. By CM, the set $\mathcal{F}$ is closed. Hence the set (4) is compact; thus the problem (1) has an optimal solution. In particular, f is bounded below on $\mathcal{F}$. Since the sequence of objective values $\{f(x^\nu, y^\nu)\}$ is strictly decreasing and bounded below; it therefore converges. Thus

$$\lim_{\nu \to \infty} (f(x^{\nu+1}, y^{\nu+1}) - f(x^\nu, y^\nu)) = 0,$$

which implies

$$\lim_{\nu \to \infty} \tau_\nu \left((df_x^\nu)^T dx^\nu + (df_y^\nu)^T (\bar{y}^\nu)'(x^\nu; dx^\nu) \right) = 0.$$

Let (x^*, y^*) be the limit of the subsequence $\{(x^\nu, y^\nu) : \nu \in \kappa\}$. There are two cases depending on whether $\liminf_{\nu(\in\kappa) \to \infty} \tau_\nu$ is positive or zero. In the former case, we must have

$$\lim_{\nu(\in\kappa) \to \infty} \left((df_x^\nu)^T dx^\nu + (df_y^\nu)^T (\bar{y}^\nu)'(x^\nu; dx^\nu) \right) = 0.$$

Since

$$(df_x^\nu)^T dx^\nu + (df_y^\nu)^T (\bar{y}^\nu)'(x^\nu; dx^\nu) + \tfrac{1}{2}(dx^\nu)^T Q_\nu dx^\nu \leq 0 \qquad (5)$$

for all ν, by BD we deduce

$$\lim_{\nu(\in\kappa) \to \infty} dx^\nu = 0. \qquad (6)$$

According to Lemma **4.2.5**, it suffices to show that for all vectors dx in $\mathcal{T}(x^*; X)$,

$$\nabla_x f(x^*, y^*)^T dx + \nabla_y f(x^*, y^*)^T (\bar{y}^*)'(x^*; dx) \geq 0. \qquad (7)$$

Let $dx \in \mathcal{T}(x^*; X)$ be arbitrary. By Lemma **6.3.3**, there exists a scalar $\varepsilon > 0$ such that $x^\nu + \varepsilon dx \in X$ for all $\nu \in \kappa$ sufficiently large and all $\varepsilon \in [0, \bar{\varepsilon}]$. For such a pair (ν, ε), we have

$$
\begin{aligned}
(df_x^\nu)^T dx^\nu + (df_y^\nu)^T (\bar{y}^\nu)'(x^\nu; dx^\nu) + \tfrac{1}{2}(dx^\nu)^T Q_\nu dx^\nu \leq \\
\varepsilon \left((df_x^\nu)^T dx + (df_y^\nu)^T (\bar{y}^\nu)'(x^\nu; dx) \right) + \tfrac{\varepsilon^2}{2} dx^T Q_\nu dx.
\end{aligned}
\qquad (8)
$$

In view of the assumption BD, we may assume without loss of generality that the sequence of matrices $\{Q_\nu\}$ converges to some limit Q_*.

For each ν, we have by the definition of the directional derivative,

$$(\bar{y}^\nu)'(x^\nu; dx) = \lim_{\tau \to 0+} \frac{\bar{y}^\nu(x^\nu + \tau dx) - y^\nu}{\tau}.$$

For each $\tau > 0$ sufficiently small, $\bar{y}^\nu(x^\nu + \tau dx)$ is the unique solution of the VI $(F(x^\nu + \tau dx, \cdot), C(x^\nu + \tau dx))$ that lies in the neighborhood U_ν of y^ν asserted by the key assumption GBIF at the pair (x^ν, y^ν). If $V_* \times U_*$ denotes the neighborhood of (x^*, y^*) asserted by GBIF at this pair, then for all $\nu \in \kappa$ sufficiently large, we have $y^\nu \in U_*$. For each such ν, since $\bar{y}^\nu(x^\nu + \tau dx) \to y^\nu$ as $\tau \downarrow 0$, it follows that there exists a $\underline{\tau}_\nu > 0$ such that $\bar{y}^\nu(x^\nu + \tau dx) \in U_*$ for all $\tau \in [0, \underline{\tau}_\nu]$. However, since $\bar{y}^*(x^\nu + \tau dx)$ is the unique solution of the VI $(F(x^\nu + \tau dx, \cdot), C(x^\nu + \tau dx))$ that lies in the neighborhood U_*, it follows that

$$\bar{y}^\nu(x^\nu + \tau dx) = \bar{y}^*(x^\nu + \tau dx)$$

for all $\tau \in [0, \underline{\tau}_\nu]$. Hence for each $\nu \in \kappa$ sufficiently large,

$$(\bar{y}^\nu)'(x^\nu; dx) = (\bar{y}^*)'(x^\nu; dx).$$

Taking limits $\nu(\in \kappa) \to \infty$ in (8) and using (6), we deduce that for each $\varepsilon \in [0, \bar{\varepsilon}]$,

$$\lim_{\nu(\in\kappa) \to \infty} \left[\varepsilon((df_x^\nu)^T dx + (df_y^\nu)^T (\bar{y}^*)'(x^\nu; dx)) + \frac{\varepsilon^2}{2} dx^T Q_\nu dx \right] \geq 0.$$

Since $\bar{y}^*$ is by assumption strongly F-differentiable at x^*, part (e) of Proposition **4.2.2** implies that

$$\lim_{\nu(\in\kappa) \to \infty} (\bar{y}^*)'(x^\nu; dx) = (\bar{y}^*)'(x^*; dx).$$

Consequently it follows that

$$(df_x^*)^T dx + (df_y^*)^T (\bar{y}^*)'(x^*; dx) + \frac{\varepsilon}{2} dx^T Q_* dx \geq 0,$$

for all $\varepsilon \in [0, \bar{\varepsilon}]$, where $df_s^* \equiv \nabla_s f(x^*, y^*)$ for $s = x, y$. Letting $\varepsilon \downarrow 0$, we obtain the desired inequality (7).

Consider the other case where $\liminf_{\nu(\in\kappa)\to\infty}\tau_\nu = 0$, or equivalently, $\limsup_{\nu(\in\kappa)\to\infty}\ell_\nu = \infty$. Without loss of generality, we may assume that

$$\lim_{\nu(\in\kappa)\to\infty}\ell_\nu = \infty.$$

Hence for all $\nu \in \kappa$ sufficiently large, the inequality (3) is violated at $\tau'_\nu \equiv \rho^{\ell_\nu - 1}$. Writing

$$u^\nu \equiv x^\nu(\tau'_\nu) \quad \text{and} \quad w^\nu \equiv \bar{y}^\nu(u^\nu),$$

we have

$$\frac{f(u^\nu, w^\nu) - f(x^\nu, y^\nu)}{\tau'_\nu} > \gamma\left((df^\nu_x)^T dx^\nu + (df^\nu_y)^T (\bar{y}^\nu)'(x^\nu; dx^\nu)\right). \quad (9)$$

We first show that $\{dx^\nu : \nu \in \kappa\}$ is bounded. As before, we have for all $\nu \in \kappa$ sufficiently large,

$$(\bar{y}^\nu)'(x^\nu; dx^\nu) = (\bar{y}^*)'(x^\nu; dx^\nu) = \lim_{\tau\to 0+}\frac{\bar{y}^*(x^\nu + \tau dx^\nu) - \bar{y}^*(x^\nu)}{\tau};$$

thus by the Lipschitz continuity of the implicit function $\bar{y}^*(x)$, it follows that there exists a constant $L > 0$ such that for all $\nu \in \kappa$ sufficiently large,

$$\|(\bar{y}^\nu)'(x^\nu; dx^\nu)\| \leq L\|dx^\nu\|. \quad (10)$$

Similarly, we can establish

$$\|w^\nu - y^\nu\| \leq L\tau'_\nu\|dx^\nu\|.$$

Since (5) holds for all ν, by BD and (10) it follows that $\{dx^\nu : \nu \in \kappa\}$ must be bounded; hence so is

$$\{(df^\nu_x)^T dx^\nu + (df^\nu_y)^T (\bar{y}^\nu)'(x^\nu; dx^\nu) : \nu \in \kappa\}.$$

By the C^1 property of f, we may write

$$f(u^\nu, w^\nu) - f(x^\nu, y^\nu) = (df^\nu_x)^T(u^\nu - x^\nu) + (df^\nu_y)^T(w^\nu - y^\nu) + o(\tau'_\nu)$$

$$= \tau'_\nu\left((df^\nu_x)^T dx^\nu + (df^\nu_y)^T (\bar{y}^\nu)'(x^\nu; dx^\nu)\right)+$$

$$(df^\nu_y)^T(w^\nu - y^\nu - \tau'_\nu(\bar{y}^\nu)'(x^\nu; dx^\nu)) + o(\tau'_\nu).$$

Consequently, by (9), we deduce

$$(1 - \gamma)\tau'_\nu \left((df^\nu_x)^T dx^\nu + (df^\nu_y)^T (\bar{y}^\nu)'(x^\nu; dx^\nu) \right) >$$

$$-(df^\nu_y)^T (w^\nu - y^\nu - \tau'_\nu (\bar{y}^\nu)'(x^\nu; dx^\nu)) + o(\tau'_\nu).$$

We have

$$\frac{w^\nu - y^\nu - \tau'_\nu (\bar{y}^\nu)'(x^\nu; dx^\nu)}{\tau'_\nu} = \frac{\bar{y}^*(x^\nu(\tau'_\nu)) - \bar{y}^*(x^\nu) - \tau'_\nu (\bar{y}^*)'(x^\nu; dx^\nu)}{\tau'_\nu},$$

and

$$\lim_{\nu(\in\kappa)\to\infty} \frac{\bar{y}^*(x^\nu(\tau'_\nu)) - \bar{y}^*(x^\nu) - \tau'_\nu (\bar{y}^*)'(x^\nu; dx^\nu)}{\tau'_\nu} = 0$$

by the strong F-differentiability of $\bar{y}^*$ at x^* and Proposition **4.2.2**(e). Consequently, since $\gamma < 1$, we deduce

$$\liminf_{\nu(\in\kappa)\to\infty} \left((df^\nu_x)^T dx^\nu + (df^\nu_y)^T (\bar{y}^\nu)'(x^\nu; dx^\nu) \right) \geq 0.$$

From (5), it follows that

$$\limsup_{\nu(\in\kappa)\to\infty} \left((df^\nu_x)^T dx^\nu + (df^\nu_y)^T (\bar{y}^\nu)'(x^\nu; dx^\nu) \right) \leq 0.$$

Thus

$$\lim_{\nu(\in\kappa)\to\infty} \left((df^\nu_x)^T dx^\nu + (df^\nu_y)^T (\bar{y}^\nu)'(x^\nu; dx^\nu) \right) = 0;$$

we are now back to the previous case. **Q.E.D.**

6.3.2 *Implementation issues*

The numerical implementation of the descent algorithm discussed in the last section is not a trivial matter. In the following discussion, we assume that the set $C(x)$ is represented by differentiable, convex inequalities as in (1.3.3) and the results in Chapter **3** are applicable. In particular, we assume that MFCQ, CRCQ, and SCOC hold at all feasible points of (1). By the proof of Theorem **1.3.4**, it follows that the solution map $\mathcal{S}$ is closed; thus assumption CM holds.

First, the computation of the search direction dx^ν in each iteration requires the solution of the piecewise quadratic program (2) which is not necessarily convex, unless the implicit function $\bar{y}^\nu(x)$ is F-differentiable at $\bar{x}^\nu$. In this case, we know from Theorem **4.2.28** that $(\bar{y}^\nu)'(\bar{x}; dx)$ is a linear function of dx; hence (2) is a strictly convex quadratic program. In general, $(\bar{y}^\nu)'(\bar{x}; dx)$ is a solution of an AVI parametrized by dx (see Theorem **4.2.25**). More specifically, (2) is equivalent to the following minimization problem in the variable $(dx, dy, d\mu_i : i \in \mathcal{I}^\nu)$:

minimize $\quad (df_x^\nu)^T dx + (df_y^\nu)^T dy + \frac{1}{2}(dx)^T Q_\nu dx$

subject to $\quad x^\nu + dx \in X$,

$$\nabla_z L(x^\nu, y^\nu, \lambda)(dx, dy) + \sum_{i \in \mathcal{I}^\nu} d\mu_i \nabla_y g_i(x^\nu, y^\nu) = 0,$$

$$\left. \begin{array}{r} dx^T \nabla_x g_i(x^\nu, y^\nu) + dy^T \nabla_y g_i(x^\nu, y^\nu) \leq 0 \\[2mm] d\mu_i \geq 0 \\[2mm] d\mu_i \left(dx^T \nabla_x g_i(x^\nu, y^\nu) + dy^T \nabla_y g_i(x^\nu, y^\nu) \right) = 0 \end{array} \right\} \; \forall i \in \mathcal{I}^\nu \text{ with } \lambda_i = 0$$

$$dx^T \nabla_x g_i(x^\nu, y^\nu) + dy^T \nabla_y g_i(x^\nu, y^\nu) = 0, \; \forall i \text{ such that } \lambda_i > 0,$$

where $\mathcal{I}^\nu \equiv \{i : g_i(x^\nu, y^\nu) = 0\}$, and λ is an arbitrary element in $M^e(x^\nu)$. Clearly the latter problem is an MP with mixed LCP constraints and a quadratic objective function. The difficulty of solving this MPAEC depends on the degeneracy of the multiplier λ chosen. If there are only a small number of indices $i \in \mathcal{I}^\nu$ with $\lambda_i = 0$, then we can solve the above direction-finding subproblem by complete enumeration of the complementarity constraints corresponding to these degenerate indices. This enumerative scheme is certainly not practical if λ is highly degenerate.

Next, after the direction dx^ν is computed, there remains the computation of the step size τ_ν. This requires the repeated evaluation of $\bar{y}^\nu(x^\nu(\tau))$ which amounts to solving the VI $(F(x^\nu(\tau), \cdot), C(x^\nu(\tau)))$ for different values of τ. As long as the number of these line searches is reasonable, and the VIs are not difficult to solve, Step 2 of the algorithm is a plausible calculation.

A third implementation issue of the descent algorithm is the choice of the matrices $\{Q_\nu\}$. Presumably, the matrix Q_ν should include some high-order derivatives of the functions f, F, and g at the current iterate (x^ν, y^ν). Due to the nonsmooth nature of the implicit function $y^\nu(x)$, a good choice for Q_ν is generally not easy. Much needs to be done on this issue.

In summary, the descent algorithm described in the last subsection remains a conceptual algorithm at this time. For it to become a practical method, many implementation issues need to be addressed. The approach described in the next section offers a promising avenue to deal with some of these issues.

6.4 A Piecewise SQP Approach

We present a piecewise programming based sequential quadratic programming method, called PSQP, for solving MP with KKT constraints. It is based on the sequential quadratic programming (SQP) method for smooth nonlinear programs [204, 30, 222] extended to a piecewise smooth setting. The PSQP method is an extension of a slightly simpler method [239] designed for MP with linear complementarity constraints.

6.4.1 A brief review of SQP methods for NLP

We begin with a sketch of SQP and some of its convergence properties. Continuing from the review of nonlinear programming in Section **5.1**, we repeat the nonlinear program (5.1.1),

$$\text{minimize} \quad f(x)$$

$$\text{subject to} \quad g_i(x) = 0, \ i = 1, ..., \ell, \tag{1}$$

$$g_i(x) \leq 0, \ i = \ell + 1, ..., \ell + m,$$

where f and $g = (g_1, ..., g_{\ell+m})$ are C^2 mappings defined on an open set in $\Re^n$ that contains the feasible region of (1), which we denote $\mathcal{F}^{\text{NLP}}$. Suppose $\bar{x} \in \mathcal{F}^{\text{NLP}}$ is a stationary point of (1). Let $M(\bar{x})$ be the set of corresponding Karush-Kuhn-Tucker multipliers $\pi \in \Re^{\ell+m}$, and $\mathcal{A}(\bar{x}) \equiv \{i : g_i(\bar{x}) = 0\}$. The Lagrangean function of (1) is

$$\mathcal{L}^{\text{NLP}}(x, \pi) \equiv f(x) + g(x)^T \pi.$$

We are particularly interested in the case $M(\bar{x}) = \{\bar{\pi}\}$; that is, when there is a unique KKT multiplier corresponding to a stationary point $\bar{x}$. As discussed in Section **3.2**, this is equivalent [148] to the strict Mangasarian-Fromovitz constraint qualification, SMFCQ, which asserts the existence of a KKT multiplier $\bar{\pi}$ such that the set $\{\nabla g_i(\bar{x}) : i \in \mathcal{A}_+(\bar{x}, \bar{\pi})\}$ consists of linearly independent vectors, where

$$\mathcal{A}_+(\bar{x}, \bar{\pi}) \equiv \{i : i = 1, ..., \ell \text{ or } \bar{\pi}_i > 0\};$$

and there exists $dx \in \Re^n$ such that

$$\nabla g_i(\bar{x})^T dx = 0, \quad i \in \mathcal{A}_+(\bar{x}, \bar{\pi}),$$

$$\nabla g_i(\bar{x})^T dx < 0, \quad i \in \mathcal{A}(\bar{x}) \setminus \mathcal{A}_+(\bar{x}, \bar{\pi}).$$

Define $\mathcal{A}_0(\bar{x}, \bar{\pi}) \equiv \mathcal{A}(\bar{x}) \setminus \mathcal{A}_+(\bar{x}, \bar{\pi})$.

Newton type methods for solving the NLP (1) originated from Wilson's dissertation [285]. Robinson [245] established the quadratic rate of convergence of these methods using the theory of perturbed KKT points. Since the main computational efforts of these methods are the solution of a sequence of quadratic programs, the term Sequential Quadratic Programming (SQP) was coined. Extensions of the SQP methods and their convergence theory to generalized equations were introduced by Josephy [127, 128], based to a large extent on the theory of strong regularity introduced by Robinson [248]. Though published a few years later, Eaves' paper [66] which considered a Newton method for solving a special class of variational inequalities arising from a market equilibrium problem was announced around the same time as Josephy's reports. Collectively, these papers have had a tremendous impact on the development of fast and robust algorithms for solving finite-dimensional variational inequalities and nonlinear complementarity problems; we refer the reader to [109, 220, 224] for a detailed historical account of these algorithms.

Just like the classical Newton method for solving a system of smooth equations [211], the early SQP methods for solving the NLP (1) were only locally convergent in the sense that their convergence could only be established if the initial iterate was assumed to be near a (strict) local minimizer. In recent years, improved local convergence results for the SQP methods were obtained using the solution stability theory of parametric NLPs and VIs; see [29, 30, 222].

In spite of the lack of global convergence, the locally convergent SQP methods have a superlinear rate of convergence which becomes quadratic under a further assumption, a principal reason why they have been the basis for many effective algorithms used in practice. S.P. Han [105] was among the pioneers who proposed a globalization scheme of the locally convergent Newton methods for solving smooth NLPs, using a penalty function approach; Burke [40, 42] and Burke and Han [43] have made many improvements to these SQP methods in terms of their globally convergent properties and practical implementation. These global SQP methods are not considered in this section.

The following is a summary of a local version of an SQP method for solving the NLP (1).

An SQP Method for NLPs

Step 0. (Initialization) Let $x^0 \in \Re^n$ and $\pi^0 \in \Re^{\ell+m}$. Set $\nu = 0$.

Step 1. Let $(dx^\nu, \pi^{\nu+1}) \in \Re^{n+m+\ell}$ be a KKT pair of the following quadratic program:

$$\begin{aligned}
\text{minimize} \quad & \nabla f(x^\nu)^T dx + \tfrac{1}{2} dx^T \nabla^2_{xx} \mathcal{L}^{\text{NLP}}(x^\nu, \pi^\nu) dx \\
\text{subject to} \quad & g_i(x^\nu) + \nabla g_i(x^\nu)^T dx = 0, \ i = 1, \ldots, \ell, \\
& g_i(x^\nu) + \nabla g_i(x^\nu)^T dx \leq 0, \ i = \ell+1, \ldots, \ell+m.
\end{aligned} \qquad (2)$$

Step 2. Set $x^{\nu+1} \equiv x^\nu + dx^\nu$ and check prescribed rule for termination. Return to Step 1 with ν replaced by $\nu + 1$ if rule is not satisfied.

For a given pair of vectors $(x, \pi) \in \Re^{n+m+\ell}$, we let $\text{QP}(x, \pi)$ denote the quadratic program (2) with (x^ν, π^ν) substituted by (x, ν). As part of the justification for the above method, we need to establish the well-definedness of each iterate $(x^{\nu+1}, \pi^{\nu+1})$; the anticipated (local) convergence of the sequence $\{(x^\nu, \pi^\nu)\}$ is the consequence of a contraction argument. In turn, both the well-definedness and the contraction property follow from a solution stability result for parametric mixed NCPs [96]. See [30, 222] for further discussion on the use of sensitivity results for the convergence analysis of Newton algorithms for solving VIs.

We explain some terminology before stating the convergence result. We say a function $\Delta(t)$ is $o(t)$, "little-oh of t", as $t \to 0+$ if $\Delta(t) \to 0$ as $t \to 0$, $t > 0$. Similarly, $\Delta(t) = O(t)$, "big-oh of t", as $t \to 0+$ if $\Delta(t)/t$ is bounded as $t \to 0$, $t > 0$. For any sequence $\{x^k\}$ which converges to a point $\bar{x}$, we say that (the rate of) convergence is Q-superlinear [211] if $x^{k+1} - \bar{x} = o(\|x^k - \bar{x}\|)$, and Q-quadratic if $x^{k+1} - \bar{x} = O(\|x^k - \bar{x}\|^2)$. From Theorem **5.1.1**, it follows that if a KKT point $(\bar{x}, \bar{\pi})$ of (1) satisfies the second-order sufficient condition that $\nabla^2_{xx} \mathcal{L}^{\text{NLP}}(\bar{x}, \bar{\pi})$ be strictly copositive on the critical cone $C(\bar{x}; \mathcal{F}^{\mathrm{NLP}})$, where

$$C(\bar{x}; \mathcal{F}^{\mathrm{NLP}}) = \{dx \in \Re^n : \nabla g_i(\bar{x})^T dx = 0, \ i \in \mathcal{A}_+(\bar{x}, \bar{\pi}),$$

$$\nabla g_i(\bar{x})^T dx \leq 0, \ i \in \mathcal{A}_0(\bar{x}, \bar{\pi})\},$$

then $\bar{x}$ is a strict local minimizer of (1). Based on this condition, we have the following convergence result for the SQP method stated above; the proof of part (a) can be found in [96]; part (b) follows from part (a) by a contraction argument.

6.4.1 Theorem. Suppose f and g are C^2, $\bar{x}$ is a stationary point of (1) with a unique KKT multiplier vector $\bar{\pi} \in \Re^{\ell+m}$, and the second-order sufficient condition holds at $\bar{x}$. Then:

(a) there exist neighborhoods U_1 and U_2 of $(\bar{x}, \bar{\pi})$ and a function Δ from $[0, \infty)$ into itself with $\Delta(t) = o(t)$ as $t \to 0+$ such that for any $(x, \pi) \in U_1$, the QP(x, π) has KKT pairs in U_2; moreover, for any KKT pair (x', π') of the QP(x, π) that belongs to U_2,

$$\|(x', \pi') - (\bar{x}, \bar{\pi})\| \leq \Delta(\|(x, \pi) - (\bar{x}, \bar{\pi})\|);$$

(b) there exists a neighborhood U of $(\bar{x}, \bar{\pi})$ such that for any $(x^0, \pi^0) \in U$, the SQP method generates a well-defined sequence $\{(x^\nu, \pi^\nu)\}$ that converges Q-superlinearly to $(\bar{x}, \bar{\pi})$; moreover, the rate of convergence is Q-quadratic if $\nabla^2 f$ and $\nabla^2 g_i$ are Lipschitz continuous near $\bar{x}$ for all i.

In the special case of nonlinear programs with affine constraints, both the algorithm and the conditions for convergence simplify; in particular, the uniqueness of the KKT vector $\bar{\pi}$ is no longer needed in the convergence result. Moreover, the second-order sufficient condition at $\bar{x}$ reduces to the strict copositivity of $\nabla^2 f(\bar{x})$ on the critical cone because for all (x, π), $\nabla^2_{xx} \mathcal{L}^{\mathrm{NLP}}(x, \pi) = \nabla^2 f(x)$. Finally, each QP$(x^\nu, \pi^\nu)$ becomes:

$$\text{minimize} \quad \nabla f(x^\nu)^T dx + \tfrac{1}{2} dx^T \nabla^2 f(x^\nu) dx$$

$$\text{subject to} \quad g_i(x^\nu + dx) = 0, \ i = 1, \dots, \ell,$$

$$g_i(x^\nu + dx) \leq 0, \ i = \ell + 1, \dots, \ell + m.$$

We let $\mathrm{QP}(x)$ denote this quadratic program with x^ν replaced by a generic vector x. Notice that $x^{\nu+1} \equiv x^\nu + dx^\nu$, where dx^ν is any stationary point of $\mathrm{QP}(x^\nu)$, must be feasible to the (now linearly constrained) NLP (1) for all $\nu \geq 0$.

In the absence of uniqueness of multipliers, the convergence result of the SQP method for solving a linearly constrained NLP is slightly different from Theorem **6.4.1** in that there is now no longer an assertion about the convergence of the multipliers; instead, convergence pertains to the primal vector x only.

6.4.2 Theorem. Suppose f is C^2 and g is affine, $\bar{x}$ is a stationary point of (1), and $\nabla^2 f(\bar{x})$ is strictly copositive on the critical cone $\mathcal{C}(\bar{x}; \mathcal{F}^{\mathrm{NLP}})$. Then:

(a) there exist neighborhoods V_1 and V_2 of $\bar{x}$ and a function Δ from $[0, \infty)$ into itself with $\Delta(t) = o(t)$ as $t \to 0+$ such that for any $x \in V_1$, the $\mathrm{QP}(x)$ has a stationary point in U_2; moreover, for any stationary point x' of the $\mathrm{QP}(x)$ that belongs to V,

$$\|x' - \bar{x}\| \leq \Delta(\|x - \bar{x}\|);$$

(b) there exists a neighborhood V of $\bar{x}$ such that for any $x^0 \in V$, the SQP method generates a well-defined sequence $\{x^\nu\}$ that converges Q-superlinearly to $\bar{x}$; moreover the rate of convergence is Q-quadratic if $\nabla^2 f$ is Lipschitz continuous near $\bar{x}$.

The (locally convergent) SQP methods for linearly constrained NLPs with C^2 objective functions were extended to linearly constrained NLPs with convex SC^1 objective functions [226]; these are convex C^1 functions with semismooth (instead of C^1) derivatives (see the discussion following Definition **4.2.8** for a brief summary and references on semismooth functions). The superlinear rate of convergence was also obtained in the generalization, but the quadratic rate is lost, due to the lack of smoothness of the derivatives. Since the class of functions with which the extended method deals is the class of convex functions, a line search was incorporated and the resulting method was shown to be globally convergent. See [237] for an alternative globalization via a "path search".

6.4.2 A PSQP method for MPEC

We now discuss the SQP approach to the KKT constrained MP using the PCP framework, resulting in the PSQP algorithm for solving this problem. We repeat the formulation of the MPEC problem (5.6.6):

$$\text{minimize}\quad f(x,y)$$

$$\text{subject to}\quad Gx + Hy + a \le 0,$$
$$F(x,y) + \sum_{i=1}^{\ell} \lambda_i \nabla_y g_i(x,y) = 0, \tag{3}$$
$$\lambda \ge 0, \quad g(x,y) \le 0, \quad \lambda^T g(x,y) = 0,$$

where $f : \Re^{n+m} \to \Re$, $F : \Re^{n+m} \to \Re^m$ are twice continuously differentiable; $g : \Re^{n+m} \to \Re^{\ell}$ is thrice continuously differentiable; $G \in \Re^{s\times n}$, $H \in \Re^{s\times m}$ and $a \in \Re^s$. Throughout, we continue to use the notation set in Subsection **3.3.1**, Section **4.3**, and Section **5.6**.

Before explaining the PSQP algorithm, we comment on a direct application of the SQP method to (3) considered as a standard nonlinear program. Consider a triple $\bar{w} \equiv (\bar{x},\bar{y},\bar{\lambda}) \in \mathcal{F}^{\text{KKT}}$. We will focus on one important requirement for the convergence of the SQP method to this triple as stipulated in Theorem **6.4.1**. This requirement is that the SMFCQ should hold at $\bar{w}$ and a corresponding KKT tuple for the constraints of (3). In turn, this SMFCQ implies that the tangent cone $\mathcal{T}(\bar{w};\mathcal{F}^{\text{KKT}})$ of (3) at $\bar{w}$ coincides with the linearized cone of this problem at the same vector. Based on the discussion at the end of Chapter **3**, we infer that in order for these two cones to be equal, we generally need $\bar{\lambda}$ to be a strictly complementary multiplier at $\bar{z}$; i.e., $\bar{\lambda} - g(\bar{z}) > 0$. Some preliminary computational experience with the SQP method applied to an LCP constrained MP suggests that the method often fails at degenerate vectors $\bar{w}$. In essence, the PSQP algorithm is a modification of the SQP method applied to (3) whose convergence does not depend on the latter strict complementarity assumption. The modification exploits the piecewise structure of the feasible set $\mathcal{F}^{\text{KKT}}$.

Throughout the following discussion, we fix the triple $(\bar{x},\bar{y},\bar{\lambda}) \in \mathcal{F}^{\text{KKT}}$. Define index sets as before, where $\mathcal{I}_0(\bar{z},\bar{\lambda})$ is the set of degenerate indices:

$$\mathcal{I}(\bar{z}) \equiv \{i : g_i(\bar{z}) = 0\},$$
$$\mathcal{I}_0(\bar{z},\bar{\lambda}) \equiv \{i : g_i(\bar{z}) = 0 = \bar{\lambda}_i\}, \quad \mathcal{I}_+(\bar{z},\bar{\lambda}) \equiv \{i : g_i(\bar{z}) = 0 < \bar{\lambda}_i\}.$$

For each partition $\mathcal{J}_1 \cup \mathcal{J}_2$ of $\{1, \ldots, \ell\}$ such that $\mathcal{J}_1 \supseteq \{i : g_i(\bar{z}) < 0\}$ and $\mathcal{J}_2 \supseteq \mathcal{I}_+(\bar{z}, \bar{\lambda})$, we consider the problem (5.6.7) which we repeat here:

$$\text{minimize} \quad f(x, y)$$

$$\text{subject to} \quad Gx + Hy + a \le 0,$$

$$L(x, y, \lambda) = 0,$$

$$\left. \begin{array}{l} \lambda_i = 0 \\[2mm] g_i(x, y) \le 0 \end{array} \right\} \quad \forall i \in \mathcal{J}_1, \tag{4}$$

$$\left. \begin{array}{l} \lambda_i \ge 0 \\[2mm] g_i(x, y) = 0 \end{array} \right\} \quad \forall i \in \mathcal{J}_2,$$

where $L(x, y, \lambda) \equiv F(x, y) + \nabla_y g(x, y)^T \lambda$. Note that in general, although the multiplier set $M(\bar{z})$ could contain infinitely many elements, there is only a finite number of NLPs of the type (4) corresponding to a given $\bar{z} \in \mathcal{F}$. As before, introducing multipliers $\zeta \in \Re^s$ and $\pi \in \Re^m$ and $\eta \in \Re^\ell$ corresponding to the constraints $z \equiv (x, y) \in Z$, the constraints $L(z, \lambda) = 0$, and the constraints defined by $g_i(x, y) \le (=)0$, respectively, we write the MPEC Lagrangean as

$$\mathcal{L}^{\text{MPEC}}(z, \lambda, \zeta, \pi, \eta) \equiv f(z) + \zeta^T (Gx + Hy - a) - \pi^T L(z, \lambda) + \eta^T g(z).$$

To apply the SQP idea, we note that

$$\{i : g_i(\bar{z}) < 0\} = \{i : g_i(\bar{z}) + \bar{\lambda}_i < 0\},$$

and

$$\{i : \bar{\lambda}_i > 0\} = \{i : g_i(\bar{z}) + \bar{\lambda}_i > 0\}.$$

Consider a given iterate $w^\nu \equiv (x^\nu, y^\nu, \lambda^\nu) \in \Re^{n+m+\ell}$, which is required to satisfy $\lambda^\nu \ge 0$ and $Gx^\nu + Hy^\nu + a \le 0$ only (these are linear constraints, so they are easy to satisfy). In particular, this iterate is not necessarily feasible to (3). Write $z^\nu \equiv (x^\nu, y^\nu)$. Also given is a triple of multipliers $(\zeta^\nu, \pi^\nu, \eta^\nu)$. Pick an arbitrary partition $\mathcal{J}_1 \cup \mathcal{J}_2$ of $\{1, \ldots, \ell\}$ such that

$$\mathcal{J}_1 \supseteq \{i : g_i(z^\nu) + \lambda_i^\nu < 0\} \quad \text{and} \quad \mathcal{J}_2 \supseteq \{i : g_i(z^\nu) + \lambda_i^\nu > 0\}. \tag{5}$$

Corresponding to this pair $(\mathcal{J}_1, \mathcal{J}_2)$, form the quadratic program associated with the NLP (4) at $(x^\nu, y^\nu, \lambda^\nu)$; the variable of this quadratic program is given by the vector $dw \equiv (dz, d\lambda)$, with $dz \equiv (dx, dy)$:

minimize $\nabla f(z^\nu)^T dz + \frac{1}{2} dw^T \nabla^2_{ww} \mathcal{L}^{\mathrm{MPEC}}(w^\nu, \zeta^\nu, \pi^\nu, \eta^\nu) dw$

subject to $G(x^\nu + dx) + H(y^\nu + dy) + a \leq 0,$

$L(w^\nu) + \nabla L(w^\nu) dw = 0,$

$$\left. \begin{array}{l} (\lambda^\nu + d\lambda)_i = 0 \\[2mm] g_i(z^\nu) + \nabla g_i(z^\nu)^T dz \leq 0 \end{array} \right\} \ \forall i \in \mathcal{J}_1, \qquad (6)$$

$$\left. \begin{array}{l} (\lambda^\nu + d\lambda)_i \geq 0 \\[2mm] g_i(z^\nu) + \nabla g_i(z^\nu)^T dz = 0 \end{array} \right\} \ \forall i \in \mathcal{J}_2.$$

A KKT tuple of this quadratic program will be used to define the next iterate.

The following is a summary of the PSQP method for solving the MP with KKT constraints (3).

A Piecewise Sequential Quadratic Programming Algorithm

Step 0. (Initialization) Let $w^0 = (z^0, \lambda^0) \in Z \times \Re^\ell_+$ and $v^0 = (\zeta^0, \pi^0, \eta^0)$ in $\Re^{s+m+\ell}$ be given. Set $\nu = 0$.

Step 1. (Main computation) Let $\mathcal{J}_1 \cup \mathcal{J}_2$ be an arbitrary partition of $\{1, \ldots, \ell\}$ satisfying (5). Let $(dw^\nu, v^{\nu+1})$ be a KKT tuple of the quadratic program (6), where $v^{\nu+1}$ is the vector of multipliers associated with the constraints in this program except for $(\lambda^\nu + d\lambda)_i = (\leq)0$.

Step 2. (Termination check and update) Set $w^{\nu+1} \equiv w^\nu + dw^\nu$ and check the prescribed rule for termination. Return to Step 1 with ν replaced by $\nu + 1$ if the rule is not satisfied.

The above PSQP algorithm is based on an idea similar to the method of Kojima and Shindo [144] for solving piecewise smooth equations $\Phi(x) = 0$, where the mapping $\Phi : \Re^n \to \Re^n$ is the continuous selection of a finite family of smooth functions $\{\Phi^i\}$ each of which maps $\Re^n$ to itself. In this reference, given x^ν, an arbitrary index $i = i(\nu)$ is chosen with $\Phi(x^\nu) = \Phi^i(x^\nu)$, and then $x^{\nu+1}$ is obtained by applying a single Newton iteration to Φ^i at x^ν; i.e., by solving the system of linear equations $\Phi^i(x^\nu) + \nabla \Phi^i(x^\nu) dx^\nu = 0$ and letting $x^{\nu+1} = x^\nu + dx^\nu$. These two methods share the same idea of solving a "linearized" subproblem on an appropriate piece of the original problem; they differ at each iteration primarily in that the subproblem of the PSQP method is a quadratic program, which is equivalent to an affine normal equation obtained from the linearization of the NLP (4) formulated as a (nonlinear) normal equation, whereas the subproblem of the Kojima-Shindo method is a system of linear equations. Superlinear or quadratic convergence, however, can be established in each case.

As with the SQP method for nonlinear programs, we only expect (superlinear) convergence of the PSQP algorithm if there is some uniqueness condition on the MPEC multipliers. Specifically, we require uniquess of the optimal multipliers associated with each nonlinear program (4) for relevant pairs of index sets $(\mathcal{J}_1, \mathcal{J}_2)$. Also we need these multipliers to be independent of the pairs $(\mathcal{J}_1, \mathcal{J}_2)$, so that no matter which nonlinear programming subproblem we apply SQP to, the current value of the multipliers is (locally) a reasonable estimate.

6.4.3 Theorem. In the context of the MPEC formulated with KKT constraints (3), let the functions f and F be twice continuously differentiable, g be thrice continuously differentiable. Suppose $\bar{w} = (\bar{x}, \bar{y}, \bar{\lambda}) \in \mathcal{F}^{\mathrm{KKT}}$ and $\bar{v} = (\bar{\zeta}, \bar{\pi}, \bar{\eta}) \in \Re^{s+m+\ell}$ are such that, for each $\mathcal{J}_1 \cup \mathcal{J}_2$ partitioning $\{1, \ldots, \ell\}$ with $\mathcal{J}_1 \supseteq \{i : g_i(\bar{z}) < 0\}$ and $\mathcal{J}_2 \supseteq \mathcal{I}_+(\bar{z}, \bar{\lambda})$, $\bar{w}$ is stationary for (4); $\bar{v}$ is the unique KKT multiplier for (4) associated with $\bar{w}$; and the second-order sufficient condition for (4) holds at $(\bar{w}, \bar{v})$. Then $\bar{w}$ is a strict local minimizer for the MPEC (3). Moreover, for each (w^0, v^0) sufficiently close to $(\bar{w}, \bar{v})$, the PSQP algorithm is well defined and produces a sequence $\{(w^\nu, v^\nu)\}$ that converges Q-superlinearly to $(\bar{w}, \bar{v})$. Finally convergence is Q-quadratic if, in addition, $\nabla^2 f$, $\nabla^2 F$ and $\nabla^3 g$ are Lipschitz near $(\bar{x}, \bar{y})$.

Proof. The claim that $\bar{w}$ is a strict local minimizer for (3) follows from Theorem **5.6.5**, because the uniqueness assumption of the multiplier $\bar{v}$ implies that the SMFCQ holds for (4) at $(\bar{w}, \bar{v})$. To prove the convergence of the PSQP algorithm, index each pair $(\mathcal{J}_1, \mathcal{J}_2)$ by subsets α of $\mathcal{I}_0(\bar{w})$, where

$$\mathcal{J}_1 \equiv (\mathcal{I}_0(\bar{w}) \setminus \alpha) \cup \{i : g_i(\bar{z}) < 0\}, \quad \text{and} \quad \mathcal{J}_2 \equiv \alpha \cup \mathcal{I}_+(\bar{w}).$$

According to Theorem **6.4.1**, for each $\alpha \subseteq \mathcal{I}_0(\bar{w})$, there is a neighborhood Θ_α of $(\bar{w}, \bar{v})$ and a function Δ_α from $[0, \infty)$ into itself such that for any

$(w, v) \in \Theta_\alpha$, one iteration of SQP applied to (4) at (w, v) yields (w', v') in Θ_α with

$$\|(w', v') - (\bar{w}, \bar{v})\| \le \Delta(\|(w, v) - (\bar{w}, \bar{v})\|) \le \tfrac{1}{2}\|(w, v) - (\bar{w}, \bar{v})\|.$$

Let

$$\Theta \equiv \bigcap_\alpha \Theta_\alpha, \quad \Delta(t) \equiv \max_\alpha \Delta_\alpha(t), \quad \forall t \ge 0,$$

where α ranges over all subsets of $\mathcal{I}_0(\bar{w})$. Since there are only finitely many such subsets α, Θ is a neighborhood of $(\bar{w}, \bar{v})$, and $\Delta(t) = o(t)$. Furthermore, for any $(w^0, v^0) \in \Theta$, it is easy to show by induction that the PSQP algorithm is well defined and generates a sequence $\{(w^\nu, v^\nu)\}$ converging to $(\bar{w}, \bar{v})$ and satisfying

$$\|(w^{\nu+1}, v^{\nu+1}) - (\bar{w}, \bar{v})\| \le \Delta(\|(w^\nu, v^\nu) - (\bar{w}, \bar{v})\|).$$

Thus convergence is Q-superlinear, and Q-quadratic if $\nabla^2 f$, $\nabla^2 F$ and $\nabla^3 g$ are Lipschitz near $\bar{z}$.   **Q.E.D.**

Using the relaxed NLP (4.3.11) corresponding to $\bar{w} = (\bar{z}, \bar{\lambda})$, that is,

$$\begin{aligned}
&\text{minimize} \quad f(x, y) \\
&\text{subject to} \quad Gx + Hy \le a, \\
&\qquad\qquad\quad L(x, y, \lambda) = 0, \\
&\qquad\qquad\quad \left.\begin{array}{l} \lambda_i = 0 \\[4pt] g_i(x, y) \le 0 \end{array}\right\} \; \forall i \notin \mathcal{I}(\bar{z}), \\
&\qquad\qquad\quad \left.\begin{array}{l} \lambda_i \ge 0 \\[4pt] g_i(x, y) \le 0 \end{array}\right\} \; \forall i \in \mathcal{I}_0(\bar{z}, \bar{\lambda}), \\
&\qquad\qquad\quad \left.\begin{array}{l} \lambda_i \ge 0 \\[4pt] g_i(x, y) = 0 \end{array}\right\} \; \forall i \in \mathcal{I}_+(\bar{z}, \bar{\lambda}),
\end{aligned} \tag{7}$$

we deduce a consequence of Theorem **6.4.3** that gives a further sufficient condition for the convergence of the PSQP algorithm.

6.4.4 Corollary. Let f, F, and g be as given in Theorem **6.4.3**. Suppose $\bar{w} = (\bar{x}, \bar{y}, \bar{\lambda}) \in \mathcal{F}^{\text{KKT}}$ is stationary for the relaxed nonlinear program (7) with the corresponding KKT multiplier $\bar{v} = (\bar{\zeta}, \bar{\pi}, \bar{\eta}) \in \Re^s \times \Re^m \times \Re^\ell$, and for this problem, the LICQ holds at $\bar{w}$ and the second-order sufficient condition holds at $(\bar{w}, \bar{v})$; i.e., the Hessian matrix $\nabla^2_{ww} \mathcal{L}^{\mathrm{MPEC}}(\bar{w}, \bar{v})$ is strictly copositive on the cone $\mathcal{C}_{\mathrm{rl}}(\bar{w})$. Then $(\bar{w}, \bar{\lambda})$ is a strict local minimizer of (3); moreover, for each (w^0, v^0) near $(\bar{w}, \bar{v})$, the PSQP algorithm generates a sequence $\{(w^\nu, v^\nu)\}$ that converges Q-superlinearly to $(\bar{w}, \bar{v})$.

Proof. Since the LICQ implies the SMFCQ, the assumptions about $(\bar{w}, \bar{v})$ imply that for each pair $(\mathcal{J}_1, \mathcal{J}_2)$ as described in the statement of Theorem **6.4.3**, $\bar{w}$ is stationary for the NLP (4) and $\bar{v}$ is the unique KKT multiplier corresponding to $\bar{w}$; see Proposition **4.3.7**. As noted in Subsection **5.6.1**, the critical cone for each nonlinear program (4) at $\bar{w}$ is contained in the critical cone for the relaxed problem at $\bar{w}$. Hence the second-order sufficient condition for the former is implied by the same for the latter, which we have assumed. Hence all the assumptions of Theorem **6.4.3** are satisfied and an application of this theorem immediately yields the desired conclusion. **Q.E.D.**

6.4.5 Remark. It can be seen that the assumptions of the above corollary actually ensure the convergence of SQP when applied to the relaxed NLP (7) at $\bar{w}$. If g is affine (e.g., in the case of the NCP constrained MP), then as stated in Proposition **5.6.8**, the strict copositivity of $\nabla^2_{ww} \mathcal{L}^{\mathrm{MPEC}}(\bar{w}, \bar{v})$ on $\mathcal{C}_{\mathrm{rl}}(\bar{w})$ is closely tied to the uniqueness of $\bar{\lambda}$ and the strict copositivity of $\nabla^2 f(\bar{z}) - \sum_{i=1}^m \bar{\pi}_i \nabla^2 F_i(\bar{z})$ on $\mathcal{C}(\bar{w})$.

Using the notation in Section **5.6**, we discuss a specialization of the PSQP algorithm for the NCP constrained MP:

$$\text{minimize} \quad f(x, y)$$
$$\text{subject to} \quad Gx + Hy + a \leq 0, \tag{8}$$
$$y \geq 0, \quad F(x, y) \geq 0, \quad y^T F(x, y) = 0.$$

At each iteration of the algorithm, we are given the tuple $(x^\nu, y^\nu, \zeta^\nu, \pi^\nu)$. We then pick an arbitrary partition $\mathcal{J}_1 \cup \mathcal{J}_2$ of $\{1, \ldots, m\}$ such that

$$\mathcal{J}_1 \supseteq \{i : y_i^\nu > F_i(y^\nu)\}, \quad \text{and} \quad \mathcal{J}_2 \supseteq \{i : y_i^\nu < F_i(y^\nu)\},$$

and solve the quadratic program in the variable $dz \equiv (dx, dy)$:

$$\text{minimize} \quad \nabla f(z^\nu)^T dz + \tfrac{1}{2} dz^T \left(\nabla^2 f(z^\nu) - \sum_{i=1}^m \pi_i^\nu \nabla^2 F_i(z^\nu) \right) dz$$

$$\text{subject to} \quad G(x^\nu + dx) + H(y^\nu + dy) + a \leq 0,$$

$$\left. \begin{array}{l} F_i(z^\nu) + \nabla F_i(z^\nu)^T dz = 0 \\[2mm] y_i^\nu + dy_i \geq 0 \end{array} \right\} \quad \forall i \in \mathcal{J}_1,$$

$$\left. \begin{array}{l} F_i(z^\nu) + \nabla F_i(z^\nu)^T dz \geq 0 \\[2mm] y_i^\nu + dy_i = 0 \end{array} \right\} \quad \forall i \in \mathcal{J}_2.$$

Let $(dz^\nu, \zeta^{\nu+1}, \eta^{\nu+1})$ be a KKT tuple of this program. We define $z^{\nu+1}$ to be $z^\nu + dz^\nu$. Specializing Theorem **6.4.3** to the case where $g(x,y) \equiv -y$ and noting Remark **6.4.5**, we obtain the local, superlinear convergence of the sequence $\{(z^\nu, \zeta^\nu, \pi^\nu)\}$ to a tuple $(\bar{z}, \bar{\zeta}, \bar{\pi})$ under the following assumptions: (i) $\bar{z} \in \mathcal{F}$; (ii) for each partition $\mathcal{J}_1 \cup \mathcal{J}_2$ of $\{1, \ldots, m\}$ such that

$$\mathcal{J}_1 \supseteq \{i : \bar{y}_i > F_i(\bar{y}) = 0\}, \quad \text{and} \quad \mathcal{J}_2 \supseteq \{i : 0 = \bar{y}_i < F_i(\bar{y})\},$$

$(\bar{z}, \bar{\zeta}, \bar{\pi})$ is a stationary point of the nonlinear program:

$$\text{minimize} \quad f(x,y)$$

$$\text{subject to} \quad Gx + Hy + a \leq 0,$$

$$F_{\mathcal{J}_1}(x,y) = 0, \quad y_{\mathcal{J}_1} \geq 0,$$

$$F_{\mathcal{J}_2}(x,y) \geq 0, \quad y_{\mathcal{J}_2} = 0;$$

(iii) $(\bar{\zeta}, \bar{\pi})$ is the unique KKT multiplier of the latter program associated with $\bar{z}$; and (iv) $\nabla^2 f(\bar{z}) - \sum_{i=1}^m \bar{\pi}_i \nabla^2 F_i(\bar{z})$ is strictly copositive on the cone $\bigcup_\beta \mathcal{C}(\bar{z}; \mathcal{F}_\beta)$, where β ranges over all subsets of $\beta(\bar{z}) \equiv \{i : \bar{y}_i = F_i(\bar{z}) = 0\}$ and $\mathcal{C}(\bar{z}; \mathcal{F}_\beta)$ is defined by (5.6.3).

Finally we turn our attention to an MPAEC where the functions F and g in the inner VIs are affine; hence the lower-level Lagrangean function $L(x, y, \lambda) = F(x,y) + \nabla_{i=1}^\ell \lambda_i \nabla_y g_i(x,y)$ is also affine. Specifically, we consider the problem:

$$\text{minimize} \quad f(x, y)$$

$$\text{subject to} \quad Gx + Hy + a \le 0, \quad \lambda \ge 0,$$

$$Dx + Ey + b \le 0, \tag{9}$$

$$Px + Qy + q + E^T\lambda = 0,$$

$$\lambda^T(Dx + Ey + b) = 0,$$

where the notation is as in Section **5.2**. Like SQP applied to nonlinear programs with linear constraints, the PSQP algorithm specialized to MPs with affine KKT constraints has locally superlinear convergence if the Hessian of the objective function $\nabla^2 f(\bar{x}, \bar{y})$ has a certain strict copositivity property, regardless of whether or not the MPEC multipliers are unique. This specialized algorithm is as follows.

The Specialized PSQP Algorithm for MPAEC

Step 0. (Initialization) Let $w^0 = (z^0, \lambda^0) \in \Re^{n+m+\ell}$ be given and let $\nu = 0$.

Step 1. (Main computation) Let $\mathcal{J}_1 \cup \mathcal{J}_2$ be an arbitrary partition of $\{1, \ldots, \ell\}$ satisfying (5). Let $(dw^\nu, v^{\nu+1})$ be a KKT tuple of the quadratic program in the variable $dw \equiv (dz, d\lambda)$:

$$\text{minimize} \quad \nabla f(z^\nu)^T dz + \tfrac{1}{2} dz^T \nabla^2 f(z^\nu) dz$$

$$\text{subject to} \quad G(x^\nu + dx) + H(y^\nu + dy) + a \le 0,$$

$$P(x^\nu + dx) + Q(y^\nu + dy) + q + E^T(\lambda^\nu + d\lambda) = 0,$$

$$\left.\begin{array}{l} \lambda_i^\nu + d\lambda_i = 0 \\ (D(x^\nu + dx) + E(y^\nu + dy) + b)_i \le 0 \end{array}\right\} \forall i \in \mathcal{J}_1,$$

$$\left.\begin{array}{l} \lambda_i^\nu + d\lambda_i \ge 0 \\ (D(x^\nu + dx) + E(y^\nu + dy) + b)_i = 0 \end{array}\right\} \forall i \in \mathcal{J}_2,$$

where $v^{\nu+1}$ is a vector of multipliers corresponding only to the functional constraints (i.e., excluding $\lambda_i^\nu + d\lambda_i = (\ge)0$ for all i).

Step 2. (Termination check and update) Let $w^{\nu+1} \equiv w^\nu + dw^\nu$ and $\nu = \nu + 1$; repeat from Step 1 if termination fails.

We recall from Theorem **5.2.3** that if $\bar{z}$ is a stationary point of the MPAEC (9) and $\nabla^2 f(\bar{z})$ is strictly copositive on the critical cone $\mathcal{C}(\bar{z};\mathcal{F})$ of this problem at $\bar{z}$, then $\bar{z}$ is a strict local minimizer of (9). It turns out that these same assumptions are sufficient for the local convergence of the above algorithm near $\bar{z}$. The proof of this convergence result is based on Theorem **6.4.2** which asserts the convergence of the primal sequence produced by an SQP method for solving a linearly constrained NLP. As a consequence, the following result asserts only the convergence of the sequence $\{(x^\nu,y^\nu)\}$ and makes no claim about the limit behavior of the sequence of KKT multipliers $\{\lambda^\nu\}$ (or about the sequence of MPEC multipliers $\{v^\nu\}$; in fact, the latter sequence plays no role in the iterations of the above algorithm).

6.4.6 Theorem. Consider the MPAEC (9) where the function f is twice continuously differentiable. Suppose that $\bar{z} = (\bar{x},\bar{y}) \in \mathcal{F}$ is stationary for the problem (9) and $\nabla^2 f(\bar{z})$ is strictly copositive on the cone $\mathcal{C}(\bar{z};\mathcal{F})$. Then for each z^0 near $\bar{z}$, the specialized PSQP algorithm is well defined and produces a sequence $\{z^\nu\}$ that converges Q-superlinearly to $\bar{z}$. Convergence is Q-quadratic if, in addition, $\nabla^2 f$ is Lipschitz near $\bar{z}$.

Proof. For each $\bar{\lambda} \in M(\bar{z})$ and each partition $\mathcal{J}_1 \cup \mathcal{J}_2$ of $\{1,\ldots,\ell\}$ satisfying

$$\mathcal{J}_1 \supseteq \{i : (D\bar{x} + E\bar{y} + b)_i < 0\}, \quad \text{and} \quad \mathcal{J}_2 \supseteq \{i : \bar{\lambda}_i > 0\}, \tag{10}$$

the problem (4) is of the form

$$\begin{aligned}
\text{minimize} \quad & f(x,y) \\
\text{subject to} \quad & Gx + Hy + a \le 0, \\
& Px + Qy + q + E^T\lambda = 0, \\
& \left.\begin{array}{l} \lambda_i = 0 \\ (Dx + Ey + b)_i \le 0 \end{array}\right\} \forall i \in \mathcal{J}_1, \\
& \left.\begin{array}{l} \lambda_i \ge 0 \\ (Dx + Ey + b)_i = 0 \end{array}\right\} \forall i \in \mathcal{J}_2.
\end{aligned}$$

Let $\mathcal{F}_{(\mathcal{J}_1,\mathcal{J}_2)} \subseteq \Re^{n+m}$ consist of all vectors (x,y) for which there exists $\lambda \in \Re^\ell$ such that (x,y,λ) is feasible to the above problem. Clearly the set

$\mathcal{F}_{(\mathcal{J}_1,\mathcal{J}_2)}$ is a polyhedron and is contained in $\mathcal{F}$; moreover the problem (4) is equivalent to

$$\begin{array}{ll}\text{minimize} & f(x,y) \\ \text{subject to} & (x,y) \in \mathcal{F}_{(\mathcal{J}_1,\mathcal{J}_2)},\end{array} \qquad (11)$$

which is a linearly constrained NLP. Since $\bar{z}$ is stationary for the MPAEC (9), it follows that $\bar{z}$ is also stationary for (11). Moreover, the strict copositivity of $\nabla^2 f(\bar{z})$ on $\mathcal{C}(\bar{z};\mathcal{F})$ implies that the second-order sufficient condition is satisfied for the problem (11) at $\bar{z}$, because $\mathcal{F}_{(\mathcal{J}_1,\mathcal{J}_2)} \subseteq \mathcal{F}$. Hence Theorem **6.4.2** is applicable to the problem (11) at $\bar{z}$. By this thereom, there is a neighborhood $V_{(\mathcal{J}_1,\mathcal{J}_2)}$ of $\bar{z}$ and a function $\Delta_{(\mathcal{J}_1,\mathcal{J}_2)} : [0,\infty) \to [0,\infty)$ such that for any $z \in V_{(\mathcal{J}_1,\mathcal{J}_2)}$, one iteration of SQP applied to (11) at z yields $z' \in V_{(\mathcal{J}_1,\mathcal{J}_2)}$ satisfying

$$\|z' - \bar{z}\| \leq \Delta(\|z - \bar{z}\|) \leq \tfrac{1}{2}\|z - \bar{z}\|.$$

Let

$$\Theta \equiv \bigcap_{(\mathcal{J}_1,\mathcal{J}_2)} V_{(\mathcal{J}_1,\mathcal{J}_2)} \quad \Delta(t) \equiv \max_{(\mathcal{J}_1,\mathcal{J}_2)} \Delta_{(\mathcal{J}_1,\mathcal{J}_2)}(t), \ \forall t \geq 0,$$

where $(\mathcal{J}_1, \mathcal{J}_2)$ ranges over all partitions of $\{1, \ldots, \ell\}$ satisfying (10). Since there are only finitely many such pairs $(\mathcal{J}_1, \mathcal{J}_2)$, Θ is a neighborhood of $\bar{z}$, and $\Delta(t) = o(t)$. The remaining proof is now similar to that of Theorem **6.4.3**. **Q.E.D.**

6.5 Computational Testing

In the previous sections, we have described several approaches for solving MPEC; two are penalty based interior point algorithms (PIPA), and the other two are based on the implicit programming and piecewise programming formulations. In this section, we shall describe our MATLAB implementations of the PIPA algorithms and report some encouraging preliminary computational testing results.

6.5.1 *Some implementation details*

We have implemented two versions of PIPA, one for the general mixed NCP constrained MP (6.1.1) and the other for the LCP constrained MP

(6.2.1). The version of PIPA we have implemented for the mixed NCP constrained MP, which we call PIPA1, differs from the one described in Section **6.1** in several minor respects. Below we highlight these differences and provide some details of our implementation.

The initialization step

In some of the computational experiments, the first-level constraints are chosen to be of the form

$$0 \le x \le b$$

where $b \in [0,1]^n$ is a vector generated randomly using the MATLAB `rand` command. Thus,

$$X = \{x \in \Re^n_+ : x \le b\}. \tag{1}$$

In this case, the initial vector x^0 is set to be the center of X, namely $x^0 = 0.5b$. The other initial values are chosen as follows

$$y^0 = w^0 = 0.5e, \quad z^0 = e,$$

where e is the vector of all ones. The parameter $\bar{\rho}$, which controls the size of the neighborhood of the central path, is set to be 0.02 in our experiments. In general, for the other initial vector (y^0, w^0), the following formula can be used:

$$\bar{\rho} = \min \left(0.02, \ \min_i \frac{m y_i^0 w_i^0}{(y^0)^T w^0} \right).$$

The penalty parameter is initially chosen to be $\alpha_{-1} = 1$. This choice differs from the requirement that $\alpha_{-1} > 1$ in the original description of the PIPA algorithms; however, we use a factor of 1.2 to scale the penalty parameter $\alpha_{\nu-1}$ in order to get the next parameter α_ν. See description below.

The direction generation step

The main computation in the direction generation step involves solving the following quadratic program QP_ν:

minimize $\quad (df_x^\nu)^T dx + (df_y^\nu)^T dy + (df_w^\nu)^T dw + (df_z^\nu)^T dz + \frac{1}{2}(dx)^T Q_\nu dx$

subject to $\quad x^\nu + dx \in X,$

$$(dx)^T dx \leq c\left(\|F^\nu\| + (w^\nu)^T y^\nu\right),$$

$$\nabla_x F^\nu dx + \nabla_y F^\nu dy + \nabla_w F^\nu dw + \nabla_z F^\nu dz = -F^\nu,$$

$$W_\nu dy + Y_\nu dw = -Y_\nu w^\nu + \sigma_\nu \mu_\nu e,$$

where we have adopted the same notations as in (6.1.11). In our MAT-LAB implementation of QP_ν, we changed the quadratic sphere constraint $(dx)^T dx \leq c\phi_\nu$ to the linear box constraint

$$-\sqrt{c\phi_\nu}\, e \leq dx \leq \sqrt{c\phi_\nu}\, e.$$

This change makes it possible to apply the MATLAB subroutine qp to the resulting linearly constrained quadratic program directly. We have set the parameter c to be 10. (Although no formal derivation is given, the change of the direction generation subproblem will not affect the convergence analysis given before.)

To solve the direction generation QP (which is the most time consuming step in each iteration), we used the last two constraints in QP_ν to express (dy, dw, dz) in terms of dx and then transformed QP_ν into the following equivalent but reduced quadratic program in dx:

minimize $\quad (\bar{d}f_x^\nu)^T dx + \frac{1}{2}(dx)^T Q_\nu dx$

subject to $\quad x^\nu + dx \in X,$ $\qquad\qquad\qquad\qquad$ (2)

$$|dx_i| \leq \sqrt{c\phi_\nu}, \quad i = 1, \ldots, n,$$

where

$$(\bar{d}f_x^\nu)^T = (df_x^\nu)^T -$$

$$\left((df_y^\nu)^T \ (df_w^\nu)^T \ (df_z^\nu)^T\right) \begin{bmatrix} \nabla_y F^\nu & \nabla_w F^\nu & \nabla_z F^\nu \\ W_\nu & Y_\nu & 0 \end{bmatrix}^{-1} \begin{pmatrix} \nabla_x F^\nu \\ 0 \end{pmatrix}.$$

Once we have computed the optimal solution dx^ν by solving (2), we can derive the corresponding solution (dy^ν, dw^ν, dz^ν) according to

$$
\begin{pmatrix} dy^\nu \\ dw^\nu \\ dz^\nu \end{pmatrix} = \begin{bmatrix} \nabla_y F^\nu & \nabla_w F^\nu & \nabla_z F^\nu \\ W_\nu & Y_\nu & 0 \end{bmatrix}^{-1} \begin{pmatrix} -F^\nu - \nabla_x F^\nu dx^\nu \\ -Y_\nu w^\nu + \sigma_\nu e \end{pmatrix}.
$$

The advantage of the above transformation is obvious: the number of variables in the original QP_ν is $n + 2m + \ell$ whereas the number of variables in the reduced quadratic program is only n. This reduction in the dimension not only significantly speeds up the direction generation step but also allows the PIPA1 to solve much larger problems than otherwise possible.

Next PIPA1 proceeds to update the penalty parameter α_ν using the update rule: choose p_ν to be the smallest integer $p \geq 1$ such that with $\alpha_\nu \equiv \alpha_{\nu-1}(1.2)^{p_\nu}$

$$
(df_x^\nu)^T dx^\nu + (df_y^\nu)^T dy^\nu - \alpha_\nu[(1-\sigma_\nu)(w^\nu)^T y^\nu + 2\|F^\nu\|^2] < -\phi_\nu.
$$

In all our experiments, the centering step size σ_ν is chosen to be 0.3. The positive definite matrix Q_ν is conveniently set to be I.

The step size determination

Once the direction $(dx^\nu, dy^\nu, dw^\nu, dz^\nu)$ is generated, the algorithm proceeds to determine a step size τ and then update the iterate $(x^\nu, y^\nu, w^\nu, z^\nu)$ accordingly. The step size determination in our implementation consists of two stages. In the first stage, we choose the smallest nonnegative integer k such that with the step size $\bar{\tau}^\nu = (0.95)^k$ the following conditions:

- $y^\nu + \bar{\tau}^\nu dy^\nu \geq 0$;

- $w^\nu + \bar{\tau}^\nu dw^\nu \geq 0$;

- for all $i = 1, \ldots, m$,

$$
(y^\nu + \bar{\tau}^\nu dy^\nu)_i (w^\nu + \bar{\tau}^\nu dw^\nu)_i \geq \bar{\rho}\frac{(y^\nu + \bar{\tau}^\nu dy^\nu)^T (w^\nu + \bar{\tau}^\nu dw^\nu)}{m};
$$

hold. In the second stage, we apply the Armijo rule to determine the smallest nonnegative integer r such that with $\tau \equiv \bar{\tau}_\nu \rho_1^r$,

$$
\phi(x^\nu(\tau), y^\nu(\tau), w^\nu(\tau), z^\nu(\tau)) \leq \phi(x^\nu, y^\nu, w^\nu, z^\nu),
$$

$$P_{\alpha_\nu}(x^\nu(\tau), y^\nu(\tau), w^\nu(\tau), z^\nu(\tau)) - P_{\alpha_\nu}(x^\nu, y^\nu, w^\nu, z^\nu)$$

$$\leq \gamma'\tau \left((df_x^\nu)^T dx^\nu + (df_y^\nu)^T dy^\nu + (df_w^\nu)^T dw^\nu + (df_z^\nu)^T dz^\nu \right.$$

$$\left. -\alpha_\nu(2\|F^\nu\|^2 + (1 - \sigma_\nu)(y^\nu)^T w^\nu) \right) \leq -\gamma'\tau\phi_\nu.$$

In our experiments, we have let $\rho_1 = 0.95$.

In some situations, the Armijo rule reduces τ repeatedly many times in one iteration; when this occurs, the step size τ will be exceedingly small and the algorithm may fail. In our implementation, we print out the following warning message

```
Armijo failed, stepsize too small
```

once the current step size τ gets smaller than 10^{-6}; when this occurs, the current step size τ is immediately accepted and the algorithm continues to the next iteration.

The termination step

We terminate the algorithm if either the condition

$$\|dx^\nu\| + (w^\nu)^T y^\nu + \|F(x^\nu, y^\nu, w^\nu, z^\nu)\| \leq \texttt{tolerance}$$

is satisfied or the number of iterations has exceeded `max_int`. In our experiments, we have set `tolerance` to be 10^{-6} and `max_int` to be 200.

We have also coded a variant of PIPA specialized to the LCP constrained MP (6.2.1). We call this implementation PIPA2. The two codes PIPA1 and PIPA2 are much the same with three differences. These are (1) PIPA2 does not permit the presence of the free variable z; (2) PIPA2 solves a different quadratic subproblem from QP_ν (cf. (6.1.11)). In particular, unlike PIPA1, PIPA2 does not use the spherical constraint $\|dx\|^2 \leq c\phi_\nu$ or the box constraint $|dx_i| \leq \sqrt{c\phi_\nu}$, $i = 1, ..., n$, in its quadratic subproblem. Instead, PIPA2 relies on the positive definite matrix Q_ν to control the size of dx. Specifically, at iteration ν, PIPA2 solves the following quadratic program

$$\text{minimize} \quad (df_x^\nu)^T dx + (df_y^\nu)^T dy + \tfrac{1}{2}(dx)^T Q_\nu dx$$

$$\text{subject to} \quad x^\nu + dx \in X,$$

$$N dx + M dy - dw = -r^\nu,$$

$$Y_\nu dw + W_\nu dy = -Y_\nu w^\nu + \sigma_\nu \mu_\nu e,$$

where we have adopted the same notations as (6.2.2). As in PIPA1, we have used the expression

$$
\begin{pmatrix} dy \\ dw \end{pmatrix} = \begin{bmatrix} M & -I \\ W_\nu & Y_\nu \end{bmatrix}^{-1} \begin{pmatrix} -r^\nu - N dx \\ -Y_\nu w^\nu + \sigma_\nu e \end{pmatrix}
$$

to eliminate the variables (dy, dw) in the above quadratic program. Using this expression, the variables (dy, dw) can easily be recovered once we know the value of dx. Thus, PIPA2 need only find dx which can be obtained by solving the following reduced quadratic program:

$$
\text{minimize} \quad (\bar{d} f_x^\nu)^T dx + \tfrac{1}{2} (dx)^T Q_\nu dx
$$

$$
\text{subject to} \quad x^\nu + dx \in X,
$$

where

$$
(\bar{d} f_x^\nu)^T = (d f_x^\nu)^T - \begin{bmatrix} (d f_y^\nu)^T & 0 \end{bmatrix} \begin{bmatrix} M & -I \\ W_\nu & Y_\nu \end{bmatrix}^{-1} \begin{pmatrix} N \\ 0 \end{pmatrix}.
$$

The third and last difference between PIPA1 and PIPA2 is that PIPA2 uses the penalty function

$$
\phi(x, y, w) \equiv \| r(x, y, w) \| + 2 \sqrt{w^T y}
$$

which is different from that used in PIPA1; here

$$
r(x, y, w) \equiv q + Nx + My - w.
$$

Accordingly, PIPA2 employs the following updating rule for penalty parameter α: choose p_ν to be the smallest integer $p \geq 1$ such that with $\alpha_\nu \equiv \alpha_{\nu-1} (1.2)^{p_\nu}$

$$
(d f_x^\nu)^T dx^\nu + (d f_y^\nu)^T dy^\nu - \alpha_\nu \left[(1 - \sigma_\nu) \sqrt{(w^\nu)^T y^\nu} + \| r^\nu \| \right] < -\phi_\nu.
$$

Similarly, we have modified the Armijo stepsize rule in PIPA2 to choose the nonnegative integer r such that with $\tau \equiv \bar{\tau}_\nu \rho_1^r$,

$$
P_{\alpha_\nu}(x^\nu(\tau), y^\nu(\tau), w^\nu(\tau), z^\nu(\tau)) - P_{\alpha_\nu}(x^\nu, y^\nu, w^\nu, z^\nu)
$$

$$
\leq \gamma' \tau \left((d f_x^\nu)^T dx^\nu + (d f_y^\nu)^T dy^\nu + (d f_w^\nu)^T dw^\nu + (d f_z^\nu)^T dz^\nu \right.
$$

$$
\left. - \alpha_\nu \left(\| r^\nu \| + (1 - \sigma_\nu) \sqrt{(y^\nu)^T w^\nu} \right) \right).
$$

The computer codes are written in MATLAB (version 4.2a) and run on an IBM RISC/6000 workstation 320. Each quadratic subproblem is solved using the MATLAB subroutine qp which implements an active-set method. We have written the MATLAB subroutines f.m, df.m, F.m, dF.m, to compute, respectively the first-level objective function, its gradient, the second-level constraint function, and its Jacobian matrix. These subroutines are called at each iteration by the main program. We have run PIPA1 and PIPA2 on several classes of problems including the LCP and general mixed NCP constrained MPs. In the results reported below, we list various statistics including the number of first-level linear constraints, the total number of iterations, and the final first-level objective function values. We do not report the total CPU times in this preliminary study since we feel that this performance measure is machine dependent; more importantly, since our goal in performing the experiments is to gain some knowledge of the algorithms' ability to solve some simple problems, we are not particularly interested in the speed of the algorithms, nor are we interested in comparing them with other algorithms. For all the experiments the CPU time ranges from seconds to several minutes (less than 5 minutes).

The following notations are used in the output tables.

ν: the iteration counter;

n: the dimension of the first-level variable x;

m: the dimension of second-level variables y and w;

ℓ: the dimension of the free variable z;

k: the number of first-level linear constraints;

α: the penalty parameter used to define P_α (cf. (6.1.9));

res: the residual defined by $\|F(x, y, w, z)\|$;

μ: the complementarity term $y^T w$;

τ: the selected step size;

$\|dx\|$: the norm of dx;

f: the value of the first-level objective function.

6.5.2 *Numerical results*

We have tested the two MATLAB codes on several classes of randomly generated problems. Below we describe the ways the random problems are generated and report the corresponding test results.

Randomly generated mixed NCP constrained MPs

For this class of problems, the first-level objective function is

$$f(x,y,w,z) = \|x\|^2 + \sum_{i=1}^{m} y_i + \sum_{i=1}^{\ell}(z_i - 1)^2.$$

The mixed NCP constraints are chosen to be

$$F(x,y,w,z) = q + Nx + M \begin{pmatrix} y \\ z \end{pmatrix} - \begin{pmatrix} w \\ 0 \end{pmatrix} + \left(\sum_{i=1}^{n}(x_i + x_i^2) \right) K \begin{pmatrix} y \\ z \end{pmatrix}$$

where q is an $(m + \ell)$-vector, N is a matrix of size $(m + \ell) \times n$, and K, M are two matrices of size $(m + \ell) \times (m + \ell)$. The vector q and the matrices N, K, and M are all randomly generated using the **rand** command in MATLAB. Specifically, the vector q is distributed uniformly in $[0, 1]^{(m+\ell)}$; the matrix N is generated with each entry randomly distributed in $[-1, 1]$; the matrices K and M must be positive semidefinite, so we let

$$K = I + 0.01 \times \bar{K}$$

with $\bar{K}$ being a random matrix in $[0, 1]^{(m+\ell) \times (m+\ell)}$, and we let

$$M = \bar{M}^T \bar{M}$$

with $\bar{M}$ a random matrix in $[-1, 1]^{(m+\ell) \times (m+\ell)}$. Notice that the nonlinearity is introduced with the last term in the definition of F.

The first-level constraint set X has the form (1), where b is a random vector in $[0, 1]^n$. In terms of the notation used in Section **6.1**, this corresponds to having $X = \{x \in \Re^n : Gx \le a\}$ with

$$G = \begin{bmatrix} I \\ -I \end{bmatrix} \quad \text{and} \quad a = \begin{pmatrix} b \\ 0 \end{pmatrix}.$$

We have run PIPA1 on many random problem instances and for many choices of m, n, and ℓ. Table 6.1 records a typical run with the choice of $m = 25$, $n = 30$, and $\ell = 5$.

ν	α	res	μ	τ	$\|dx\|$	f
1	5.1598	27.8328	0.099450	0.3585	1.65811	10.6549
2	5.1598	13.5041	0.051845	0.6302	1.3517	5.88693
3	5.1598	7.15587	0.025337	0.5133	0.650417	4.05877
4	5.1598	3.45617	0.012509	0.5404	0.444869	3.09953
5	5.1598	0.718247	4.126e-3	0.8145	0.242834	2.53425
6	6.1917	0.0239396	1.2946e-3	1	0.134858	2.42355
7	12.839	3.22941e-3	4.3879e-4	1	0.0563915	2.40197
8	12.839	2.14233e-4	1.4243e-4	1	0.014690	2.38201
9	12.839	7.41164e-6	4.4817e-5	1	1.88917e-3	2.3789
10	12.839	2.35526e-7	1.388e-5	1	2.81262e-4	2.37704
11	12.839	5.66328e-8	3.5304e-6	1	7.60135e-5	2.37639
12	12.839	3.46917e-9	7.3902e-7	1	1.93169e-5	2.37623
13	12.839	1.2447e-10	1.2841e-7	1	3.99843e-6	2.3762
14	12.839	3.53969e-12	1.858e-8	1	6.92168e-7	2.37619

Table 6.1

In Table 6.1, it can be seen that the penalty parameter α is increased only modestly, and that the step size $\tau = 1$ is eventually used. Also, the f value is seen to decrease monotonically. However, this need not be the case in general; in fact it has been observed that f can increase during the course of computation. The small value of $\|dx\|$ at termination indicates (see the convergence proof in Subsection **6.1.3**) that the last iterate approximately satisfies the stationarity condition of MPEC.

Table 6.2 reports the numerical results of PIPA1 for different choices of m, n, and ℓ. For each random instance, we record the various parameters from the last iteration when the algorithm terminates. In all instances the algorithm eventually selects the step size τ to be 1. Also, the total number of iterations ν remains relatively small even as the dimension of the problem instances increases. Of course, each iteration will take longer as the dimension of the quadratic subproblem increases. Finally, we note that PIPA1 successfully solved all but one randomly generated problems (about 30 of them with varying sizes). In this failed instance, the warning message "Armijo rule failed, the step size too small" was given and the algorithm could not converge.

m,n,ℓ	ν	α	res	μ	τ	$\|dx\|$	f
5,3,1	11	15.4	1.3e-13	3.2e-8	1	3.9e-7	0.4678
7,5,2	13	12.8	4.0e-14	1.5e-8	1	1.7e-7	0.6629
9,8,4	13	15.4	3.4e-14	7.8e-9	1	1.8e-7	1.8771
13,8,5	13	15.4	1.2e-13	1.4e-8	1	2.6e-7	2.2256
19,11,7	13	15.4	3.3e-12	4.7e-8	1	9.2e-7	2.7251
25,15,9	16	15.4	2.4e-12	3.6e-8	1	8.3e-7	3.8939
30,20,10	16	15.4	1.3e-12	3.2e-8	1	8.2e-7	7.3195
40,25,15	18	15.4	1.3e-13	5.6e-9	1	1.9e-7	6.0394
50,30,15	18	15.4	1.4e-12	2.1e-8	1	7.0e-7	6.5302
60,40,25	19	15.4	3.8e-12	2.4e-8	1	9.6e-7	12.479
70,50,30	21	15.4	2.7e-13	7.8e-9	1	2.8e-7	13.047
80,55,35	22	15.4	1.0e-12	1.3e-8	1	5.9e-7	17.204
100,70,40	24	15.4	4.2e-13	7.6e-09	1	3.6e-7	17.790

Table 6.2

Dempe's example problem

We have also applied PIPA1 to the solution of the following bilevel program considered by Dempe [59]:

$$\text{minimize} \quad (x - 3.5)^2 + (z + 4)^2$$
$$\text{subject to} \quad z \in \text{argmin}\{(z-3)^2 : z^2 - x \le 0\}.$$

By writing the second-level optimization problem in terms of its KKT condition, we obtain the following mixed NCP constrained MP:

$$\text{minimize} \quad (x - 3.5)^2 + (z + 4)^2$$
$$\text{subject to} \quad z - 3 + 2zw = 0,$$
$$z^2 - x + y = 0,$$
$$yw = 0, \quad y \ge 0, \quad w \ge 0.$$

We choose the initial values of (x, y, w, z) to be $(1, 1, 1, 1)$. After 13 iterations, PIPA1 terminates at the point $(1.7296, 0.000, 0.6406, 1.3151)$ with the first-level function value being 31.385. It can be verified that this point is a stationary point of the above bilevel program.

Randomly generated LCP constrained MPs

For this class of problems, the first-level objective function is

$$f(x, y, w) = \frac{1}{2}\|x\|^2 + \sum_{i=1}^{m} y_i.$$

The first-level linear constraint set

$$X = \{x \in \Re^n : Gx \le a\}$$

is generated randomly. Specifically, we let $G = G_1 - G_2$ with G_1 and G_2 two uniformly distributed random matrices over $[0, 1]^{k \times n}$, and we let $a = 0.5Ge + 0.5 * \text{abs}(Ge)$ where $e \in \Re^n$ is the vector of ones and $\text{abs}(\cdot)$ denotes the entry-wise absolute value operation. In this way, we know $x^0 = 0.5e$ and $x^* = 0$ are both feasible vectors of X. The parametric LCP constraint is given by

$$q + Nx + My - w = 0, \quad y \ge 0, \quad w \ge 0, \quad y^T w = 0,$$

where the vector $q \in \Re^m$ and the matrices $N \in \Re^{m \times n}$ and $M \in \Re^{m \times m}$ are generated randomly. More specifically, N is chosen to equal the difference of two uniformly distributed random matrices over $[0, 1]^{m \times n}$, whereas M is chosen to equal $\bar{M}^T \bar{M}$ with $\bar{M}$ a uniformly distributed random matrix over $[0, 1]^{m \times m}$. The matrix M is thus symmetric positive semidefinite.

By the nonnegativity of q, the vector $(x^*, y^*) = (0, 0)$ is feasible for the above LCP constraint. Since $f(x, y, w)$ is nonnegative for all (x, y, w) and $f(x^*, y^*, w) = 0$ for all w, we see that the (globally) optimal value of this class of MPEC is always equal to zero. Thus, the algorithm, if it converges, should generate a sequence of function values converging to zero.

We used PIPA2 to solve this class of randomly generated LCP constrained MP. The initialization is given by

$$(x^0, y^0, w^0) = (0.5e, 0.5e, 0.4 \times 10^{-2}e).$$

Table 6.3 records a typical run of PIPA2 on a random problem instance with dimensions $m = 15$, $n = 10$, and $k = 5$.

ν	α	res	μ	τ	$\|dx\|$	f
1	1	35.264	0.0192	0.046	2.497	8.151
4	1	17.164	8.1e-3	0.166	1.392	5.228
7	1	12.923	3.8e-3	0.122	0.697	2.462
10	1	4.682	1.4e-3	0.488	0.351	0.831
13	1	1.443	5.6e-4	0.463	0.155	0.231
16	1.44	0.416	1.8e-4	0.105	0.138	4.7e-2
19	2.07	0.372	1.7e-4	0.015	0.059	4.0e-2
22	2.07	0.217	9.8e-5	0.105	0.029	1.4e-2
25	2.49	0.064	2.2e-5	0.513	0.049	1.5e-3
27	2.99	6.0e-4	2.0e-7	0.95	7.1e-3	8.7e-6
28	2.99	1.1e-16	5.6e-9	1	4.4e-4	3.2e-7
30	2.99	2.8e-17	8.0e-12	1	5.8e-7	4.6e-10

Table 6.3

Table 6.4 reports the numerical results of PIPA2 for different choices of m, n, and k. As for PIPA1, for each random instance, we record the various parameters from the last iteration when the algorithm terminates.

m,n,k	ν	α	res	μ	τ	$\|dx\|$	f
5,3,1	14	6.19	1.1e-16	2.1e-9	1	6.8e-7	1.5e-8
7,5,2	17	5.16	1.1e-16	3.3e-9	1	9.6e-8	8.0e-11
9,8,4	22	4.3	1.1e-16	1.5e-13	1	7.6e-7	1.5e-11
13,8,5	35	2.99	1.1e-16	2.4e-12	1	7.1e-8	6.1e-11
19,11,7	26	2.49	1.1e-16	2.3e-16	0.95	8.5e-8	1.8e-13
25,15,9	27	2.07	1.1e-16	1.3e-14	0.95	5.0e-7	5.0e-12
30,20,10	26	2.07	1.1e-16	1.6e-13	0.95	5.3e-8	1.7e-11
40,25,15	34	1.73	1.1e-16	9.9e-15	0.95	6.9e-8	2.1e-12
50,30,15	34	1.44	1.3e-12	1.1e-15	0.95	7.2e-7	3.1e-12

Table 6.4

A bilevel linear program

We have applied PIPA2 to the bilevel linear program considered by Hansen, Jaumard, and Savard [106]. In this reference, a branch-and-bound approach was used. This BLP is defined by the the variables $x = (x_1, x_2)$

and $y = (y_1, y_2, y_3)$:

maximize $f(x, y) \equiv 8x_1 + 4x_2 - 4y_1 + 40y_2 + 4y_3$

subject to $x_1 + 2x_2 - y_3 \leq 1.3$,

$\qquad x_1 \geq 0, \; x_2 \geq 0$,

and

$$y \in \operatorname{argmin}\{2y_1 + y_2 + 2y_3 : (y_1, y_2, y_3) \geq 0,$$

$$-y_1 + y_2 + y_3 \leq 1,$$

$$4x_1 - 2y_1 + 4y_2 - y_3 \leq 2,$$

$$4x_2 + 4y_1 - 2y_2 - y_3 \leq 2\}.$$

To solve this BLP by PIPA2, we first cast it into the MPEC framework by using the primal-dual formulation of the second-level linear program:

minimize $-f(x, y) = -8x_1 - 4x_2 + 4y_1 - 40y_2 - 4y_3$

subject to $x_1 + 2x_2 - y_3 \leq 1.3$,

$\qquad x_1 \geq 0, \; x_2 \geq 0$,

and $w_1 = 2 - y_4 - 2y_5 + 4y_6$,

$\qquad w_2 = 1 + y_4 + 4y_5 - 2y_6$,

$\qquad w_3 = 2 + y_4 - y_5 - y_6$,

$\qquad w_4 = 1 + y_1 - y_2 - y_3$,

$\qquad w_5 = 2 - 4x_1 + 2y_1 - 4y_2 + y_3$,

$\qquad w_6 = 2 - 4x_2 - 4y_1 + 2y_2 + y_3$,

$\qquad y_i \geq 0, \; w_i \geq 0, \; y_i w_i = 0, \; i = 1, ..., 6$,

where (y_4, y_5, y_6) is the dual vector for the second-level linear program. Notice that this is an LCP constrained MP to which PIPA2 can be directly applied. (Strictly speaking, PIPA2 assumes the first-level constraint to be independent of the second-level variables; this is not the case here. Nonetheless, PIPA2 can still be applied; see the discussion in the next subsection.)

We have initialized PIPA2 as follows:

$$x^0 = (1, 1); \quad y^0 = e; \quad w^0 = 0.4e.$$

In 17 iterations, PIPA2 generates the following solution

$$\bar{x} = (0, 0.65); \quad \bar{y} = (0, 0.3, 0, 0, 0, 0.5); \quad \bar{w} = (4, 0, 1.5, 0.7, 0.8, 0)$$

with the first-level objective value equal to $f = 14.6$. This is a stationary point of the BLP. With a different set of initial values

$$x^0 = (0.5, 1); \quad y^0 = (0.5, 0.5, 1, 1, 1, 1); \quad w^0 = (1, 0.1, 0.1, 0.1, 0.1, 0.1),$$

PIPA2 finds the following solution

$$x^* = (0.5, 0.8); \quad y^* = (0, 0.2, 0.8)$$

in 22 iterations. This solution provides a better objective value $f = 18.4$.

It is interesting to note that the reference [106] obtains the same optimal solution (x^*, y^*) by using a branch-and-bound technique and claims that this is the optimal solution of the BLP.

6.5.3 *Discussion*

Although the computational results reported in the previous section are encouraging, they are still very preliminary at this stage. For one thing, many of the test problems used in the experiments are randomly generated and thus are likely to be nondegenerate. Hence they may not represent the usual degenerate problems arising from realistic applications. We have, however, conducted some limited testing of PIPA, PSQP, and the standard SQP algorithms on several artificially contrived degenerate problems. The results indicate that the three algorithms can all handle the degeneracy in these problems, with the standard SQP requiring 5 times more iterations than the other two algorithms.

We have not yet addressed many efficiency issues in our implementations. For example, in order to solve large-scale real problems, we must incorporate ways to handle sparse data. In all of our randomly generated problem instances, the matrices are dense and no attention has been given to the ways of exploiting any sparsity structure of the matrices. Another issue that requires further investigation is the choice of the various algorithm parameters. We have found that a good set of parameters may speed up the convergence substantially. So far, we have not yet found a good way to "optimize" the choices of these parameters.

In principle, both PIPA1 and PIPA2 can only find a stationary point of an MPEC. This is similar to the case of traditional nonconvex programming where all the iterative descent algorithms are capable of generating only local minima at best. To find a globally optimal solution, one must rely on such techniques as branch-and-bound or implicit enumerations. We have not incorporated these globalization techniques in PIPA1 or PIPA2. One easy way of combining PIPA1 or PIPA2 with a branch-and-bound process is as follows: since the solutions computed by PIPA1 or PIPA2 are always feasible solutions of the MPEC, we can use them as upper bounds to help navigate through the branch-and-bound tree.

Another point worth noting is that both PIPA1 and PIPA2 assume that the first-level constraint set Z is independent of the second-level variables. That is, Z is of the form $X \times \Re^{m+\ell}$. This assumption is tactically made so that the direction generation step is well defined (i.e., the subproblem QP_ν is always solvable). If Z contains coupling constraint, we can modify QP_ν by replacing the constraint $x^\nu + dx \in X$ with $(x^\nu, y^\nu) + (dx, dy) \in Z$. With this modification, QP_ν is no longer guaranteed to have a solution. Nonetheless, assuming it does, so the algorithm is well defined at each iteration, the convergence results (Theorems **6.1.13**, **6.1.17**, and **6.2.3**) can be shown to remain valid. In fact, the proof of the original convergence results can easily be modified to accommodate the modified constraint in QP_ν: $(x^\nu, y^\nu) + (dx, dy) \in Z$. The only thing that needs to be changed is the KKT condition for QP_ν and a few minor modifications in the remaining proof. What this entails in practice is significant: we can still apply PIPA1 and PIPA2 to solve an MPEC containing coupled first-level constraints so long as all the iterates are well defined. In these cases, if the algorithm terminates at some point, then we can still use Theorems **6.1.13** and **6.1.17** (or Theorem **6.2.3**) to deduce that this point must be a stationary solution of MPEC. The last numerical problem in this section is an example to this effect; we have successfully applied PIPA2 to solve this BLP which contains one coupled first-level linear constraint.

We have not yet tested the implicit programming algorithm described in Section **6.3**. To implement and test this algorithm, a subroutine is needed to solve the direction generation subproblem at each iteration. Recall that this subproblem is given as an MPAEC. Thus, the penalty based algorithm PIPA2 appears well suited for this task. Moreover, since PIPA2 does not require a feasible starting point, this makes it possible to use the previous iterate as the starting point for PIPA2 in each iteration.

To conclude, we believe that it would be extremely useful to expand our experiments and perform a systematic computational study of the algorithms presented in this chapter (as well as others) in solving some realistic applications of MPEC, such as those discussed in Section **1.2**. To facilitate such a study and further algorithmic development, it would be particularly valuable to build and maintain a library of challenging test problems with numerical data representing realistic models of MPEC.

Bibliography

[1] M. Abdulaal and L.J. LeBlanc, "A continuous equilibrium network design model," *Transportation Research* 13B (1979) 19–32.

[2] E. Aiyoshi and K. Shimizu, "A solution method for the static constrained Stackelberg problem via penalty method," *IEEE Transactions on Automatic Control* AC-29 (1984) 1111–1114.

[3] A. Ambrosetti and G. Prodi, *A Primer of Nonlinear Analysis*, Cambridge University Press, Cambridge, England (1993).

[4] G. Anandalingam and T. Friesz, Editors, "Hierarchical optimization," *Annals of Operations Research* 34 (1992).

[5] G. Anandalingam and D.J. White, "A solution method for the linear static Stackelberg problem using penalty function," *IEEE Transactions on Automatic Control* 35 (1990) 1170–1173.

[6] P. Armstrong, "Quality control in services," Ph.D. thesis, Department of Decision Sciences, The Wharton School, University of Pennsylvania, Philadelphia (1993).

[7] J.P. Aubin and H. Frankowska, *Set-Valued Analysis*, Birkäuser, Boston (1990).

[8] G. Auchmuty, "Variational principles for variational inequalities", *Numerical Functional Analysis and Optimization* 10 (1989) 863–874.

[9] A. Auslender, *Optimization: Méthodes Numériques*, Masson, Paris (1976).

[10] A. Auslender, "Convergence of stationary sequences for variational inequalities with maximal monotone operators," *Applied Mathematics and Optimization* 28 (1993) 161–172.

[11] A. Auslender, R. Cominetti, and J.-P. Crouzeix, "Convex functions with unbounded level sets and applications to duality theory," *SIAM Journal on Optimization* 3 (1993) 669–687.

[12] A. Auslender and J.-P. Crouzeix, "Well behaved asymptotical convex functions," *Analyse Non-linéare* (1989) 101–122.

[13] T. Başr and G.J. Olsder, *Dynamic Noncooperative Game Theory*, Academic Press, New York (1982).

[14] E. Balas, "Disjunctive programming," *Annals of Discrete Mathematics* 5 (1979) 3–51.

[15] B. Bank, J. Guddat, D. Klatte, B. Kummer, and K. Tammer, *Non-Linear Parametric Optimization*, Birkhäuser Verlag (1982).

[16] D. Baraff, "Issues in computing contact forces for non-penetrating rigid bodies," *Algorithmica* 10 (1993) 292–352.

[17] J.F. Bard, "An algorithm for solving the general bilevel programming problem," *Mathematics of Operations Research* 8 (1983) 260–272.

[18] J.F. Bard, "Optimality conditions for the bilevel programming problem," *Naval Research Logistics Quarterly* 31 (1984) 13–26.

[19] J.F. Bard, "Convex two-level optimization," *Mathematical Programming* 40 (1988) 15–27.

[20] O. Ben-Ayed and C.E. Blair, "Computational difficulties of bilevel linear programming," *Operations Research* 38 (1990) 556–559.

[21] O. Ben-Ayed, C.E. Blair, D.E. Boyce, and L.J. LeBlanc, "Construction of a real-world bilevel linear programming model of the highway network design problem," *Annals of Operations Research* 34 (1992) 219–254.

[22] C. Bergthaller and I. Singer, "The distance to a polyhedron," *Linear Algebra and its Applications* 169 (1992) 111–129.

[23] Z. Bi, P. Calamai, and A. Conn, "An exact penalty function approach for the linear bilevel programming problem," Technical Report #167-0-310789, Department of Systems Design and Engineering, University of Waterloo, Waterloo (1989).

[24] W.F. Bialas and M.H. Karwan, "On two-level linear optimization," *IEEE Transaction on Automatic Control* AC-27 (1982) 211–214.

[25] W.F. Bialas and M.H. Karwan, "Two-level linear programming," *Management Science* 30 (1984) 1004–1020.

[26] E. Bierstone and P.D. Milman, "Semianalytic and subanalytic sets," *Institut des Hautes Etudes Scientifiques, Publications Mathématiques* 67 (1988) 5–42.

[27] S.C. Billups, "Algorithms for complementary problems and generalized equations," Mathematical Programming Technical Report 95-14, Computer Sciences Department, University of Wisconsin, Madison (August 1995).

[28] J. Bisschop, W. Candler, J.H. Duloy, and G.T. O'Mara, "The Indus basin model: a special application of two-level linear programming," *Mathematical Programming Study* 20 (1982) 30–38.

[29] J.F. Bonnans, "Local study of Newton type algorithms for constrained problems," in S. Dolecki, ed., *Lecture Notes in Mathematics*, Springer-Verlag (1989), pp. 13–24.

[30] J.F. Bonnans, "Local analysis of Newton-type methods for variational inequalities and nonlinear programming," *Applied Mathematics and Optimization* 29 (1994) 161–186.

[31] J.F. Bonnans, A.D. Ioffe, and A. Shapiro, "Développment de solutions exactes et approchées en programmation non linéaire," *Comptes Rendus Academie de Sciences, Paris* 315 (1992) 119–123.

[32] J.F. Bonnans, A.D. Ioffe, and A. Shapiro, "Expansion of exact and approximate solutions in nonlinear programming," in W. Oettli and

D. Pallaschke, eds., *Advances in Optimization, Proceedings of 6th French-German Conference on Optimization*, Lecture Notes in Economics and Mathematical Systems No. 382, Springer Verlag, Berlin (1992).

[33] J.F. Bonnans and A. Sulem, "Pseudopower expansion of solutions of generalized equations and constrained optimization problems," *Mathematical Programming* 70 (1995) 123–148.

[34] J. Bracken and J.T. McGill, "Mathematical programs with optimization problems in the constraints," *Operations Research* 21 (1973) 37–44.

[35] J. Bracken and J.T. McGill, "Defense applications of mathematical programs with optimization problems in the constraints," *Operations Research* 22 (1974) 1086–1096.

[36] J. Bracken and J.T. McGill, "A method for solving mathematical programs with nonlinear programs in the constraints," *Operations Research* 22 (1974) 1097–1101.

[37] J. Bracken and J.T. McGill, "Equivalence of two mathematical programs with optimization problems in the constraints," *Operations Research* 22 (1974) 1102–1104.

[38] J. Bracken and J.T. McGill, "Production and marketing decisions with multiple objectives in a competitive environment," *Journal of Optimization Theory and Applications* 24 (1978) 449–458.

[39] P.S. Bradley, O.L. Mangasarian, and W.N. Street, "Feature selection via mathematical programming," Mathematical Programming Technical Report 95-21, Department of Computer Sciences, University of Wisconsin, Madison (December 1995).

[40] J.V. Burke, "A sequential quadratic programming method for potentially infeasible mathematical programs," *Journal of Mathematical Analysis and Applications* 139 (1989) 319–351.

[41] J.V. Burke, "An exact penalization viewpoint of constrained optimization," *SIAM Journal on Control and Optimization* 29 (1991) 968–998.

[42] J.V. Burke, "A robust trust region method for constrained nonlinear programming problems," *SIAM Journal on Optimization* 2 (1992) 325–347.

[43] J.V. Burke and S.P. Han, "A robust sequential quadratic programming method," *Mathematical Programming* 43 (1989) 277–303.

[44] R.G. Cassidy, M.J.L. Kirby, and W.M. Raike, "Efficient distribution of resources through three levels of government," *Management Science* 17 (1971) B462–B473

[45] R.W. Chaney, "Piecewise C^k functions in nonsmooth analysis," *Nonlinear Analysis, Theory, Methods & Applications* 15 (1990) 649–660.

[46] Y. Chen and M. Florian, "The nonlinear bilevel programming problem: formulations, regularity and optimality conditions," *Optimization* (1994).

[47] Y. Chen and M. Florian, "O-D demand adjustment problem with congestion: part I. Model analysis and optimality conditions," Publication CRT-94-56, Centre de Recherche sur les Transports, Université de Montréal, Montréal, Canada (December 1994).

[48] S.C. Choi, W.S. Desarbo, and P.T. Harker, "Product positioning under price competition," *Management Science* 36 (1990) 175–199.

[49] C.C. Chou, K.F. Ng, and J.S. Pang, "Minimizing and stationary sequences of optimization problems," *SIAM Journal on Control and Optimization*, submitted (September 1995).

[50] P.A. Clark and A.W. Westerberg, "A note on the optimality conditions for the bilevel programming problem," *Naval Research Logistics Quarterly* 35 (1988) 413–418.

[51] F.H. Clarke, *Optimization and Nonsmooth Analysis*, John Wiley & Sons, New York (1983).

[52] L.B. Contesse, "Une caractérisation complète des minima locaux en programmation quadratique," *Numerische Mathematik* 34 (1980) 315–332.

[53] W. Cook, A.M.H. Gerards, A. Schrijver, and E. Tardös, "Sensitivity theorems in integer linear programming," *Mathematical Programming* 34 (1986) 251–264.

[54] R.W. Cottle, J.S. Pang, and R.E. Stone, *The Linear Complementarity Problem*, Academic Press, Boston (1992).

[55] I. Constantin and M. Florian, "A method for optimizing the frequencies in a transit network: a special case of nonlinear bilevel programming", Technical report TRISTAN 1, Centre de recherche sur les transports, University of Montréal (1991).

[56] G.B. Dantzig, R.P. Harvey, Z.F. Lansdowne, D.W. Robinson, and S.F. Maier, "Formulating and solving the network design problem by decomposition," *Transportation Research* 13B (1979) 5–17.

[57] J.P. Dedieu, "Penalty functions in subanalytic optimization," *Optimization* 26 (1992) 27–32.

[58] T. De Luca, F. Facchinei and C. Kanzow, "A semismooth equation approach to the solution of nonlinear complementarity problems, *Mathematical Programming* (1997), forthcoming.

[59] S. Dempe, "A necessary and a sufficient optimality condition for bilevel programming problems," *Optimization* 25 (1992) 341–354.

[60] S. Dempe, "Directional differentiability of optimal solutions under Slater's condition," *Mathematical Programming* 59 (1993) 49–69.

[61] A.H. de Silva, "Sensitivity formulas for nonlinear factorable programming and their application to the solution of an implicitly defined optimization model of U.S. crude oil production," Ph.D. thesis, Department of Operations Research, The George Washington University (January 1978).

[62] D. De Wolf and Y. Smeers, "Mathematical properties of formulations of the gas transmission problem," manuscript, Center for Operations Research and Econometrics, Unviersité Catholique de Louvain, Louvain, Belgium (June 1994).

[63] S.P. Dirkse and M.C. Ferris, "The PATH solver: a non-monotone stabilization scheme for mixed complementarity problems," *Optimization Methods & Software* 5 (1995) 123–156.

[64] B.C. Eaves, "On quadratic programming," *Management Science* 17 (1971) 698–711.

[65] B.C. Eaves, "On the basic theorem of complementarity," *Mathematical Programming* 1 (1971) 68–75.

[66] B.C. Eaves, "Where solving for stationary points by LCPs is mixing Newton iterates," in B.C. Eaves, F.J. Gould, H.O. Peitgen, and M.J. Todd, eds., *Homotopy Methods and Global Convergence*, Plenum Press, New York (1983), pp. 63–78.

[67] F. Facchinei and C. Kanzow, "On (un)constrained and constrained stationary points of the implicit Lagrangian," *Journal of Optimization Theory and Applications* 92 (1997).

[68] F. Facchinei, H. Jiang and L. Qi, "A smoothing method for mathematical programs with equilibrium constraints," Technical Report, Dipartimento di Informatica e Sistemistica, Università di Roma "La Sapienza," Rome, Italy (March 1996).

[69] J.E. Falk and J. Liu, "On bilevel programming, part I: general nonlinear case," *Mathematical Programming* 70 (1995) 47–72.

[70] M.C. Ferris, *Weak Sharp Minima and Penalty Functions in Mathematical Programming*, Ph.D. thesis, University of Cambridge, England (1988).

[71] M.C. Ferris and J.S. Pang, "Nondegenerate solutions and related concepts in affine variational inequalities," *SIAM Journal on Control and Optimization* 34 (1996) 244–263.

[72] M.C. Ferris and D. Ralph, "Projected gradient methods for nonlinear complementarity problems via normal maps," in D.Z. Du, L. Qi, and R.S. Womersley, eds., *Recent Advances in Nonsmooth Optimization*, World Scientific, Singapore (1995), pp. 57–87.

[73] A.V. Fiacco, *Introduction to Sensitivity and Stability Analysis in Nonlinear Programming*, Academic Press, New York (1993).

[74] A.V. Fiacco and G.P. McCormick, *Nonlinear Programming: Sequential Unconstrained Minimization Technique*, John Wiley and Sons, Inc., New York (1968).

[75] A. Fischer, "A special Newton-type optimization method," *Optimization* 24 (1992) 269–284.

[76] A. Fischer, "A Newton-type method for positive semidefinite linear complementarity problems," *Journal of Optimization Theory and Applications* 86 (1995) 585–608.

[77] A. Fischer, "An NCP-function and its use for the solution of complementarity problems," in D.-Z. Du, L. Qi, and R.S. Womersley, eds., *Recent Advances in Nonsmooth Optimization*, World Scientific, Singapore (1995), pp. 88–105.

[78] M. Florian, "Mathematical programming applications in national, regional and urban planning," in M. Iri and K. Tanabe, eds., *Mathematical Programming: Recent Developments and Applications*, Kluwer Academic Publishers, Tokyo (1989), pp. 57–82.

[79] C.A. Floudas and I.E. Grossmann, "Synthesis of flexible heat exchanger networks with uncertain flowrates and temperatures," *Computers and Chemical Engineering* 11 (1987) 319–336.

[80] J. Fortuny-Amart and B. McCarl, "A representation and economic interpretation of a two-level programming problem," *Journal of Operational Research Society* 32 (1981) 783–792.

[81] M. Frank and P. Wolfe, "An algorithm for quadratic programming," *Naval Research Logistics Quarterly* 3 (1956) 95–110.

[82] T.L. Friesz, H.J. Cho, N.J. Mehta, R.L. Tobin, and G. Anandalingam, "A simulated annealing approach to the network design problem with variational inequality constraints," *Transportation Science* 28 (1992) 18–26.

[83] T.L. Friesz and P.T. Harker, "Freight network equilibrium: a review of the state of the art," in A. Daughety, ed., *Analytical Studies in Transportation Economics*, Cambridge University Press, Cambridge (1985), pp. 161–206.

[84] T.L. Friesz, R.L. Tobin, T.E. Smith, and P.T. Harker, "A nonlinear complementarity formulation and solution procedure for the general derived demand network equilibrium problem," *Journal of Regional Science* 23 (1983) 337–359.

[85] T. Fujisawa and E.S. Kuh, "Piecewise-linear theory of nonlinear networks," *SIAM Journal of Applied Mathematics* 22 (1972) 307–328.

[86] M. Fukushima, "Equivalent differentiable optimization problems and descent methods for asymmetric variational inequality problems," *Mathematical Programming* 53 (1992) 99–110.

[87] M. Fukushima, "Merit functions for variational inequality and complementarity problems," in G. Di Pillo and F. Giannessi, eds., *Nonlinear Optimization and Applications*, Plenum Publishing Corporation, New York, forthcoming.

[88] M. Fukushima, Z.-Q. Luo, and J.S. Pang, "A globally convergent sequential quadratic programming algorithm for mathematical programs with linear complementarity constraints," manuscript, Department of Mathematical Sciences, The Johns Hopkins University (1996), forthcoming.

[89] M. Fukushima and J.S. Pang, "Minimizing and stationary sequences of merit functions for complementarity problems and variational inequalities," manuscript, Department of Mathematical Sciences, The Johns Hopkins University, Baltimore (November 1995).

[90] S.A. Gabriel and J.S. Pang, "An inexact NE/SQP method for solving the nonlinear complementarity problem," *Computational Optimization and Applications* 1 (1992) 67–92.

[91] S.A. Gabriel and J.S. Pang, "A trust region method for constrained nonsmooth equations," in W.W. Hager, D.W. Hearn, and P. Parda-

los, eds., *Large-Scale Optimization: State of the Art*, Kluwer Academic Publishers, Boston (1994), pp. 159–186.

[92] M. Garey and D. Johnson, *Computers and Intractability*, W.H. Freeman, San Francisco (1979).

[93] J. Gauvin, "A necessary and sufficient regularity condition to have bounded multipliers in nonconvex programming," *Mathematical Programming* 12 (1977) 136–138.

[94] R. Gibbons, *Game Theory for Applied Economists*, Princeton University Press, Princeton (1992).

[95] M.S. Gowda "An analysis of zero set and global error bound properties of a piecewise affine function via its recession function," *SIAM Journal on Matrix Analysis* 17 (1996).

[96] M.S. Gowda and J.S. Pang, "Stability analysis of variational inequalities and nonlinear complementarity problems, via the mixed linear complementarity problem and degree theory," *Mathematics of Operations Research* 19 (1994), 831–879.

[97] M.S. Gowda and R. Sznajder, "On the pseudo-Lipschitzian behavior of the inverse of a piecewise affine function," *Mathematical Programming*, forthcoming.

[98] A. Graham, "Aspects of bilevel programming," Honors thesis, Department of Mathematics, University of Melbourne, Parkville, Australia (1994).

[99] I.E. Grossmann and C.A. Floudas, "Active constraint strategy for flexibility analysis in chemical processes," *Computers and Chemical Engineering* 11 (1987) 675–693.

[100] J. Guddat, F. Guerra Vasquez, and H.Th. Jongen, *Parametric Optimization: Singularities, Pathfollowing and Jumps*, John Wiley & Sons, Chichester (1990).

[101] M. Guignard, "Generalized Kuhn-Tucker conditions for mathematical programming problems in a Banach space," *SIAM Journal on Control* 7 (1969) 232–241.

[102] O. Güler, A.J. Hoffman, and U.G. Rothblum, "Approximations to solutions to systems of linear inequalities," *SIAM Journal on Matrix Analysis and Applications* 16 (1995) 688–696.

[103] C.D. Ha, "Application of degree theory in stability of the complementarity problem," *Mathematics of Operations Research* 31 (1985) 327–338.

[104] W.W. Hager, "Lipschitz continuity for constrained process," *SIAM Journal on Control and Optimization* 17 (1979) 321–338.

[105] S.P. Han, "A globally convergent method for nonlinear programming," *Journal of Optimization Theory and Applications* 22 (1977) 297–309.

[106] P. Hansen, B. Jaumard, and G. Savard, "New branch-and-bound rules for linear bilevel programming," *SIAM Journal on Scientific and Statistical Computing*, 13 (1992) 1194–1217.

[107] P.T. Harker, *Predicting Intercity Freight Flows*, VNU Science Press, Utrecht, The Netherlands (1987).

[108] P.T. Harker and J.S. Pang, "On the existence of optimal solutions to mathematical program with equilibrium constraints," *Operations Research Letters* 7 (1988) 61–64.

[109] P.T. Harker and J.S. Pang, "Finite-dimensional variational inequalities and complementarity problems: a survey of theory, algorithms and applications," *Mathematical Programming* 60 (1990) 161–220.

[110] D.W. Hearn, "The gap function of a convex program," *Operations Research Letters* 1 (1982) 67–71.

[111] R. Hirabayashi, H.Th. Jongen, and M. Shida, "Stability for linearly constrained optimization problems," *Mathematical Programming* 66 (1994) 351–360.

[112] H. Hironaka, *Introduction to real-analytic sets and real-analytic maps*, Instituto Matematico "L. Tonelli" dell'Università de Pisa, Italy (1973).

[113] H. Hironaka, "Subanalytic sets," *Number Theory, Algebraic Geometry and Commutative Algebra* (1973) 453–493.

[114] B.F. Hobbs and K.A. Kelly, "Using game theory to analyze electric transmission pricing policies in the United States," *European Journal of Operational Research* 56 (1992) 154–171.

[115] B.F. Hobbs and S.K. Nelson, "A nonlinear bilevel model for analysis of electric utility demand-side planning issues," *Annals of Operations Research* 34 (1992) 255–274.

[116] A.J. Hoffman, "On approximate solutions of systems of linear inequalities," *Journal of Research of the National Bureau of Standards* 49 (1952) 263–265.

[117] L. Hörmander, "On the division of distributions by polynomials," *Arkiv for Mathematik* 3 (1958) 555–568.

[118] T. Ibaraki, "Complementary programming," *Operations Research* 19 (1971) 1523–1528.

[119] V.I. Istratescu, *Fixed Point Theory*, D. Reidel Publishing Co., Boston (1981).

[120] R. Janin, "Directional derivative of the marginal function in nonlinear programming," *Mathematical Programming Study* 21 (1984) 110–126.

[121] R.G. Jeroslow, "Cutting planes for complementarity constraints," *SIAM Journal on Control and Optimization* 16 (1978) 56–62.

[122] R.G. Jeroslow, "The polynomial hierarchy and a simple model for competitive analysis," *Mathematical Programming* 32 (1985) 146–164.

[123] H. Jiang and L. Qi, "A new nonsmooth equations approach to nonlinear complementarity problems," *SIAM Journal on Control and Optimization*, forthcoming.

[124] K. Jittorntrum, "Solution point differentiability without strict complementarity in nonlinear programming," *Mathematical Programming Study* 21 (1984) 127–138.

[125] H.Th. Jongen, D. Klatte, and K. Tammer, "Implicit functions and sensitivity of stationary points," *Mathematical Programming* 49(1990) 123–138.

[126] H.Th. Jongen, T. Mobert, J. Rückmann, and K. Tammer, "On inertia and Schur complement in optimization," *Linear Algebra and its Applications* 95 (1987) 97–109.

[127] N.H. Josephy, "Newton's method for generalized equations," Technical report No. 1965, Mathematics Research Center, University of Wisconsin, Madison (1979).

[128] N.H. Josephy, "Quasi-Newton methods for generalized equations," Technical report No. 1966, Mathematics Research Center, University of Wisconsin, Madison (1979).

[129] J.J. Júdice and A.M. Faustino, "A sequential LCP method for bilevel linear programming," *Annals of Operations Research* 34 (1992) 89–106.

[130] T.J. Kim and S. Suh, "Toward developing a national transportation planning model: a bilevel programming approach for Korea," *Annals of Regional Science* 22 (1988) 65–80.

[131] A. Klarbring, "A mathematical programming approach to three-dimensional contact problems with friction," *Computational Methods in Applied Mechanics and Engineering* 58 (1986) 175–200.

[132] A. Klarbring, "Mathematical programming and augmented Lagrangian methods for frictional contact problems," in A. Curnier, ed., *Proceedings Contact Mechanics International Symposium*, Presses Polytechniques et Universitaires Romandes (1992), pp. 409–422.

[133] A. Klarbring, "Mathematical programming in contact problems," in M.H. Aliabadi and C.A. Brebbia, eds., *Computational Methods in Contact Mechanics*, Computational Mechanics Publications, Southampton (1993), Chapter 7, pp. 233–263.

[134] A. Klarbring and G. Björkman, "A mathematical programming approach to contact problems with friction and varying contact surface," *Computers & Structures* 30 (1988) 1185–1198.

374   Bibliography

[135] A. Klarbring, J. Petersson, and M. Rönnqvist, "Truss topology optimization involving unilateral contact," Journal of Optimization Theory and Applications 87 (1995) 1–31.

[136] A. Klarbring and M. Rönnqvist, "Nested approach to structural optimization in non-smooth mechanics," Structural Optimization 10 (1995) 79–86.

[137] D. Klatte and K. Tammer, "Strong stability of stationary solutions and Karush-Kuhn-Tucker points in nonlinear optimization," Annals of Operations Research 27 (1990) 285–310.

[138] M. Kočvara and J.V. Outrata, "On optimization systems governed by implicit complementarity problems," Numerical Functional Analysis and Optimization 15 (1994) 869–887.

[139] M. Kočvara and J.V. Outrata, "On the solution of optimum design problems with variational inequalities," in D.Z. Du, L. Qi, and R.S. Womersley, eds., Recent Advances in Nonsmooth Optimization, World Scientific, Singapore (1995), pp. 172–192.

[140] M. Kojima, "Strongly stable stationary solutions in nonlinear programs," in S.M. Robinson, ed., Analysis and Computation of Fixed Points, Academic Press, New York (1980), pp. 93–138.

[141] M. Kojima and P. Hirabayashi, "Continuation deformation of nonlinear programs," Mathematical Programming Study 21 (1984) 150–198.

[142] M. Kojima, N. Megiddo, T. Noma, and A. Yoshise, A Unified Approach to Interior Point Algorithms for Linear Complementarity Problems, Lecture Notes in Computer Science 538 Springer Verlag, Berlin (1991).

[143] M. Kojima and R. Saigal, "A study of PC^1 homeomorphisms on subdivided polyhedrons," SIAM Journal on Mathematical Analysis 10 (1979) 1299–1312.

[144] M. Kojima and S. Shindo, "Extensions of Newton and quasi-Newton methods to systems of PC^1 equations," Journal of Operations Research Society of Japan 29 (1986) 352–374.

[145] D. Kuhn and R. Löwen, "Piecewise affine bijections of $\mathbf{R}^n$ and the equation $Sx^+ - Tx^- = y$," *Linear Algebra and its Applications* 96 (1987) 109–129.

[146] L. Kuntz and S. Scholtes, "Structural analysis of nonsmooth mappings, inverse functions, and metric projections," *Journal of Mathematical Analysis and Applications* 188 (1994) 346–386.

[147] L. Kuntz and S. Scholtes, "A nonsmooth variant of the Mangasarian-Fromovitz constraint qualification," *Journal of Optimization Theory and Applications* 82 (1994) 59–75.

[148] J. Kyparisis, "On uniqueness of Kuhn-Tucker multipliers in nonlinear programming," *Mathematical Programming* 32 (1985) 242–246.

[149] J. Kyparisis, "Sensitivity analysis framework for variational inequalities," *Mathematical Programming* 38 (1987) 203–213.

[150] J. Kyparisis, "Sensitivity analysis for nonlinear programs and variational inequalities with nonunique multipliers," *Mathematics of Operations Research* 15 (1990) 286–298.

[151] J. Kyparisis, "Solution differentiability for variational inequalities," *Mathematical Programming* 48 (1990) 285–302.

[152] J. Kyparisis, "Parametric variational inequalities with multilevel solution sets," *Mathematics of Operations Research* 17 (1992) 341–364.

[153] W. Li, "The sharp Lipschitz constants for feasible and optimal solutions of a perturbed linear program," *Linear Algebra and its Applications* 187 (1993) 15–40.

[154] W. Li, "Remarks on convergence of the matrix splitting algorithm for the symmetric linear complementarity problem," *SIAM Journal on Optimization* 3 (1993) 155–163.

[155] W. Li, "Error bounds for piecewise convex quadratic programs and applications," *SIAM Journal on Control and Optimization* 33 (1995) 1510–1529.

[156] W. Li, "Linearly convergent descent methods for unconstrained min-
      imization of convex quadratic splines," *Journal of Optimization The-
      ory and Applications* 86 (1995) 145-172.

[157] M.B. Lignola and J. Morgan, "Topological existence and stability for
      Stackelberg problems," *Journal of Optimization Theory and Appli-
      cations* 84 (9915) 145-169.

[158] M.B. Lignola and J. Morgan, "Stability of regularized bilevel pro-
      gramming problem," Preprint n. 14, Dipartimento di Matematica e
      Applicazioni, Università di Napoli "Federico II," Italy (1995).

[159] J. Liu, "Sensitivity analysis in nonlinear programs and variational
      inequalities via continuous selections," *SIAM Journal on Control and
      Optimization* 33 (1995) 1040-1061.

[160] J. Liu, "Strong stability in variational inequalities," *SIAM Journal
      on Control and Optimization* 33 (1995) 725-749.

[161] J. Liu, "Perturbation Analysis in Nonlinear Programs and Varia-
      tional Inequalities," Ph.D. dissertation, Department of Operations
      Research, The George Washington University, Washington, D.C.
      (1995).

[162] N.G. Lloyd, *Degree Theory*, Cambridge University Press, Cambridge
      (1978).

[163] M.S. Lojasiewicz, "Sur le problème de la division," *Studia Mathe-
      matica* 18 (1959) 87-136.

[164] M.S. Lojasiewicz, "Ensembles semi-analytiques," Institut des Hautes
      Etudes Scientifiques, Bures-sur-Yvette (1964).

[165] P. Loridan and J. Morgan, "Weak via strong Stackelberg problem:
      new results," Preprint n. 30, Dipartimento di Matematica e Appli-
      cazioni, Università di Napoli "Federico II," Italy (1994).

[166] P. Lötstedt, "Coulomb friction in two-dimensional rigid body sys-
      tems," *Zeitschrift Angewandte Mathematik und Mechanik* 61 (1981)
      605-615.

[167] X.-D. Luo and Z.-Q. Luo, "Extension of Hoffman's error bound to polynomial systems," *SIAM Journal on Optimization* 4 (1994) 383–392.

[168] X.-D. Luo and P. Tseng, "Conditions for a projection-type error bound for the linear complementarity problem to be global," *Linear Algebra and its Applications* (1996), forthcoming.

[169] Z.-Q. Luo, "Convergence analysis of primal-dual interior point algorithms for convex quadratic programs," in R.P. Agarwal, ed., *Recent Trends in Optimization Theory and Applications*, World Scientific, Singapore (1995), pp. 255–270.

[170] Z.-Q. Luo, O.L. Mangasarian, J. Ren, and M. Solodov, "New error bounds for the linear complementarity problem," *Mathematics of Operations Research* 19 (1994) 880–892.

[171] Z.-Q. Luo and J.S. Pang, "Error bounds for analytic systems and their applications," *Mathematical Programming* 67 (1995) 1–28.

[172] Z.-Q. Luo, J.S. Pang, D. Ralph, and S.Q. Wu, "Exact penalization and stationarity conditions of mathematical programs with equilibrium constraints," *Mathematical Programming*, forthcoming.

[173] Z.-Q. Luo and P. Tseng, "A decomposition property for a class of square matrices," *Applied Mathematics Letters* 4 (1991) 67–69.

[174] Z.-Q. Luo and P. Tseng, "On the linear convergence of descent methods for convex essentially smooth minimization," *SIAM Journal on Control and Optimization* 30 (1992) 408–425.

[175] Z.-Q. Luo and P. Tseng, "Error bound and convergence analysis of matrix splitting algorithms for the affine variational inequality problem," *SIAM Journal on Optimization* 2 (1992) 43–54.

[176] Z.-Q. Luo and P. Tseng, "On global error bound for a class of monotone affine variational inequality problems," *Operations Research Letter* 11 (1992) 159–165.

[177] Z.-Q. Luo and P. Tseng, "Perturbation analysis of a condition number for linear systems," *SIAM Journal on Matrix Analysis and Applications* 15 (1994) 636–660.

[178] Z.-Q. Luo and P. Tseng, "Error bounds and convergence analysis of feasible descent methods: a general approach," *Annals of Operations Research* 46 (1993) 157–198.

[179] L. Mallozzi and J. Morgan, "Existence of feedback equilibrium for two-stage Stackelberg games," Preprint n.13, Dipartimento di Matematica e Applicazioni, Università di Napoli "Federico II," Italy (1994).

[180] O.L. Mangasarian, *Nonlinear Programming*, McGraw-Hill Book Company, New York (1969); Japanese Edition (1971); SIAM Classics in Applied Mathematics 10, Philadelphia (1994).

[181] O.L. Mangasarian, "A condition number of linear inequalities and equalities," in G. Bamber and O. Optiz, eds., *Methods of Operations Research 43*, Proceedings of 6th. Symposium über Operations Research, Universität Augsburg, September 7–9, 1981, Verlagsgruppe Athennäum/Hain/Scriptor/Hanstein, Konigstein (1981), pp. 3–15.

[182] O.L. Mangasarian, "Sufficiency of exact penalty minimization," *SIAM Journal on Control and Optimization* 23 (1985) 30–37.

[183] O.L. Mangasarian, "Simple computable bounds for solutions of linear complementarity problems and Linear Programs," *Mathematical Programming Study* 25 (1985) 1–12.

[184] O.L. Mangasarian, "A condition number for differentiable convex inequalities," *Mathematics of Operations Research* 10 (1985) 175–179.

[185] O.L. Mangasarian, "A simple characterization of solution sets of convex program," *Operations Research Letters* 7 (1988) 21–26.

[186] O.L. Mangasarian, "Global error bounds for monotone affine variational inequality problems," *Linear Algebra and its Applications* 174 (1992) 153–163.

[187] O.L. Mangasarian, "Misclassification minimization," *Journal of Global Optimization* 5 (1994) 309–323.

[188] O.L. Mangasarian, "The ill-posed linear complementarity problem," Mathematical Programming Technical Report 95-15, Computer Sciences Department, University of Wisconsin, Madison (revised November 1995).

[189] O.L. Mangasarian, "Machine learning via polyhedral concave minimization," in H. Fischer, B. Riedmuller, and S. Schaeffler, eds., *Applied Mathematics and Parallel Computing – Festchrift for Klaus Ritter*, Physica-Verlag, Berlin (1996), pp. 175–188.

[190] O.L. Mangasarian and R. De Leone, "Error bounds for strongly convex programs and (super)linearly convergent iterative schemes for the least 2–norm solution of linear programs," *Applied Mathematics & Optimization* 17 (1989) 1–14.

[191] O.L. Mangasarian and S. Fromovitz, "The Fritz John optimality necessary conditions in the presence of equality and inequality constraints," *Journal of Mathematical Analysis and Applications* 17 (1967) 37–47.

[192] O.L. Mangasarian and R.R. Meyer, "Nonlinear perturbation of linear programs," *SIAM Journal on Control and Optimization* 17 (1979) 745–752.

[193] O.L. Mangasarian and J.S. Pang, "Exact penalty functions for mathematical programs with linear complementarity constraints", manuscript, forthcoming.

[194] O.L. Mangasarian and J. Ren, "New improved error bounds for the linear complementarity problem," *Mathematical Programming* 66 (1994) 241–256.

[195] O.L. Mangasarian and T.-H. Shiau, "Lipschitz continuity of solutions of linear inequalities, programs and complementarity problems," *SIAM Journal on Control and Optimization* 25 (1987) 583–595.

[196] O.L. Mangasarian and M.V. Solodov, "Nonlinear complementarity as unconstrained and constrained minimization," *Mathematical Programming* 62 (1993) 277–297.

[197] P. Marcotte, "Network optimization with continuous control parameters," *Transportation Science* 17 (1983) 181–197.

[198] P. Marcotte, "A new algorithm for solving variational inequalities with application to the traffic assignment problem," *Mathematical Programming* 33 (1985) 339–351.

[199] P. Marcotte, "Network design problem with congestion effects: a case of bilevel programming," *Mathematical Programming* 34 (1986) 142–162.

[200] P. Marcotte and J.P. Dussault, "A note on a globally convergent Newton method for solving monotone variational inequalities," *Operations Research Letters* 6 (1987) 35–42.

[201] P. Marcotte and D.L. Zhu, "Exact and inexact penalty methods for the generalized bilevel programming problems," *Mathematical Programming* 74 (1996), forthcoming.

[202] R. Mathias and J.S. Pang, "Error bounds for the linear complementarity problem with a P-matrix," *Linear Algebra and its Applications* 132 (1990) 123–136.

[203] B.A. McCarl and T.H. Spreen, "Price endogenous mathematical programming as a tool for sector analysis," *American Journal of Agricultural Economics* (1980) 87–102.

[204] G.P. McCormick, *Nonlinear Programming: Theory, Algorithms, and Applications*, John Wiley & Sons, New York (1983).

[205] R. Mifflin, "Semismooth and semiconvex functions in constrained optimization," *SIAM Journal on Control and Optimization* 15 (1977) 957–972.

[206] T. Miller, T.L. Friesz, and R.L. Tobin, "Heuristic algorithms for delivered price spatially competitive network facility location problems," *Annals of Operations Research* 34 (1992) 177–202.

[207] J.F. Nash, "Non-cooperative games," *Annals of Mathematics* 54 (1951) 286–295.

[208] M.G. Nicholls, "Aluminum production modeling–A nonlinear bilevel programming approach," *Operations Research* 43 (1995) 208–218.

[209] M.G. Nicholls, "The application of non-linear bi-level programming to the aluminium industry," manuscript, School of Information Systems, Swinburne University of Technology (1995).

[210] K. Okuguchi, *Expectations and Stability in Oligopoly Models*, Lecture Notes in Economics and Mathematical Systems, No. 138, Springer-Verlag, Berlin (1976).

[211] J.M. Ortega and W.C. Rheinboldt, *Iterative Solution of Nonlinear Equations in Several Variables*, Academic Press, New York (1970).

[212] J.V. Outrata, "On the numerical solution of a class of Stackelberg problems," *Zeitschrift für Operations Research* 4 (1990) 255–278.

[213] J.V. Outrata, "On necessary optimality conditions for Stackelberg problems," *Journal of Optimization Theory and Applications* 76 (1993) 305–320.

[214] J.V. Outrata, "On optimization problems with variational inequality constraints," *SIAM Journal of Optimization* 4 (1994) 340–357.

[215] J.V. Outrata and J. Zowe, "A numerical approach to optimization problems with variational inequality constraints," *Mathematical Programming* 68 (1995) 105–130.

[216] G. Owen, *Game Theory*, 2nd edition, Academic Press, New York (1982).

[217] J.S. Pang, "A posteriori error bounds for the linearly–constrained variational inequality Problem," *Mathematics of Operations Research* 12 (1987) 474–484.

[218] J.S. Pang, "Solution differentiability and continuation of Newton's method for variational inequality problems over polyhedral sets," *Journal of Optimization Theory and Applications* 66 (1990) 121–135.

[219] J.S. Pang, "Newton's method for B-differentiable equations," *Mathematics of Operations Research* 15 (1990) 311–341.

[220] J.S. Pang, "Complementarity problems," in R. Horst and P. Pardalos, eds., *Handbook on Global Optimization*, Kluwer Academic Publishers, B.V., Dordrecht (1994), pp. 271–338.

[221] J.S. Pang, "A degree-theoretic approach to parametric nonsmooth equations with multivalued perturbed solution sets," *Mathematical Programming, Series B* 62 (1993) 359–384.

[222] J.S. Pang, "Convergence of splitting and Newton methods for complementarity problems: an application of some sensitivity results," *Mathematical Programming* 58 (1993) 149–160.

[223] J.S. Pang, "Necessary and sufficient conditions for solution stability in parametric nonsmooth equations," in D.Z. Du, L. Qi, and R.S. Womersley, eds., *Recent Advances in Nonsmooth Optimization*, World Scientific, Singapore (1995), pp. 261–288.

[224] J.S. Pang and S.A. Gabriel, "NE/SQP: a robust algorithm for nonlinear complementarity problems," *Mathematical Programming, Series A* 60 (1993) 295–338.

[225] J.S. Pang, S.P. Han, and N. Rangaraj, "Minimization of locally Lipschitzian functions," *SIAM Journal of Optimization* 1 (1991) 57–82.

[226] J.S. Pang and L. Qi, "A globally convergent Newton method for convex SC^1 minimization problems," *Journal of Optimization Theory and Applications* 85 (1995) 633–648.

[227] J.S. Pang and D. Ralph, "Piecewise smoothness, local invertibility, and parametric analysis of normal maps," *Mathematics of Operations Research* 21 (1996).

[228] J.S. Pang and J.C. Trinkle, "Complementarity formulations and existence of solutions of dynamic multi-rigid-body contact problems with Coulomb friction," *Mathematical Programming*, forthcoming.

[229] J.S. Pang and J.M. Yang, "Two-stage parallel iterative methods for the symmetric linear complementarity problem," *Annals of Operations Research* 14 (1988) 61–75.

[230] J.M. Peng, "Equivalence of variational inequality problems to unconstrained optimization," Technical report, State Key Laboratory of Scientific and Engineering Computing, Academia Sinica, Bejing, China (December 1994).

[231] J. Petersson, *Optimization of Structures in Unilateral Contact*, Linköping Studies in Science and Technology, Ph.D. dissertation, No. 397, Division of Mechanics, Department of Mechanical Engineering, Linköping University, Linköping (1995).

[232] L. Qi, "Convergence analysis of some algorithms for solving nonsmooth equations," *Mathematics of Operations Research* 18 (1993) 227–244.

[233] L. Qi and J. Sun, "A nonsmooth version of Newton's method," *Mathematical Programming* 58 (1993) 353–368.

[234] Y. Qiu and T.L. Magnanti, "Sensitivity analysis for variational inequalities," *Mathematics of Operations Research* 17 (1992) 61–76.

[235] H. Rademacher, "Über partielle und totale Differenzierbarkeit von Funktionen mehrerer Variabler. I," *Mathematical Annals* 79 (1919) 340–359.

[236] D. Ralph, "A new proof of Robinson's homeomorphism theorem for piecewise linear maps," *Linear Algebra and its Applications* 178 (1993) 249–260.

[237] D. Ralph, "Global convergence of damped Newton's method for nonsmooth equations, via the path search," *Mathematics of Operations Research* 19 (1994) 352–389.

[238] D. Ralph, "On branching numbers of normal manifolds," *Journal of Nonlinear Analysis: Theory, Methods & Applications* 22 (1994) 1041–1050.

[239] D. Ralph, "Sequential quadratic programming for mathematical programs with linear complementarity constraints," in A. Easton and R. May, eds., *Computational Techniques and Applications (CTAC95)*, World Scientific, Singapore (1996), pp. 663–669.

[240] D. Ralph and S. Dempe, "Directional derivatives of the solution of a parametric nonlinear program," *Mathematical Programming* 70 (1995) 159–172.

[241] D. Ralph and S. Scholtes, "Sensitivity analysis and Newton's method for composite piecewise smooth equations," *Mathematical Programming*, forthcoming.

[242] E. Rasmusen, *Games and Information: An Introduction to Game Theory*, Basil Blackwell (1994).

[243] A. Reinoza, "The strong positivity conditions," *Mathematics of Operations Research* 10 (1985) 54–62.

[244] S.M. Robinson, "Bounds for error in the solution set of a perturbed linear program," *Linear Algebra and its Applications* 6 (1973) 69–81.

[245] S.M. Robinson, "Perturbed Kuhn-Tucker points and rates of convergence of a class of nonlinear-programming problems," *Mathematical Programming* 7 (1974) 1–16.

[246] S.M. Robinson, "An application of error bounds for convex programming in a linear space," *SIAM Journal on Control* 13 (1975) 271–273.

[247] S.M. Robinson, "Generalized equations and their applications, part I: basic Theory," *Mathematical Programming Study* 10 (1979) 128–141.

[248] S.M. Robinson, "Strongly regular generalized equations," *Mathematics of Operations Research* 5 (1980) 43–62.

[249] S.M. Robinson, "Some continuity properties of polyhedral multifunctions," *Mathematical Programming Study* 14 (1981) 206–214.

[250] S.M. Robinson, "Generalized equations and their applications, part II: applications to nonlinear programming," *Mathematical Programming Study* 19 (1982) 200–221.

[251] S.M. Robinson, "Local structure of feasible sets in nonlinear programming, part III: stability and sensitivity," *Mathematical Programming Study* 30 (1987) 45–66. Corrigenda, *Mathematical Programming* 49 (1987) 143.

[252] S.M. Robinson, "An implicit-function theorem for a class of nonsmooth functions," *Mathematics of Operations Research* 16 (1991) 292–309.

[253] S.M. Robinson, "Normal maps induced by linear transformations," *Mathematics of Operations Research* 17 (1992) 691–714.

[254] S.M. Robinson, "Homeomorphism conditions for normal maps of polyhedra," in A. Ioffe, M. Marcus, and S. Reich, eds., *Optimization and Nonlinear Analysis*, Longman, London (1992) 691–714.

[255] S.M. Robinson, "Nonsingularity and symmetry for linear normal maps," *Mathematical Programming, Series B* 62 (1993) 415–426.

[256] R.T. Rockafellar, *Convex Analysis*, Princeton University Press, Princeton (1970).

[257] H. Scheel, "Ein straffunktionsansatz für optimierungsprobleme mit gleichgewichtsrestriktionen," Diploma thesis, Operations Research und Wirtschaftsinformatik, Universität Dortmund (1995).

[258] S. Scholtes, "Introduction to piecewise differentiable equations," Habilitation thesis, Institut für Statistik und Mathematische Wirtschaftstheorie, Universität Karsruhe, Germany (1994).

[259] S. Scholtes, "Homeomorphism conditions for coherently oriented piecewise affine mappings," Research report, Institut für Statistik unde Mathematische Wirtschaftstheorie, Universität Karlsruhe, Germany (1994).

[260] R. Schramm, "On piecewise linear functions and piecewise linear equations," *Operations Research* 5 (1980) 510–522.

[261] A. Shapiro, "Sensitivity analysis of nonlinear programs and differentiability properties of metric projections," *SIAM Journal on Control and Optimization* 26 (1988) 628–645.

[262] A. Shapiro, "On concepts of directional differentiability," *Journal of Optimization Theory and Applications* 66 (1990) 477–487.

[263] H.D. Sherali, A.L. Soyster, and F.H. Murphy, "Stackelberg-Nash-Cournot equilibria: characterizations and computations," *Operations Research* 31 (1983) 253–276.

[264] M.J. Smith, "The existence, uniqueness and stability of traffic equilibrium," *Transportation Research* 13B (1979) 295–394.

[265] H. Van Stackelberg, *The Theory of Market Economy*, Oxford University Press, Oxford (1952).

[266] G.E. Stavroulakis, "Optimal prestress of cracked unilateral structures: finite element analysis of an optimal control problem for variational inequalities," *Computer Methods in Applied Mechanics and Engineering* (1995), forthcoming.

[267] G.E. Stavroulakis, "Optimal prestress of structures with frictional unilateral contact interfaces," *Archives of Applied Mechanics* (1996), forthcoming.

[268] S. Suh and T.J. Kim, "Solving nonlinear bilevel programming models of the equilibrium network design problem: a comparative review," *Annals of Operations Research* 34 (1992) 203–218.

[269] C. Suwansirikul, T.L. Friesz, and R.L. Tobin, "Equilibrium decomposed optimization: a heuristic for the continuous equilibrium network design problem," *Transportation Science* 21 (1987) 254–263.

[270] R.E. Swaney and I.E. Grossmann, "An index for operational flexibility in chemical process design, part I: formulation and theory," *Journal of the American Institute of Chemical Engineers* 31 (1985) 621–630.

[271] R.E. Swaney and I.E. Grossmann, "An index for operational flexibility in chemical process design, part II: computational algorithms," *Journal of the American Institute of Chemical Engineers* 31 (1985) 631.

[272] R. Sznajder and M.S. Gowda, "Generalizations of P_0- and P-properties; extended vertical and horizontal LCPs," *Linear Algebra and its Applications* 223/224 (1995) 695–715.

[273] R. Sznajder and M.S. Gowda, "Nondegeneracy concepts for zeros of piecewise affine functions," *Mathematics of Operations Research*, forthcoming.

[274] K. Taji, M. Fukushima, and T. Ibaraki, "A globally convergent Newton method for solving strongly monotone variational inequalities," *Mathematical Programming* 58 (1993) 369–383.

[275] R.L. Tobin, "Uniqueness results and algorithm for Stackelberg-Cournot-Nash equilibria," *Annals of Operations Research* 34 (1992) 21–36.

[276] M.J. Todd and Y. Ye, "A centered projective algorithm for linear programming," *Mathematics of Operations Research* 15 (1990) 508–529.

[277] P. Tseng, "Growth behavior of a class of merit functions for the nonlinear complementarity problem," *Journal of Optimization Theory and Applications* 89 (1996) 17–37.

[278] L.N. Vicente and P.H. Calamai, "Bilevel and multilevel programming: a bibliography review," *Journal of Global Optimization* 5 (1994) 291–306.

[279] V. Visweswaran, C.A. Floudas, M.G. Ierapetritou, and E.N. Pistikopoulos, "A decomposition-based global optimization approach for solving bilevel linear and quadratic programs," *Global Optimization*, Kluwer Academic Publishers, Boston (1995).

[280] T. Wang, R.D.C. Monteiro, and J.S. Pang, "An interior point potential reduction method for constrained equations," *Mathematical Programming* (1996).

[281] T. Wang and J.S. Pang, "Global error bounds for convex quadratic inequality systems," *Optimization* 31 (1994) 1–12.

[282] J.G. Wardrop, "Some theoretical aspects of road traffic research," *Proceedings of the Institute of Civil Engineers, Part II* 1 (1952) 325–378.

[283] J. Warga, "A necessary and sufficient condition for a constrained minimum," *SIAM Journal on Optimization* 2 (1992) 665–667.

[284] D.J. White and G. Anandalingam, "A penalty function approach for solving bi-level linear programs," *Journal of Global Optimization* 3 (1993) 397–419.

[285] A.B. Wilson, "A simplicial algorithm for concave programming," Ph.D. dissertation, Graduate School of Business Administration, Harvard University (1963).

[286] A.N. Wilson, Jr., "A useful generalization of the P_0-matrix concept," *Numerische Mathematique* 17 (1971) 62–70.

[287] S. Wright and D. Ralph, "A superlinear infeasible interior point algorithm for monotone complementarity problems," *Mathematics of Operations Research*, forthcoming.

[288] N. Yamashita and M. Fukushima, "On stationary points of the implicit Lagrangian for nonlinear complementarity problems," *Journal of Optimization Theory and Applications* 84 (1995) 653–663.

[289] N. Yamashita and M. Fukushima, "Modified Newton methods for solving semismooth reformulations of monotone complementarity problems," *Mathematical Programming*, forthcoming.

[290] N. Yamashita, K. Taji, and M. Fukushima, "Unconstrained optimization reformulations of variational inequality problems," *Journal of Optimization Theory and Applications*, forthcoming.

[291] J.J. Ye and D.L. Zhu, "Optimality conditions for bilevel programming problems," *Optimization*, forthcoming.

[292] J.J. Ye, D.L. Zhu, and Q. Zhu, "Generalized bilevel programming problem," manuscript, Department of Mathematics and Statistics, University of Victoria, Victoria, B.C., Canada (1993).

[293] A. Yezza, "First-order necessary optimality conditions for general bilevel programming problems," *Journal of Optimization Theory and Applications* 89 (1996) 189–219.

[294] R. Zhang, "Problems of hierarchical optimization: nonsmoothness and analysis of solutions," Ph.D. dissertation, Department of Applied Mathematics, University of Washington, Seattle (1990).

[295] R. Zhang, "Problems of hierarchical optimization in finite dimensions," *SIAM Journal on Optimization* 4 (1994) 521–536.

[296] Y. Zhang, "On the convergence of an infeasible interior-point algorithm for linear programming and other problems," *SIAM Journal on Optimization* 4 (1994) 208–227.

Index